PHYSICS AND CHEMISTRY OF FISSION

Proceedings of the XVIIIth International Symposium on Nuclear Physics

Devoted to the Fiftieth Anniversary of the Discovery of Nuclear Fission

PHYSICS AND CHEMISTRY OF FISSION

Edited by H. Märten and D. Seeliger

Organized by the Technical University Dresden in Cooperation with the International Atomic Energy Agency

Nova Science Publishers, Inc.
New York

Nova Science Publishers, Inc.
6080 Jericho Turnpike, Suite 207
Commack, New York 11725

Book Production Manager: Janet Glanzman White
Graphics: Elenor Kallberg, Christopher Concannon, and Michael A. Masotti

ISBN 1-56072-023-9

International Symposium on Nuclear Physics (18th : 1988 : Gaussig, Germany)
Physics and chemistry of fission : proceedings of the XVIIIth International Symposium on Nuclear Physics, Gaussig (GDR), 21-25 November 1988 / organized by the Technical University Dresden in cooperation with the International Atomic Energy Agency ; edited by H. Märten and D. Seeliger.
p. cm.
"Devoted to the fiftieth anniversary of the discovery of nuclear fission."
Includes bibliographical references and indexes.
ISBN 1-56072-023-9 : $98.00
1. Nuclear fission--Congresses. 2. Nuclear chemistry--Congresses. I. Marten, H. II. Seeliger, D. (Dieter) III. Technische Universität Dresden. IV. International Atomic Energy Agency. V. Title.
QC789.7.I58 1988 92-16061
539.7'62--dc20 CIP

Printed in the United States of America

CONTENTS

OPENING ADDRESS

Dieter Seeliger

Chairman of the Organizing Committee

On behalf of the President of the Technical University Dresden and as the chairman of the Organizing Committee, I welcome you at the old castle Gaussig at the beginning of the XVIIIth International Symposium on Nuclear Physics - Physics and Chemistry of Fission.

Since 1971 every year scientists from many countries have come here for participation in the International Symposia on Nuclear Physics organized by the Technical University Dresden. In this year our meeting has been organized in co-operation with the International Atomic Energy Agency and the Physical Society of the German Democratic Republic. It is devoted to the fiftieth anniversary of the discovery of nuclear fission - a milestone in the history of nuclear physics. Just fifty years ago, in autumn of 1938, Hahn and Staβmann investigated by radiochemical methods the reaction and decay products of uranium irradiated by neutrons. In a first publication in the journal "Naturwissenschaft" on 6th January 1939 they indicated that Barium was found among the reaction products. This was assumed as a possible result of neutron induced fission of uranium nuclei. In a second publication submitted on 28th January 1939 this assumption was clearly stated - a new physical phenomenon was discovered. Immediately after first information of this work, very active research work started in many laboratories all over the world. Within a few months between December 1938 and March 1939 the most essential features of this surprising nuclear process were found.

Today we know that this discovery was one of the major steps in the history of physical sciences in the 20th century. It opened the window for the development of nuclear energetics. It had a large influence on the whole society. The discovery of nuclear fission - this was one of the crucial conclusions by Otto Hahn in the fall of 1945 - also raised the question about the responsibility of scientists for the kind of use of their scientific results.

Dear colleagues, starting to organize this symposium, we were forced to select between two possibilities: either the arrangement of a broad-profile conference including historical aspects as well as the whole range of nuclear technologies or to concentrate on a more scientific meeting on current physical problems in nuclear fission research. We selected the second way, having in mind that there are several other conferences in the next year, which have a much broader scope. The advantage of this decision is that many distinguished scientists and actively working specialists in nuclear fission came here. I hope this will result in many fruitful discussions during this week in the old Gaussig castle.

In co-operation with the IAEA we also invited scientists from developing countries to this meeting. As a matter of fact, a fast development of nuclear energetics is now observed in those countries. Therefore, we hope that contacts and discussions at this meeting will stimulate also further fission-related research and its application in developing countries.

I wish you a pleasant stay at Gaussig and many interesting scientific discussions, but also exciting sightseeing and cultural events in Dresden and its surroundings. With your help the XVIIIth International Symposium on Nuclear Physics certainly will become a successful scientific meeting stimulating the further international contacts and co-operation.

I. FISSION FRAGMENT DISTRIBUTIONS: EXPERIMENT AND THEORY

FISSION BARRIERS, FISSION CHANNELS, FISSION VALLEYS

A. Michaudon
Institut Laue - Langevin, Grenoble, France

ABSTRACT

Many important properties of fission are determined by the potential energy surface in the space of deformation parameters. This paper presents an attempt to understand some aspects of fission in terms of this surface properties at various deformations: in the barrier region where two or even three humps may appear, on tops of these barriers where fission channels may play a role and a larger deformations where fission valleys seem to be present. The interplay between the properties of the potential energy at these various deformations is also discussed.

1. INTRODUCTION

Fission is a violent collective phenomenon whereby a heavy nucleus undergoes a series of oscillations until it is driven to a very large deformation culminating in scission, i.e., the breaking of this nucleus into two or more fragments.

This process is difficult to understand mainly because the properties of nuclear matter at such extreme deformations are not well known. To understand fission, one needs to know i) the statics, i.e., the potential energy surface (P.E.S.) at all relevant deformations in the space $\{s\}$ of shape parameters s_i and ii) the dynamics i.e. the motion of the nuclear system on this P.E.S. along the fission path. The statics are now fairly well known thanks to improvements in nuclear models at large deformations: this is in contrast to the dynamics which are poorly understood because of the complex effects of dissipation and inertia.

Experimentally one has access to fission cross sections, governed by fission barrier penetrability and to the properties of the fission fragments and of their de-excitation products at infinity (not at scission).

This talk presents some aspects of fission, in terms of the statics only, with special consideration to the fission barriers, the fission channels and the fission valleys. Also the interplay between these various concepts is discussed.

2. FISSION BARRIERS

In the early days of fission, only macroscopic models could be used for the determination of the P.E.S. and they were very approximate. They resulted in single-hump fission barriers (SHFB) that proved very useful indeed at the beginning (1) but failed later to explain several puzzling fission phenomena which first appeared uncorrelated: fission isomers, vibrational resonances and intermediate structure in subthreshold fission cross sections (2). These phenomena were subsequently interpreted in a coherent manner, at least qualitatively, with double-hump fission barriers (DHFB) that resulted from calculations using an ingenious macroscopic-microscopic (Mμ) method first put forward by STRUTINSKY (3). In this method, the macroscopic energy $V_M(\{s\})$ is modulated by shell-energy and pairing corrections $\Delta E_{sh}(\{s\})$ oscillating as a function of deformation.

Fission barrier heights obtained in this manner are in good agreement with experimental results (2,4) with a few exceptions, for example for Thorium isotopes. This so-called Thorium anomaly can be best illustrated by the 720 keV resonance observed below threshold in the ^{230}Th (n,f) cross section. The calculated inner barrier of the DHFB is too low to accommodate a class II vibrational level at the resonance energy. But a closer examination of shell effects near the outer barrier can provide an explanation. When the mass asymmetry parameter s_3 is taken into account, two shallow wells appear in the neighbourhood of the second saddle point. This 720 keV ^{230}Th neutron resonance is now interpreted as caused by vibrational levels in these so-called class III wells. Coupling between such levels of the same energy through the barrier in the s_3 plane results in levels of opposite parities slightly splitted in energy. The fine structure in this resonance that appears in high resolution measurements can now be interpreted in terms of two rotational bands built on two class III vibrational levels of opposite parities. Results obtained in (d,pf) measurements substantiate this interpretation up to higher spin states with similar parameters (5).

In conclusion, the existence of the triple-hump fission barrier (THFB) seems now well established for this ^{230}Th resonance for several reasons:

- Calculations show that the inner barrier of the DHFB is too low to accommodate a class II vibrational state at the resonance energy;
- The fission cross section cannot be fitted with one rotational band only, thus excluding the class II vibrational resonance hypothesis;
- The (n,f) and (d,pf) cross section and anisotropy data can be fitted with $K^{\pi} = 1/2^{+}$ <u>and</u> $1/2^{-}$ rotational bands having the same parameters;
- The inertia parameter $\frac{h^2}{2\mathfrak{J}}$ of these rotational bands is about 2 keV as compared to 3keV for fission isomers, thus demonstrating that the deformation of the states responsible for the ^{230}Th neutron resonance is greater than that of a class II state.

The fission barriers (DHFB <u>and</u> THFB) calculated with the Mμ method now give a good description of actinide fission cross sections.

3. FISSION CHANNELS

The traditional concept of exit channel in a nuclear reaction consists in a given final state (nature and quantum states) of the products of the reaction. If applied to fission, the extreme variety of fission fragments (both in nuclear species and excited states) would lead to a very large number of fission exit channels (of the order of 10^{10}). This is in violent disagreement with the large fluctuations of the fission widths, as observed in the neutron resonances for fissile nuclei (2). These fluctuations are compatible with a few fission exit channels only.

This apparent paradox was solved with the fission channel theory of A. BOHR (6). According to this theory, the fission properties are partially determined at the saddle-point of the SHFB by transition states whose spectrum resembles that at ground-state deformation. Since in low-energy fission, the fissioning nucleus is cold at the saddle point, only a few transition states are available and make a substantial contribution to the fission process. This theory proved very successful in explaining many fission properties, such as the large fluctuations in the fission widths. It could explain also the clear separation in the ^{239}Pu fission widths for $J^{\pi} = 0^{+}$ and $J^{\pi} = 1^{+}$ neutron resonances in terms of two 0^{+} and 1^{+} transition states having different energies. But this theory was less successful in interpreting the fission properties in the resonances, for example the mass distribution and the kinetic energy of the fission fragments and the average number $\overline{\nu}$ of the fission neutrons.

This theory hat do be modified later with the advent of the DHFB. Instead of one set of transition states, one has to consider two sets of such states at the two saddle points. In addition, one has to take into account also the influence of class II states acting as intermediate states in the fission exit channels. Therefore, the fission channel theory of A. BOHR, first proposed in the frame of the SHFB, is more difficult to apply for a DHFB. In this case, the theory must take into account the interplay between transition states at the two barriers and the class II states. This interplay depends on the relative heights of the two barriers and the depth of the second well.

But, in addition to the difficulties associated with the DHFB, the fission theory of A. BOHR should also consider the influence of the last phase of fission, the passage from the last saddle point to scission. This phase being very rapid, its effect was ignored for a long time but recent experimental and theoretical results seem now to show that it has an influence, as illustrated in the next chapter.

4. FISSION VALLEYS

Experimental results and P.E.S. calculations at very large deformations demonstrate that the transition from the last saddle point to scission is complex. For example some Mμ calculations of the P.E.S. show the existence of several valleys (or fission paths) beyond the last saddle point and these valleys can help interpret several experimental results as illustrated below.

4.1. Bimodal Symmetric Fission for Heavy Actinides (7)

Correlated measurements of the yields Y(M) and total kinetic energies TKE(M) of the fission fragments as a function of their mass M show interesting features for actinides between Z = 100 and Z = 104. The Y(M) distribution is composed of a sharp symmetric peak superimposed on a broad one. The sharp (broad) peak is associated with a high (low) TKE component. These results can be interpreted in terms of two fission modes: mode 1 with a compact scission configuration (high TKE) and mode 2 with a more elongated one (low TKE). This so-called bimodal fission appears also in Mμ calculations, for example ^{258}Fm (8). Two valleys appear at large deformations:

1) a "fusion valley" leading to a compact configuration of two touching magic spherical fragments having Z = 50 and therefore a large Coulomb energy at scission and 2) the more conventional "fission valley" with a more elongated configuration at scission, hence with a lower Coulomb energy. Clearly the fusion and the fission valleys can be associated with modes 1 and 2 respectively. In addition, the P.E.S. leaves provision for a so-called "switchback" path whereby nuclear matter in the fusion valley can leak into the fission valley at extreme deformations.

4.2. Fission properties of the $^{235}U(n,f)$ reaction

Some calculations with the Mµ method show that 3 fission valleys appear at large deformations in the P.E.S. for ^{236}U (9). The fission paths associated with these valleys are called: superlong, standard I and standard II and they can help understand several experimental results described below.

a) Fission induced in ^{235}U by thermal neutrons (10)

The yields Y(M), average total kinetic energies ($\overline{TKE}(M)$) and standard deviation $\sigma_{TKE}(M)$ of the TKE distributions were measured as a function of M. These data were fitted simultaneously using a superposition of the three fission modes referred to above. For each mode i, 5 free parameters were used in the fitting procedure: the weight Wi, the average mass Mi, the standard deviation σ_{Mi} of the mass distribution, the average distance D_i between charge centers at scission and the standard deviation σ_{Di} for the distribution of this distance. The parameters Mi, σ_{Mi}, Di and $\overline{TKE}$ derived for each mode from these fits are in good agreement with the calculations, thus supporting the hypothesis of these 3 fission modes.

b) Fission induced in ^{235}U by resonance neutrons (11)

Fluctuations in some properties such as Y(M) and TKE(M) have been observed in the ^{235}U neutron resonances but they are weak and not correlated with the spins and parities (J^{π} = 3^- or 4^-) of these resonances (2). The small amplitude of these fluctuations is expected from the A. BOHR theory because of the similarit between 3^- and 4^- transition states, but the absence of correlation with J^{π} casts some doubt on the validity of the theory. Precise measurements of Y and TKE were made as a function of M in many isolated ^{235}U neutron resonances below 130 eV using the CBMN linac as a pulsed neutron source (11). The data analysis was made for each resonance in terms of the three fission modes referred to above. The variations in Y(M) could be accounted for by those of Wi from resonance to resonance. The variations in $\overline{TKE}(M)$ as high as 500 keV can also be explained by those of Wi, in the same manner as for the Y data, but no correlation with J^{π} can be seen. The variations in $\overline{\nu}$, as obtained by other groups, are correlated with those of $\overline{TKE}$, as expected.

Therefore the three fission paths observed in the ^{236}U P.E.S. seem to correspond to three fission modes that provide a good interpretation of the fission properties in the $^{235}U(n,f)$ reaction whereas these properties could not be explained satisfactorily in terms of the A. BOHR channel theory of fission.

4.3. Fission properties in the ^{252}Cf spontaneous fission

Calculations with the Mμ method, of the same type as those mentioned above for ^{236}U, show the existence of four valleys in the ^{252}Cf P.E.S. at large deformations: standard St (asymmetric), super-short (symmetric), super-long SL (symmetric) and super-asymmetric. These valleys terminate at elongations ranging from 14fm to 20.7fm at scission and their access is controlled by barriers with heights ranging from 10 to 19 MeV (12). Measurements of Y(M), TKE(M) and $\sigma_{TKE}(M)$ were made by two groups (13, 14) and they can be interpreted in terms of these four fission modes. Whereas the St path accounts for most of the yields in the asymmetric peaks in Y(M), the symmetric contribution of the SL path is visible in the dip at symmetry of the St one, with a value of TKE lower by 6 MeV. The super-short contribution should fall in the wings of the St yields and therefore cannot be seen. On the other hand the very low yields of the super-asymmetric component can be seen because they lie in the far end of the wings of the St yields (for $M > 176$). This contribution explains the change of slope of the Y(M) curve beyond $M = 176$. One can notice also in that region fluctuations in Y(M) that are interpreted as shell effects at scission.

Therefore experimental results observed in the ^{252}Cf spontaneous fission can be explained in terms of the fission valleys in the P.E.S. of this nucleus at large deformations.

4.4. Cold fragmentation

The energy balance in fission can be expressed of each fragmentation M in the following manner:

$$Q(M) = TKE(M) + E_{exc}(M)$$

When the excitation energy $E_{exc}(M)$ for each fragment lies below its emission threshold $S_n(M)$, there cannot be prompt neutron emission. Fission without neutron emission is a few % probable and is called "cold fragmentation" (CF). Primary and secondary fission fragments are the same nuclei, though in different states, and this simplifies the analysis of fission data. The low value of E_{exc}, which is the sum of deformation and excitation energies at scission, implies that the nascent fragments are formed close to their ground state, therefore in a compact configuration. This configuration is not far from the saddle-point on the fission path and results in a large TKE (since E_{exc} is small) with only a small component of precession-kinetic energy. The yields decrease rapidly with increasing TKE and consequently CF can be observed only with fission fragment spectrometers of high sensitivity. This is the case of the Lohengrin spectrometer installed at the Grenoble ILL High Flux Reactor. This spectrometer can separate the ionized fission fragments with an excellent resolution by combining electrostatic and magnetic deflections. Another very promising method consists in using a high resolution back-to-back double-grid ionisation chamber in which a very thin deposit of fissil material on a very thin backing can be inserted (16).

The yields of the fission of ^{235}U by thermal neutrons are measured with these sophisticated instruments with a clear separation in proton and neutron numbers, and show that CF is observed for practically all types of fragmentation except for symmetric and very asymmetric mass splits. Very cold fragmentation for which TKE is very close to Q (E_{exc} is then very small) is also observed in a few cases, for example when $100 < A_L < 105$ and $40 < Z_L < 42$ (the subscript L stands for light fragment). This corresponds to nascent fission fragments stabilized by shell effects near ^{132}Sn (spherical) and ^{104}Mo (deformed). Even-odd (e-o) effects are clearly visible in Q(M) but not so much in $\overline{TKE}$(M) (17).

These results can be interpreted in terms of microscopic calculations of the P.E.S using the Hartree-Fock-Bogolyubov (HFB) method (18). These calculations made for ^{240}Pu can also be used for ^{236}U and show the existence of two valleys, separated by a barrier in the Q_{40} deformation coordinate. The fission valley, called N_1, describes the conventional descent from saddle-point to scission, whereas the steeper fusion valley N_2 corresponds to two fragments approaching each other. The so-called Q_{40} barrier between those two valleys decreases with increasing deformation until it vanishes at scission. The most probable fission path is therefore the descent down N_1 until the nucleus reaches the elongation at scission where it falls down into the fusion valley with neck breaking. But shape transitions from N_1 to N_2 at shorter elongations can also occur though with a smaller probability.

Cold fragmentation is interpreted as one of these transitions and is not typical of conventional fission proceeding down N_1. Viscosity plays a minor role in CF since the transition from N_1 to N_2 occurs very early in the descent from the saddle-point. Also superfluidity, if present in N_1, may be destroyed when the nucleus tunnels through the Q_{40} barrier. Lastly some o-e effects can be explained using simple statistical arguments (17).

Therefore, these two valleys obtained in microscopic P.E.S. calculations provide a coherent description of cold fragmentation observed in the $^{235}U(n,f)$ reaction with thermal neutrons.

These high sensitivity fission fragment spectrometers mentioned above give access to exotic very neutron rich nuclei (produced with yields as low as 10^{-9}) such as $^{70-74}Ni$ and $^{73-77}Cu$, as observed with Lohengrin at the ILL (19). The variations of these yields with proton and neutron numbers can be used to explore fission valleys in the P.E.S. Also, the properties of these neutron rich nuclei can help to give a more precise description of the r-process in astrophysics.

5. CONCLUSION

A thorough understanding of fission requires a detailed knowledge of the P.E.S. at all deformations. In the barrier region, the modulation of this P.E.S. by shel effects leads to the famous double-hump barrier for actinides and even to a tripl hump barrier in the Thorium region. The calculations made with the macroscopic-microscopic method are quite precise and can explain qualitatively and quantitati vely the behaviour of fission cross sections and phenomena such as: fission isome vibrational resonances and intermediate structure in subthreshold fission cross-sections.

The concept of A. BOHR's fission channels needs to be adapted to this new situati in order to take into account transition states on top of both barriers of the DH and also the intermediate class II states in the second well. But in addition, th influence of the last fission phase between the saddle point and scission cannot be ignored any longer.

Shell effects seem to play a role in that region of large deformation also, givin rise to fission valleys of various sorts depending on the fissioning nucleus. Motion along and access to these valleys can be complex with possible cross-over between them. Some of them can be seen in fission events of low probability, henc the interest of very sensitive fission detectors. These low fission yields are not typical of gross fission properties and they must be interpreted with caution especially as far as dissipation is concerned. A critical assessment of P.E.S. calculations must be made in terms of the method, the parameterization and the models that are used, in order to have a coherent and consistent description of the fission valleys.

Lastly the concept of fission channel needs to be revisited. It must take into account all possible paths up to scission. In particular it must consider the interplay between the A. BOHR transition states on top of all barriers, the class-II states, the fission valleys at large deformations and possibly shell effects at scission. The interplay may vary according to the type of fission events being considered and this complicates dissipation studies.

AKNOWLEDGEMENTS

The author is extremely grateful to the organisers and in particular to Professor Dr. Seeliger for having had the opportunity to present this overview of some aspects of fission which would need more developments than those presented briefly here. The author apologises to the reader for not having shown in the text any of the numerous figures projected for illustration during the oral presentation, because of the lack of space. But they can be sent upon request to those who are interested.

REFERENCES

(1) N. Bohr and J.A. Wheeler, Phys. Rev. 56:426 (1939).

(2) A. Michaudon, Advances in Nuclear Physics, Vol. 6, 1, (1973) Plenum Press

- International Conference on the Interactions of Neutrons with Nuclei (Lowell, USA, July 1976) CONF-760715, P2, p. 641.
- Neutron Interlab Seminar (Oxford, July 1-3, 1981). Note CEA-N-2232 NEANDC (E) 224 (Dec. 1981).

(3) V.M. Strutinsky, Nucl. Phys., A95:420 (1967) Nucl. Phys., A122:1 (1968).

(4) S. Bjørnholm and J.E. Lynn, Rev. Mod. Phys. 52, 725 (1980).

(5) J. Blons, C. Mazur, D. Paya, M. Ribrag and H. Weigmann, Nuclear Physics, A414, 1, (1984).

(6) A. Bohr, Proc. Int. Conf. Peaceful Uses of Atomic Energy (Geneva 1955), Vol. II, United Nations, New York (1986), p. 220.

(7) E.K. Hulet, J.F. Wild, R.J. Dougan, R.W. Lougheed, J.H. Landrum, A.D. Dougan, M. Schädel, R.L. Hahn, P.A. Baisden, C.M. Henderson, R.J. Dupzyk, K. Sümmerer and G.R. Bethune, Phys. Rev. Let., 56 Nb 4, 313 (1986).

(8) P. Möller, J.R. Nix and W.J. Swiatecki, submitted to Nuclear Physics.

(9) U. Brosa, S. Grossmann and A. Müller, Z. Naturforsch., 41a, 1341, (1986) and several other papers quoted in (10).

(10) H.H. Knitter, F.J. Hambsch, C. Budtz-Jørgensen and J.P. Theobald, Z. Naturforsch., 42a, 786 (1987).

(11) F.J. Hambsch, H.H. Knitter and C. Budtz-Jørgensen, submitted to Nuclear Physics.

(12) U. Brosa, S. Grossmann and A. Müller, Z. Naturforsch., 41a, 1341 (1986).

(13) G. Barreau, A. Sicre, F. Caïtucoli, M. Asghar, T.P. Doan, B. Leroux, G. Martinez and T. Benfoughal, Nucl. Phys., A432, 411 (1985).

(14) C. Budtz-Jørgensen and H.H. Knitter, Proceedings of the Fifteenth International Symposium on Nuclear Physics, Gaussig (1985).

(15) E. Moll, H. Schrader, G. Siegert, M. Asghar, J.P. Bocquet, G. Bailleul, J.B. Gautheron, J. Greif, G. Crawford, C. Chauvin, H. Ewald, H. Wollnik, P. Armbruster, G. Fiebig, H. Lawin and K. Fistemich, Nucl. Inst. Methods, 123, 615 (1975).

(16) C. Signarbieux, G. Simon, J. Trochon and F. Brisard, J. Physique Lett., 46, 1095 (1985).

(17) F. Gönnenwein, Nuclear Data for Basic and Applied Science (Proceedings of a Conference, Santa Fe, New Mexico, USA, 13-17 May 1985), Gordon and Breach Publishers.

(18) J.F. Berger, M. Girod and D. Gogny, Nuclear Physics, A428, 23c, (1984).

(19) P. Armbruster, M. Bernas, J.P. Bocquet, R. Brissot, H.R. Faust and P. Roussel Europhys. Lett., 4 (7), 793 (1987).

FRAGMENT CHARGE DISTRIBUTIONS IN LOW ENERGY FISSION

F. Gönnenwein
Physikalisches Institut der Universität Tübingen,
Morgenstelle 14, D 7400 Tübingen, FRG

Abstract

Improved and new techniques for measuring the nuclear charges of fission fragments have been put into use in recent years. Before all studies in low energy neutron induced fission of actinides ranging from Th to Cf have brought about a comprehensive and reliable set of data on fragment yields $Y(A, Z, E)$, with A the mass, Z the nuclear charge and E the kinetic energy of fragments. The present review surveys the available data with special emphasis put on the nuclear charges. The mean charges $\bar{Z}(A)$ and the variances $\sigma^2(Z|A)$ of charge distributions for a given mass chain A, and the global charge distributions $Y(Z)$ are discussed and confronted with theoretical predictions. The odd-even effect in the charge yields is evaluated to yield the intrinsic excitation energy of fragments at scission.

1. INTRODUCTION

In binary fission a heavy nucleus decays into two fragments, with mass and charge numbers A^* and Z, respectively. The excitation energy of the "primary" fragments being carried away very fast (typ. 10^{-15}s) by neutrons and gammas, the fragments being detected by nowadays techniques are the "secondary" fragments or "products" after prompt neutron evaporation. With $\nu(A)$ the number of neutrons being emitted from fragments to yield the product mass A, and with $\langle\nu(A)\rangle$ the average neutron number one has

$$A' = A + \langle\nu(A)\rangle \tag{1}$$

for the average primary fragment mass A'. One expects A' to be a good estimate for A^*.

The charge numbers Z of fragments undergo changes through β-decay. Hence, the time scales are much slower (typ. > 100 ms) than for neutron evaporation.

Most of the data on fragments yields $Y(A, Z)$ have been obtained by radiochemical methods. Evidently, much work had to go into developing fast chemical separation techniques [HER 82] and into establishing the decay characteristics of isotopes in order to cope with β-decay. Even for the technically important fission reactions comprehensive data on independent yields $Y(Z, A)$ have only become available in recent years [WAH 88].

In parallel to radiochemistry also physical methods to measure nuclear charges of fission fragments have been put forward. The most reliable and most general one is the $\Delta E - E$ technique and its variants. Here the nuclear charges are identified from an inspection of the Bragg-curve (stopping power) for mass separated fragments. The different techniques are reviewed in [DEN 86].

The chemical and physical methods to measure fragment charges are complementary to one another. Radiochemistry covers all nuclides through the periodic table. However, the information on the kinetic energy of fragments is lost. In contrast, the $\Delta E - E$ technique is limited to charge numbers $Z \lesssim 42-44$, which means that the charges in the heavy mass group can not be determined. On the other hand, first the identification of Z is fast (typ. $1\mu s$) and one is not hampered by β-decay, and second the yields $Y(Z, A)$ are measured for known fragment kinetic energies. This latter piece of information proves to be very valuable for the interpretation of experimental results.

For the present review on charge distribution in fission the results obtained by the two methods outlined above have been put together. The reactions covered are mainly neutron-induced fission of nuclei ranging from ^{229}Th(n,f) up to ^{249}Cf(n,f). The neutron energies lie between thermal and a few MeV, i.e. we limit ourselves to low energy fission. This means that the (Z_F, A_F) values of the fissioning compound nucleus are uniquely determined. Needless to stress that this choice of reactions is merely dictated by the availability of comprehensive data on $Y(Z, A)$.

In the following we present the experimental yields $Y(Z, A')$ not directly but, instead, give the first and second moments of the isobaric charge distributions, i.e. the average charges $\langle Z(A')\rangle$ and the variances $\sigma^2(Z|A')$. The experimental data are confronted with theoretical model predictions. The element yields $Y(Z)$, and here before all the odd-even effects in the charge yields, are compared for the nuclei studied and their dependence on the excitation energy of the compound nucleus and on the total kinetic energy release of the fragments is discussed. It is shown that from the systematic study of the odd-even effect one may learn something on the energy dissipated into intrinsic degrees of freedom already at the scission point. This quantity has a direct bearing on the viscosity of nuclear matter in dynamic fission models.

2. THE AVERAGE CHARGES OF MASS CHAINS

2.1 EXPERIMENTAL RESULTS

It is observed experimentally that the charge distribution $Y(Z|A')$ of any fixed isobaric mass chain A' is to first approximation rather well represented by a model Gaussian

$$Y_m(Z|A') \sim exp\{Z(A') - Z_p(A')\}^2/2\sigma_z^2(A') \quad . \tag{2}$$

For most mass chains A', however, a distinct odd-even staggering shows up, with the yields for even Z being larger than for odd ones. This has to be accounted for in eq. (2) by a modulating factor $F_z \geq 1$: all even Z-yields have to multiplied by F_z, while all odd Z-yields have to be divided by F_z. Within a mass chain A' the factor F_z is taken to be constant.

It is therefore customary to discuss charge distributions in terms of the average isobaric charge $\bar{Z}(A') = \langle Z(A') \rangle$, the isobaric charge variance $\sigma^2(Z|A') = \langle \{Z(A') - \bar{Z}(A')\}^2 \rangle$ and odd-even effects. We keep to this scheme in the following.

For the average charge $\bar{Z}(A')$ it is convenient not to plot $\bar{Z}(A')$ itself, but instead the deviation ΔZ of $\bar{Z}(A')$ from an "unchanged charge distribution"(UCD). If there were no charge polarization of the fragments upon breaking into pieces a fissioning nucleus (Z_F, A_F), one would expect the charge Z_{UCD} of a fragment with mass A' to follow the rule

$$Z_{UCD}/A' = Z_F/A_F \quad . \tag{3}$$

In all of the figures shown in the following the charge deviation ΔZ defined by

$$\Delta Z = (\bar{Z} - Z_{UCD})_H = (Z_{UCD} - \bar{Z})_L \tag{4}$$

is plotted. As usual, H and L designate the heavy and light fragment, respectively. The change of sign in the definition of eq. (4) keeps track of the conservation of charge numbers.

A compilation of mainly radiochemical data for thermal neutron fission of ^{233}U, ^{235}U and ^{239}Pu, and the spontaneous fission of ^{252}Cf has recently been published by A.C. Wahl [WAH 88]. Results obtained for ^{235}U(n,f) are shown in Fig. 1, where the first and second moments of isobaric charge distributions are depicted as a function of fragment mass A'. The open symbols are experimental data and the continuous and broken lines come from Gaussian model fits, with and without taking odd-even effects into consideration, respectively. Deferring the discussion of variances (termed $RMS(A')$ in Fig. 1) to the next chapter we summarize the main features of $\Delta Z(A')$ to be read from the figure as:

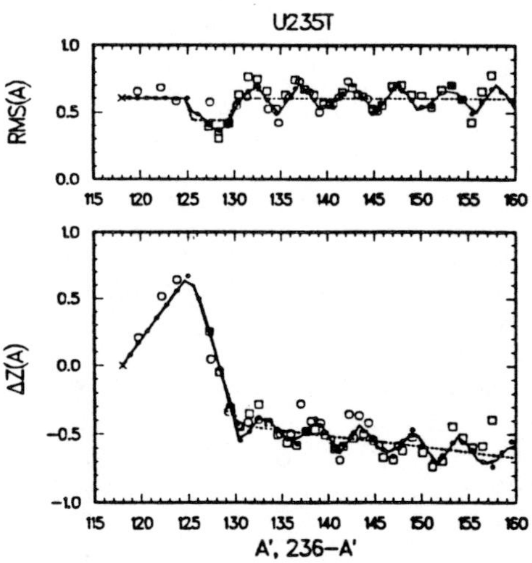

FIGURE 1
Characteristics of isobaric charge distributions in thermal neutron fission of 235 U.
Bottom: avarage charge plotted as $\Delta Z = (\bar{Z} - Z_{UCD})_H$
Top: variance of charge distribution
Open symbols: experimental data
Full and dashed line: model fits with and without odd-even effects, respectively.

a) for asymmetric fission ($A'_H \gtrsim$ 128 amu) ΔZ is negative and typically $\Delta Z = -0.5$ ch. units; there is a pronounced fluctuation of $\Delta Z(A')$ with a spacing of $\Delta A' \approx 5$ amu.

b) for symmetric fission ($A'_H \lesssim$ 128 amu) ΔZ becomes positive and is not fluctuating. It should be underlined that the measurement of independent yields $Y(Z, A')$ close to mass symmetry is a recent major achievement of radiochemistry [WAH 85]. The spectacular change of sign for ΔZ being not anticipated by theory has come as a surprise.

As already pointed out, fission yields studied by the $\Delta E - E$ technique contain the additional information on fragment kinetic energies. An example of such data is given in Fig. 2

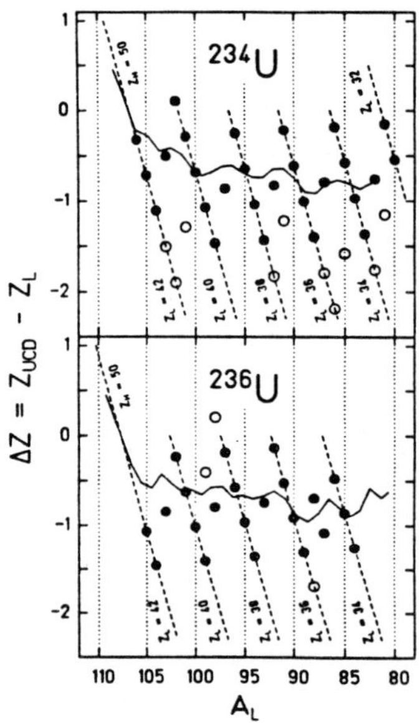

FIGURE 2
Charges dominating the isobaric yield for thermal neutron fission of ^{233}U (bottom) at fixed kinetic energies E_L of the light fragment: E_L = 110.55 MeV (^{234}U) and E_L = 108.0 MeV (^{236}U). The charges are plotted as the deviation of ΔZ (full points). For the open points and lines s. text.

for the reactions ^{233}U(n,f) and ^{235}U(n,f) [CLE 86]. For fixed kinetic energies E_L of the light fragment, the charges Z (viz. the deviations ΔZ) carrying in a mass chain the highest yields are plotted as full points. The energies E_L chosen are E_L = 110.55 MeV and E_L = 108.0 MeV for ^{234}U and ^{236}U, respectively. These energies are well above average and, in fact, one is approaching cold fission, where all of the available reaction energy $Q(A, Z)$ goes into the kinetic energies of the fragments. One finds that upon coming close to cold fission, in most cases the charge number $Z(A)$ maximizing the yield $Y(Z|A)$ is identical to the charge maximizing the Q-value $Q(Z|A)$. For mass chains where this is not the case, open points in Fig. 2 indicate the ΔZ-values corresponding to the maximum isobaric $Q(Z|A)$. The dashed lines refer to charge splits with even fragment charges. The full lines are the $\Delta Z's$ being averaged over fragment kinetic energies. It is evident from Fig. 2 that the ΔZ fluctuations become much more pronounced for high kinetic energies, with a tendency for the charges to stick to even charge numbers (dashed lines). The fluctuations being closely linked to odd-even effects in the charge yields will be discussed further in chapter 4.

2.2 DISCUSSION OF ΔZ

For the interpretation of isobaric charge distributions one has to calculate the fragment charges going with a given fragment mass split. A model serving this purpose and having been proposed very early in the history of fission is the Minimum Potential Energy (MPE) model [PRE 47]. In its most simple version one visualizes the scission configuration by two almost touching spheres and computes the potential energy V of this configuration from

$$V = E_{LD}(Z_1, A_1) + E_{LD}(Z_2, A_2) + Z_1 Z_2 e^2/R \tag{5}$$

for given A_1 and A_2. In eq. (5) E_{LD} are the selfenergies of the two fragments calculated from a Liquid Drop mass formula with the constraints $Z_1 + Z_2 = Z_F$ and $A_1 + A_2 = A_F$, while the last term on the RHS of eq. (5) is the Coulomb interaction of two homogeneously charged spheres. Obviously, in a more realistic approach one has to include deformed fragment shapes, shell and pairing corrections, nuclear interactions and eventually charge polarization inside the fragments. The most probable charge Z_p for given A_1 is then found by minimizing the potential energy V, i.e. from setting

$$\left. \frac{\partial V}{\partial Z_1} \right|_{A_1} = 0 \quad . \tag{6}$$

From the interplay between Coulomb and asymmetry energy it turns out that energetically it is advantageous for half a proton, say, to move from the heavy to the light fragment as compared to an unchanged charge distribution with $Z_1/A_1 = Z_2/A_2 = Z_F/A_F$. This is in rough agreement with experiment for asymmetric fission with $\Delta Z \approx -0.5$ ch. units. However, the MPE model fails to predict correctly the trends and fluctuations of ΔZ with mass and, perhaps even more serious, there is no indication from the model for the change in sign of ΔZ.

Some authors have minimized the potential energy all along the way down from the saddle to the scission point, in order to find out at which point in deformation space theoretical and experimental results agree best. From the comparison it is inferred that fragment charges are not determined at the "geometrical" scission point with neck radius zero. Instead, it appears that the charges are fixed earlier when the neck radius is still 1 - 2 fm [ADE 76, ADE 77, VOL 79]. This point in deformation space coincides roughly with the "exit point", where a rapid necking-in leading to neck rupture takes place. While on the average the agreement with experiment is quite good, the points which have already been criticized in connection with the MPE-model are not resolved either.

Similar objections hold for other theoretical prescriptions how to calculate charge distributions: statistical models [FON 56, IGN 69], scission point models [WIL 76], fragmentation theory [GUP 75, GUP 84]. In none of these theories the manifest change of sign of ΔZ (s. Fig. 1) has been foreseen.

The theoretical models are more fully covered in recent reviews [FREI 84, OGA 85].

3. THE WIDTH OF ISOBARIC CHARGE DISTRIBUTIONS

3.1 EXPERIMENTAL RESULTS

Before coming to some experimental results one should draw attention to the fact that the physical variance $\sigma^2(Z|A') = \langle\{Z(A') - \bar{Z}(A')\}^2\rangle$ pertains to integer charge numbers, while the width parameter $\sigma_z^2(A')$ in a Gaussian model approximation to $Y(Z|A')$ in eq. (2) is calculated for a continuous Z variable. For comparison with the physical variance one has to introduce a grouping correction to the width parameter. In the notation of A.C. Wahl [WAH 88] one finds for the variance

$$\{RMS(A')\}^2 = \sigma_z^2(A') + \frac{1}{12} \quad . \tag{7}$$

The top part of Fig. 1 shows $RMS(A')$ for thermal neutron fission of ^{235}U. The meaning of the symbols has already been given in chapter 2.1. On the average one observes for this reaction $\langle RMS(A')\rangle = 0.63$ ch. units, or $\langle\sigma^2(Z|A')\rangle = 0.4$ (ch. units)^2. Similar to the mean charge, a marked fluctuation shows up in the case of asymmetric fission, while close to mass symmetry the fluctuation disappears. It is also noteworthy that the change in sign of ΔZ correlates with a minimum in the variance.

Again for the reaction $^{235}U(n,f)$, Fig. 3 presents results from a physical measurement for the square root $\sigma(Z|A)$ of the conditional variance at fixed kinetic energies E_L of the light product indicated in the inserts [LAN 80]. Instead of giving on the abscissa the mass number A of the isobaric chain, the corresponding average charge $\bar{Z}_L$ of the light product has been plotted. This brings into evidence that the minima of $\sigma(Z|A)$ correspond to even average charges. It is seen from the figure that similar to ΔZ (s. Fig. 2) the amplitude of the odd-even staggering of $\sigma(Z|A)$ gets larger for increasing kinetic energies, but in contrast to this trend the average value of $\sigma(Z|A)$ is pretty independent from the fragment energies.

This feature is brought about more convincingly in Fig. 4 where $\langle\sigma^2(Z|A)\rangle$ averaged over all measured product mass chains is plotted as a function of the light product kinetic energy [SCH 84]. The full and open points stand for thermal neutron fission of ^{235}U and ^{239}Pu, respectively. It is seen that $\langle\sigma^2(Z|A)\rangle$ slightly decreases the larger the kinetic energies are. However, it has been shown by a Monte-Carlo simulation of neutron evaporation that a constant average variance of the isobaric charge distribution for the primary fragments is in perfect agreement with the observed energy dependence of the variance for the products [CLE 79]. In fact, for low kinetic product energies the corresponding excitation energies and, hence, the number of evaporated neutrons will be large; for large neutron numbers ν also their variance $\sigma^2(\nu)$ is large and, therefore, a given primary mass A' (together with

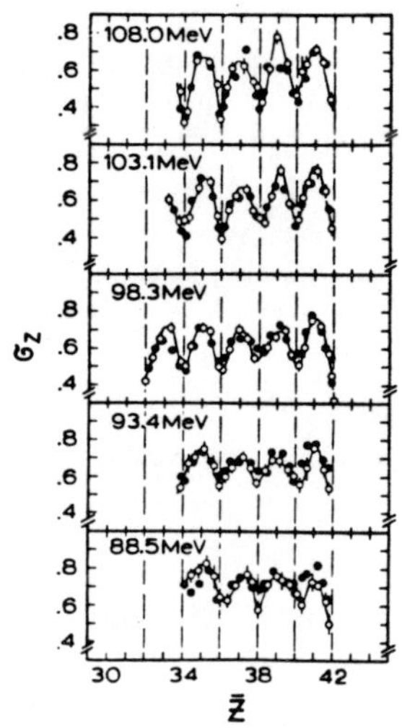

FIGURE 3
Standard deviation $\sigma(Z|A)$ of isobaric product charge distribution for thermal neutron fission of ^{235}U at fixed kinetic energies (s. insert) of light product plotted as a function of average product charge $\bar{Z}(A)$. Open points: experimental data.
Full points: Monte Carlo simulation.

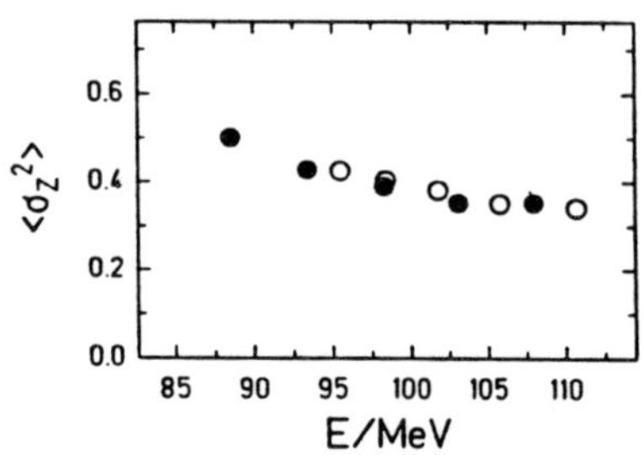

FIGURE 4
Variance $\langle\sigma^2(Z|A)\rangle$ averaged over product mass as a function of light product kinetic energy for thermal neutron fission of ^{235}U (full points) and ^{239}Pu (open circles)

its intrinsic charge distribution) will be spread over a wider range of product masses A. The results of the Monte Carlo calculation are given in Fig. 3 as full points.

The result then is that independent from fragment energy the mass averaged isobaric charge variance in ^{235}U(n,f) is

$$\langle\sigma^2(Z|A')\rangle = 0.40 \pm 0.05 \ \text{(ch. units)}^2 \quad .$$

For the highest kinetic energies studied in Fig. 4 the total excitation energy TXE left to the fragments is just $TXE \approx 12$ MeV. Nevertheless, it is interesting to pursue the evolution of fragment charge distributions up to the highest feasible kinetic energies, that is to say in Cold Fission. It is a notable progress in experimental techniques that has made possible the study of fragment charge data at the low yields prevailing in this limiting case of nuclear fission. It was found that, for all mass chains where events with the total kinetic energy release coming close to the maximum Q-value $Q_{max}(Z|A')$ are to be observed at all, only one charge for given A' survives. As already pointed out in chapter 2.1, these charges are in most cases identical to the charges maximising $Q(Z|A')$. This sounds trivial, but with only one charge left per mass chain the variance $\sigma^2(Z|A')$ vanishes. It means that in Fig. 4 the otherwise constant $\langle\sigma^2(Z|A')\rangle$ abruptly drops to zero upon pushing the fragment energies a few MeV higher than shown. This behaviour has been discovered for all three reactions studied so far: thermal neutron fission of ^{233}U, ^{235}U and ^{239}Pu [SIM 86].

3.2 DISCUSSION OF $\langle\sigma^2(Z/A')\rangle$

In heavy ion physics the charge exchange between reaction partners has become a cornerstone for testing theoretical models against experimental evidence. This has revived the interest in fission fragment charge distributions.

It is rather straightforward to generalize the MPE-model, expounded in chapter 2., for calculating the most probable charge Z_p of a given mass chain A, to yield also the width or variance of isobaric charge distributions. To this purpose one expands the potential $V(Z_1, A_1)$ of eq. (5) around the conditional minimum $\partial V/\partial Z_1 = 0$ for fixed A_1 in terms of a harmonic oscillator and obtains

$$V(Z_1|A_1) = V(Z_p|A_1) + \frac{1}{2}\frac{\partial^2 V}{\partial Z^2}\bigg|_{A_1} (Z - Z_p)^2 \quad . \tag{8}$$

Depending on the liquid drop mass parameters used for the calculations one finds for the stiffness constant c_z of the charge equilibration oscillator

$$c_z \equiv \frac{\partial^2 V}{\partial Z_1^2}\bigg|_{A_1} \approx 3.2 \pm 0.3 \;\; \text{MeV/(ch. units)}^2 \quad . \tag{9}$$

To introduce a width to the charge distribution one places the oscillator in a heat bath of temperature T. Assuming classical statistics to hold one has for the probability $P(Z_1|A_1)$ to observe the charge Z_1 in the mass chain A_1

$$\begin{aligned} P(Z_1|A_1) &\sim exp(-V(Z_1|A_1)/T) \\ &\sim exp(-\frac{1}{2}c_z(Z - Z_p)^2/T) \quad . \end{aligned} \tag{10}$$

Thus, $P(Z_1|A_1)$ is given by a Gausssian

$$P(Z_1|A_1) \sim exp(-(Z - Z_p)^2/2\sigma^2) \tag{11}$$

with the variance

$$\sigma^2(Z_1|A_1) = T/c_z \tag{12}$$

increasing linearly with temperature. However, experiment tells that $\sigma^2(Z_1|A_1)$ is both, independent from the excitation energy of the compound nucleus up to energies of about 50 MeV [MCH 68, FRE 82], and independent from the excitation energies of the fragments [CLE 79].

Therefore, it is felt that one should look for another explanation for the constancy of $\sigma^2(Z_1|A_1)$ as a function of excitation energy. A similar observation in heavy ion collisions has led to the idea that the constant variance is due to the zero-point oscillation of a quantal oscillator [BER 79]. For a quantal harmonic oscillator with the characteristic frequency ω being coupled to a heat bath T, the formula of eq. (12) is only valid in the high temperature limit, i.e. $T \gg \hbar\omega$. In the other extreme of very low temperatures, viz. $T \ll \hbar\omega$, one has a purely quantal fluctuation with

$$\sigma^2(Z_1|A_1) = \frac{1}{2}\hbar\omega/c_z \quad . \tag{13}$$

Evidently, the quantal variance of eq. (13) is independent from T or the excitation energy. Adopting this formula for the fission problem at hand one calculates with $\sigma^2 = 0.4$ and $c_z = 3.2$ the phonon energy $\hbar\omega$ to be $\hbar\omega = 2.56$ MeV.

The next question is whether the above phonon energy lends itself to a physical interpretation. Early in the history of fission it was conjectured [HIL 53] that collective isovector dipole oscillations may be the mechanism behind charge equilibration. This idea has been reconsidered, but there is an obvious difficulty. From the systematics of the phonon energy $\hbar\omega$ for giant $E1$ resonances it is known that $\hbar\omega$ is well reproduced by

$$\hbar\omega = 78/A^{1/3} \text{ MeV} \quad . \tag{14}$$

For a dinuclear system, viz. a fission-prone nucleus close to scission, one would hence expect

$$\hbar\omega = 78/(A_1^{1/3} + A_2^{1/3}) \text{ MeV} \quad . \tag{15}$$

For symmetric fission of ^{236}U one calculates $\hbar\omega \approx 8$ MeV which is clearly at variance with the experimental findings.

A possible way out of the difficulty is the suggestion that mass flow through a neck, getting the narrower the more fission proceeds, should be hindered [UPD 71, BRO 78, BRO 80, NIF 80, MYE 81, HER 81]. The equivalent increase in inertia will bring down the oscillator frequency and hence the charge variance (s. eq. (13)). At the point in the course of fission, however, where the neck gets rapidly pinched the charges no longer adapt and the charge width is frozen out. Quantitative figures for the variance σ^2 are obtained by studying an oscillator with time-dependent mass. The predictions from theory agree reasonably well with experiment indicating that the freezing out point of charge coincides with the exit point where the neck is still rather thick.

A different approach, chosen by several authors, is to consider again a collective charge oscillator in a heat bath provided by the intrinsic degrees of freedom, but to calculate charge fluctuations from transport theory [HOF 79, MAR 82, ADE 85, POP 88].

As an alternative to the explanation of charge distributions in terms of collective modes, a mechanism based on the stochastic exchange of individual nucleons has been proposed for heavy ion reactions [NOR 74, MER 81, SCH 81, DE 82]. In heavy ion collisions the question is still open whether collective or independent nucleon motions are the appropriate way to understand charge equilibration. Unfortunately, to our knowledge, there are no calculations available having applied the nucleon exchange model to tackle charge distributions in fission.

A final comment should address isobaric charge distributions in Cold fission where we have seen that the conditional variance decreases down to zero. This appears to be in contradiction with the idea of zero-point oscillations leading to a constant variance. One should not overlook, however, that in Cold Fission one is pushing the system to a limiting case where all mass-charge fragmentations but one well defined are energetically ruled out. Therefore, if Cold Fission is attained at all, for given masses the charges are uniquely fixed, i.e. the conditional charge variance $\sigma^2(Z|A)$ vanishes.

4. ODD-EVEN EFFECTS IN THE CHARGE YIELDS

4.1 EXPERIMENTAL RESULTS

Repeatedly we have pointed out that the fluctuations of $\bar{Z}(A')$ and $\sigma^2(Z|A')$ are correlated with the odd or even character of the fragment charge Z dominating in a given mass chain A'. We now want to scrutinize in more detail these odd-even fluctuations.

The comprehensive data set of independent yields $Y(Z,A')$ compiled in [WAH 88] allows for a presentation differing from the one in Fig. 1 and bringing into focus more clearly odd-even aspects. Instead of taking together yield data as isobaric charge distributions $Y(Z|A')$ one may reshuffle the yields and consider instead isotopic mass distributions $Y(A'|Z)$. As a result of this analysis Fig. 5 shows the deconvolution of the global product mass yield $Y(A)$ for ^{233}U(n,f) into its isotopic mass distributions $Y(A|Z)$ [WAH 87]. It is obvious from the figure that for asymmetric fission there is a conspicuous odd-even staggering of the charge yields (with even-Z yields being favored as compared to odd-Z yields), while in symmetric fission odd and even Z contribute in equal weight. Evidently this odd-even effect in the charge yields is responsible for the fluctuations observed in $\Delta Z(A')$ and $\sigma^2(Z|A')$ (s. Fig. 1): with no odd-even effect both, $\bar{Z}(A')$ and $\sigma^2(Z|A')$, would vary smoothly with mass number A', while for example for mass chains coinciding with a maximum of the isotopic distribution $Y(A|Z)$ with even Z (s. Fig. 5), it is the preponderant even-Z yield which will control $\bar{Z}(A)$ and bring down the conditional variance $\sigma^2(Z|A)$.

Of course, the odd-even effects in the charge yields are most readily seen in the global charge distributions $Y(Z)$. To have a quantitative measure for the odd-even effect it is customary to introduce

$$\delta := (Y_e - Y_o)/(Y_e + Y_o) \tag{16}$$

as the difference between the even and odd total charge yields Y_e and Y_o, respectively, normalized to their sum. The proton odd-even effect δ (given as a percentage) for thermal

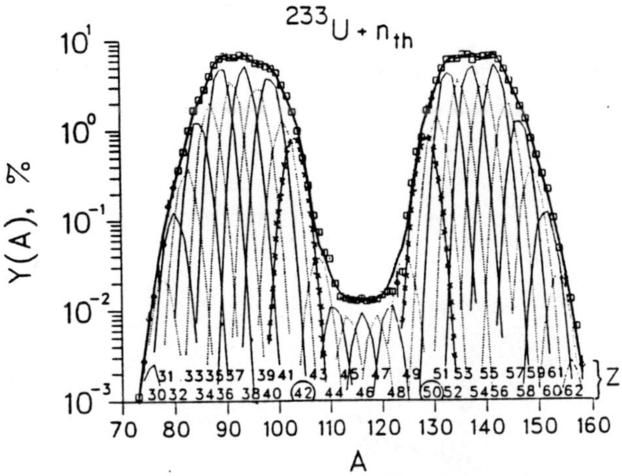

FIGURE 5
Global (squares) and isotopic (lines) product mass distributions for thermal neutron fission of ^{233}U. The mass distributions for Z_L = 42 and Z_H = 50 are enhanced by crosses.

neutron fission of all even-Z_F actinides having been studied so far in reasonable detail is shown in Fig. 6 as a function of fissility Z_F^2/A_F of the fissioning nucleus [LAN 80, DJE 84, SCH 84, DJE 88, QUA 88, SCH 88]. There is an obvious correlation between the odd-even effect and the fissility. No such smooth correlation is found between δ and other variables being characteristic for the fissioning nucleus, like the fission barrier height B_f, the neutron binding energy B_n, or the excitation energy at the saddle point $(B_n - B_f)$.

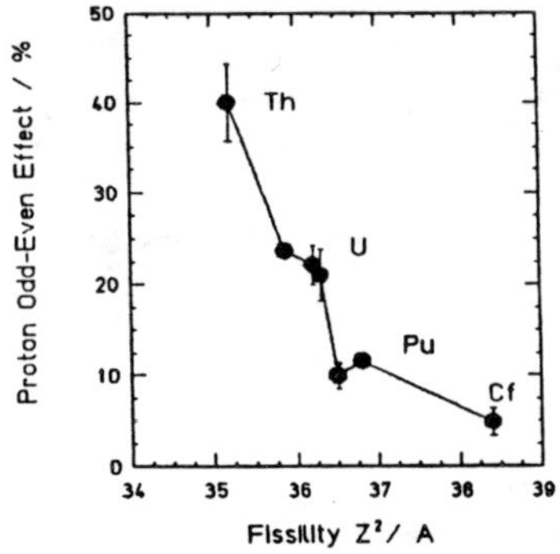

FIGURE 6
The odd-even effect in the charge yields (averaged over kinetic energy) as a function of fissility Z^2/A of the compound nucleus for thermal neutron fission.

For a given compound nucleus the odd-even effect strongly depends on the excitation energy. Experimental results for neutron induced fission of ^{235}U with different incoming energies are depicted in Fig. 7 [AMI 77, LAN 80, MAR 81]. The counting statistics for MeV neutron fission is low and, therefore, the error bars are large, but the decrease of the odd-even effect with increasing compound excitation is manifest. A similar observation has been reported for photofission of U-isotopes [FRE 84], where the odd-even effect has virtually vanished to zero for Bremsstrahlung photons of maximum energy of 12 MeV.

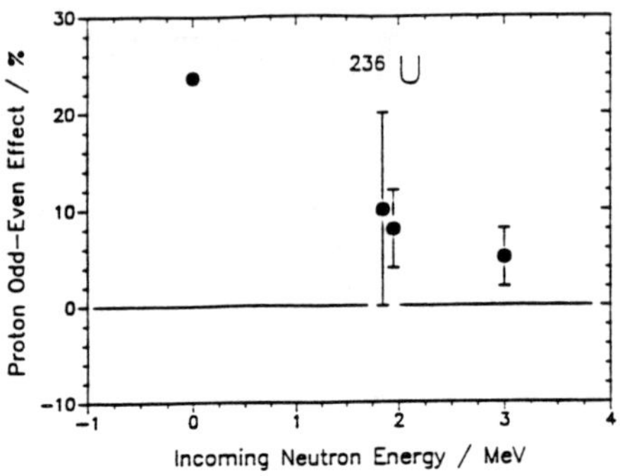

FIGURE 7
The odd-even effect in the charge yield (averaged over kinetic energy) for ^{235}U(n,f) as a function of incoming neutron energy.

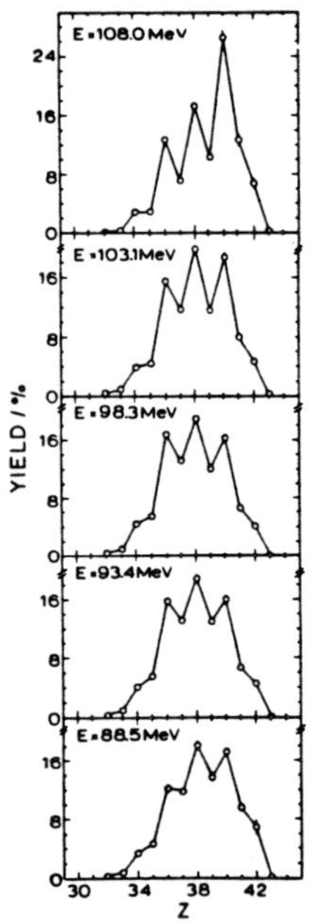

Finally, it is interesting to study charge yields at fixed kinetic energies of the fragments. Measurements for thermal neutron fission of ^{235}U are shown in Fig. 8 with the energies in the inserts giving the kinetic energy of the light fragment having been kept constant. The odd-even effect increases markedly with growing kinetic fragment energy. If, as argued above, odd-even effects in the yields $Y(Z)$ are the origin of the fluctuations in $\Delta Z(A')$ and $\sigma^2(Z|A')$, one should observe similar trends for the latter quantities. In fact, in Figs. 2 and 3 the fluctuations in $\Delta Z(A')$ and $\sigma^2(Z|A')$ are seen to become more pronounced with an increase of the fragment kinetic energy.

FIGURE 8: charge distribution at fixed light product energy (s. insert) for thermal neutron fission of ^{233}U.

4.2 DISCUSSION OF ODD-EVEN EFFECTS

Starting with an even-Z parent nucleus one has to conclude from the mere existence of odd-Z fragments that in fission some pairbreaking always occurs. Since energy has to be supplied to break a nucleon pair it is to be expected that the odd-Z yield will increase with excitation energy and thereby diminish the odd-even effect δ. Indeed, as Fig. 7 tells, by increasing the energy of neutrons inducing fission and hence the excitation energy of the compound nucleus, the odd-even effect δ decreases dramatically. The question then is whether the odd-even effect could not serve as a tool to measure intrinsic excitation energies.

The dependence of δ on the fissility parameter Z^2/A as shown in Fig. 6 is fully in line with this idea. In fact, as borne out by all calculations of the potential energy surface between saddle and scission, the gain in potential energy ΔE_{pot} rises with fissility. With nuclear viscosity coming into play, part of the energy gain will go into intrinsic excitation energy at scission. The increase of excitation energy with increasing fissility will then bring down the odd-even effect δ, as observed experimentally.

For a more quantitative discussion one has to rely on a model, since the large error bars in Fig. 7 prevent an experimental calibration of the δ-thermometre. Several models have been proposed how to link the odd-even effect to the internal excitation energy [GIN 81, MAN 82, NIF 82]. For the present purpose we follow the approach having been put forward by H. Nifenecker [NIF 82]. There the odd-even effect δ is calculated by a combinatorial analysis of pair-breaking. One finds

$$\delta = (1 - 2p\varepsilon q)^{N_{max}} \tag{17}$$

with N_{max} the maximum number of broken pairs (depending on the energy gain ΔE_{pot}), q the probability to break a pair, ε the probability for the broken pair to be a proton pair and p the probability for the nucleons of a broken pair to end up in different fragments. The probability ε is simply set equal to the relative numbers of protons , i.e. $\varepsilon = Z/A$, which for the actinides under study reads $\varepsilon \approx 0.39$. The probability p is set equal to $p = 1/2$. The average internal excitation energy at scission E^*_{SCI} is then obtained from

$$E^*_{SCI} = 2\Delta\langle N\rangle = 2\Delta q N_{max} \tag{18}$$

with Δ the gap parameter. Evidently, in eq. (18) it is assumed that the gap parameter Δ is independent from excitation energy or temperature. This is a simplification which tends to overestimate E^*_{SCI}, but the error introduced is balanced through the factor N_{max} which would increase if the decrease of Δ at high temperatures were taken properly into account. For the gap parameter we adopt the value recommended at the saddle point $2\Delta = 1.7$ MeV [BRI 74]. Combining eqs. (17) and (18) one obtains

$$E^*_{SCI} \approx -4 ln\delta \text{ MeV} \tag{19}$$

where fortunately enough the numerical factor in eq. (19) is rather insensitive to the choice of q (q is set q $\approx 1/2$ in eq. (19)). Bearing in mind that both, the excitation energy at the saddle point E^*_{SAD} and the dissipated energy E_{DIS} contribute to the excitation energy at scission, one has

$$E^*_{SCI} = E^*_{SAD} + E_{DIS} \quad . \tag{20}$$

The excitation energy at the saddle is straightforward to calculate from the incoming neutron binding and kinetic energy B_n and E_n, respectively, and the fission barrier height B_f:

$$E^*_{SAD} = B_n - B_f + E_n \quad . \tag{21}$$

The interesting quantity is E_{DIS}, the energy dissipated between saddle and scission; it is evaluated from eqs. (19) – (21). It should be stressed that in the approach chosen any pair-breaking occuring at the moment of scission, viz. the rapid non-adiabatic necking-in phase, is neglected. Since from studies of Cold Fission, corresponding to the most compact scission configurations with ΔE_{pot} coming close to zero, it is known that odd-Z or odd-N fragments show up as soon as energetically allowed by the Q-value, this should be interpreted to mean that at scission at least one nucleon pair has a good chance to get broken. If this pair-breaking mechanism were always to hold, the present formalism would overestimate the number of pairs broken and hence the energy dissipated through viscosity.

All available data for the odd-even effect δ as measured in thermal and MeV-neutron fission have been analyzed according to the above scheme to yield the dissipated energy E_{DIS} [AMI 77, LAN 80, MAR 81, HAM 83, MEI 83, DJE 84, SCH 84, DJE 88, SCH 88, SRI 88, QUA 88, WAH 88]. The results from the analysis are plotted in Fig. 9 as a function of Z^2/A of the fissioning nucleus (full points: thermal neutron data; open squares: MeV neutron data).

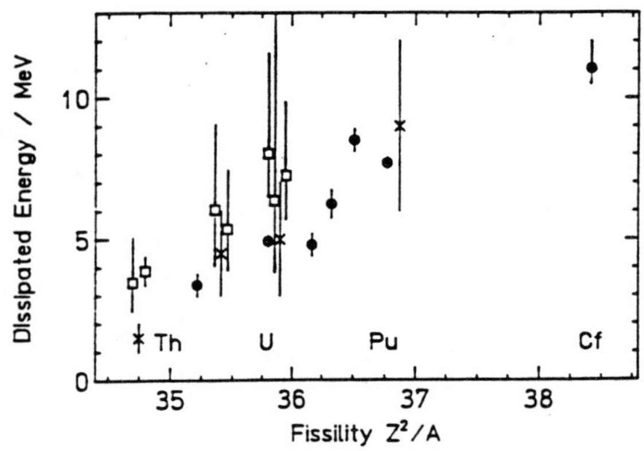

FIGURE 9
The energy dissipated between saddle and scission point for neutron induced fission as a function of z^2/A of the compound nucleus. Explanation of symbols s. text.

A clear correlation between the energy dissipated in the course of fission and the fissility parameter of the compound nucleus shows up. The thermal and fast neutron data fit together rather satisfactorily showing that treating the excitation energy at the saddle point E^*_{SAD} and the dissipated energy E_{DIS} on an equal footing, as expressed in eq. (20), is a sensible approach. The error bars in the figure have been calculated from the experimental error bars of the input data for δ. They do not include the uncertainties due to the model behind the evaluation. Taking the experimental dependence of δ on compound nucleus excitation shown in Fig. 7 at face value the dissipated energies depicted in Fig. 9

appear to be upper limits. It is comforting, however, that dissipated energies coming from an analysis of the symmetric/asymetric mass yield ratio and its dependence on compound nucleus excitation energy [GIN 79, WIL 86] are compatible with the energies from the present study.

In spite of the above systematic uncertainties it is safe to state that the energy E_{DIS} dissipated in the course of fission between saddle and scission is small. From Fig. 9 one concludes that this energy ranges between about 2 MeV for Th and 11 MeV for Cf. To link E_{DIS} to the question of nuclear viscosity one has to compare the figures quoted for E_{DIS} with the available energy, i.e. the potential energy gain ΔE_{pot} between saddle and scission. The energies ΔE_{pot} have to be calculated from theory. An example is provided in [ASG 84]. Relying on these latter calculations the ratio $E_{DIS}/\Delta E_{pot}$ amounts to about 30 %. But again it should be stressed that the various model calculations give quite different values for ΔE_{pot}, with the proper definition of the scission point being an especially subtle point.

The dependence of δ on the kinetic energy release in fission (s. Fig. 8) may be understood from the same physical picture as the dependence on fissility. For a given fissioning species the width in the kinetic energy distribution comes mainly about through fluctuations in the location of the scission point. High kinetic energies correspond to compact scission configurations and correlate to low potential energy gains, and vice versa. Therefore, high kinetic energy events should carry low intrinsic excitation energy and in experiment this becomes observable as a pronounced odd-even effect. The trend to be read from Fig. 8 with δ getting larger for increasing kinetic energy is thus at least qualitatively easy to explain. A more detailed discussion of odd-even effects in connection with the kinetic energy release may be found in [GON 87].

As a last comment we come back to Fig. 5. It was remarked that close to mass symmetry there is no measurable odd-even effect δ in the charge yields. In the light of the foregoing discussion and assuming the observational limit for δ to be $\delta = 3\%$ the dissipated energy at scission must be superior to $E_{DIS} \approx 14$ MeV. This value is for the case of ^{234}U, shown in Fig. 5, much larger than the average $E_{DIS} \approx 5$ MeV typical for asymmetric fission. Yet, this unexpectedly large value for E_{DIS} should be seen together with two other well-known facts: the kinetic energy dip and the increased number of evaporated neutrons for symmetric mass divisions in the fission of actinides. Compared to asymmetric fission, all these observations consistently point to large deformations (small TKE), which due to the easy deformability of symmetric fragments will lead to large potential energy gains (hence large E_{DIS}), while finally the unusually large dissipated energy and the potential energy of deformation will give rise to high excitation energies at infinity (large neutron numbers).

5. SUMMARY

For many years fragment charge distributions from nuclear fission were considered to bear not much interesting information. The charges were known to closely follow the fragment masses and the slight polarization of charge $\Delta Z \neq 0$ was understood from simple models.

However, recent progress in the experimental techniques to measure nuclear charges has brought forward a wealth of new and surprising data:

a) the charge deviation ΔZ changes sign close to mass symmetry,

b) the variance $< \sigma^2(Z|A') >$ of isobaric charge distributions averaged over mass is independent from the fragment excitation energy,

c) the charge yields show a pronounced odd-even effect, with even Z fragmentations being favored, depending on both, the fissility and the excitation energy of the compound nucleus, and the kinetic (or excitation) energy of the fragments.

From theory it is probably the constancy of $< \sigma^2(Z|A') >$ which is best understood, though it is still unclear whether collective oscillations or stochastic exchanges of individual nucleons have to be invoked. The behaviour of ΔZ as a function of fragment mass is predicted by no theory at all, while for the odd-even effect a systematic study is missing.

The odd-even effect δ is analyzed in terms of a simple model allowing to interpret the dependence of δ on fissility and fragment kinetic energy within one single picture. The main result concerns the energy E_{DIS} dissipated as fission proceeds between saddle and scission. It is shown that the odd-even effect is directly linked to the sum of the excitation energy at the saddle point and the dissipated energy. The energy dissipated is found to be low for the (n,f) reactions studied in the lighter actinides: about 2 MeV for Th up to about 11 MeV for Cf. For symmetric mass splits, however, the disappearance of the odd-even effect indicates that here the dissipated energies are large. On the other hand, it should be stressed that even for Cold Fission, where energy dissipation is ruled out by energy conservation, no perfect proton (or neutron) pairing is observed. This could mean that in the very act of scission some pair-breaking occurs.

REFERENCES

[ADE 76] G. D. Adeev, et al., Sov. J. Nucl. Phys. **23** (1976) 15

[ADE 77]. G. D. Adeev, and T. Dossing, Phys. Lett. **66B** (1977) 11

[ADE 85]. G. D. Adeev, et al., Sov. J. Nucl. Phys. **42** (1985) 25

[AMI 77]. S. Amiel, et al. Phys. Rev. **C15** (1977) 2119

[ASG 80a]. M. Asghar, Z. Physik. **A296** (1980) 79

[ASG 84]. M. Asghar, and R. W. Hasse, J. de Physique **C6** (1984) 455

[BER 79] M. Berlanger, et al., Z. Physik. **A291** (1979) 133

[BRI 74]. H. C. Britt, and J. R. Huizenga, Phys. Rev. **C9** (1974) 435

[BRO 78]. U. Brosa, and H. J. Krappe, Z.Phys. **A284** (1978) 65

[BRO 80] U. Brosa, and D. H. E. Gross, Z. Physik. **A294** (1980) 217

[CLE 79]. H.-G. Clerc et al., Proc."Physics and Chemistry of Fission" Jülich 1979, Vol.2, p65

[CLE 86] H.-G. Clerc et al., Nucl. Phys. **A452** (1986) 277

[DE 82]. J. N. De, Phys. Lett. **113B** (1982) 455

[DJE 84]. M. Djebara, et al., Nucl. Phys. **A425** (1984) 120

[DJE 88]. M. Djebara, et al., Nucl. Phys. to be published

[DEN 86] H. O. Denschlag, Nucl. Science and Eng. **94** (1986) 337

[FON 56] P. Fong, Phys. Rev. **102** (1956) 434

[FRE 82]. D. De Frenne, et al., Phys. Rev. **C26** (1982) 1356

[FRE 84]. D. De Frenne, et al., Phys. Rev. **C29** (1984) 1908

[FREI 84]. H. Freiesleben, and J. V. Kratz, Physics Reports **106** (1984) 1

[GIN 79] J. E. Gindler, et al., Proc. "Physics and Chemistry of Fission" Jülich 1979, Vol. 2, p 111

[GIN 81] J. E. Gindler, et al., Lect. Notes in Physics 158, Springer 1982, p 145

[GON 87] F. Gönnenwein, et al., Proc. "XVIIth Int.Symp.on Nuclear Physics" Gaussig/GDR

[GUP 75] R. K. Gupta, et al., Phys. Rev. Lett. **35** (1975) 353

[GUP 84]. R. K. Gupta, and D. R. Saroha, Phys. Rev. **C30** (1984) 395

[HAM 83]. C. Hamelin, PHD thesis, Univ. of Grenoble, 1983

[HER 81] E. S. Hernandez, et al., Nucl. Phys. **A361** (1981) 483

[HER 82]. G. Herrmann, and N. Trautmann, Ann.Rev.Nucl. and Part.Sci **32** (1982) 117

[HIL 83] D. L. Hill, and J. A. Wheeler, Phys. Rev. **89** (1953) 1102

[HOF 79]. H. Hofmann, et al., Z. Phys. **A293** (1979) 229

[IGN 69] A. V. Ignatyuk, Sov. J. Nucl. Phys. **9** (1969) 208

[LAN 80]. W. Lang, et al., Nucl. Phys. **A345** (1980) 34

[MAN 82b]. G. Mantzouranis, and J. R. Nix, Phys. Rev. **C25** (1982) 918

[MAR 81]. G. Mariolopoulos, et al., Nucl. Phys. **A361** (1981) 213

[MAR 82]. E. Martschev, and K. Pomorski, Acta Phys. Pol. **B13** (1982) 747

[MCH 68] J. A. McHugh, and M. C. Michel, Phys. Rev. **172** (1968) 1160

[MEI 83]. H. H. Meixler, et al., Can. Journ. Chem. **61** (1983) 665

[MER 81] A. C. Merchant, and W. Nörenberg, Phys. Lett. **104B** (1981) 15
and. Z.Phys. **A308** (1982) 315

[MYE 81] W. D. Myers, et al., Phys. Lett. **98B** (1981) 1

[NIF 80]. H. Nifenecker, Journal de Physique Lettres **41** (1980) L47

[NIF 82] H. Nifenecker, et al., Z. Phys. **A308** (1982) 39

[NOR 74] W. Nörenberg, Phys. Lett. **53B** (1974) 289

[OGA 85]. Yu. Ts. Oganessian and Yu. A. Lazarev, Treatise on Heavy-Ion Science, D.A. Bromley ed., Plenum Press 1985, Vol. 4, p 3

[POP 88] A. Pop, et al., Z. Phys. **A329** (1988) 357

[PRE 47] R. D. Present, Phys. Rev. **72** (1947) 7

[QUA 88]. U. Quade, et al., Nucl. Phys. **A487** (1988) 1

[SCH 81]. W. U. Schröder, et al., Phys. Lett. **98B** (1981) 355

[SCH 84]. C. Schmitt, et al., Nucl. Phys. **A430** (1984) 21

[SCH 88]. P. Schillebeeckx, PHD thesis, Univ. of Gent/Belgium 1988

[SIM 86] G. Simon, et al., Proc. "Journées d'Études sur la Fission", Arcachon France 1986, B. Leroux ed., CENBG 8722, B21

[SRI 88] A. Srivastava, and H. O. Denschlag, Radiochim. Acta **46** (1989) 17

[UPD 71] W. E. Updegraff, and D. S. Onley, Nucl. Phys. **A161** (1971) 191

[VOL 79]. N. G. Volkov, et al., Sov. J. Nucl. Phys. **29** (1979) 603

[WAH 85]. A. C. Wahl, Phys. Rev. **C32** (1985) 184

[WAH 87]. A. C, Wahl "New Directions in Physics" Acad.Press, New York 1987, p 163

[WAH 88] A. C. Wahl, At. Data and Nucl. Data Tables **39** (1988) 1

[WIL 76] B. D. Wilkins, et al., Phys. Rev. **C14** (1976) 1832

[WIL 86]. B. D. Wilkins, and J. E. Gindler, Proc. Int. Symp. "Nuclear Fission and Heavy-Ion-Induced Reactions", Rochester 1986, Harwood Acad. Publ., p 101

DOUBLE-ENERGY, DOUBLE-VELOCITY MEASUREMENT OF FISSION FRAGMENTS FROM THERMAL NEUTRON INDUCED FISSION

I. Kimura
Department of Nuclear Engineering, Kyoto University,
Yoshida, Kyoto 606, Japan
Y. Nakagome
Research Reactor Institute, Kyoto University,
Kumatori-cho, Osaka 590-04, Japan
I. Kanno
Japan Atomic Energy Research Institute, Tokai-mura, Ibaraki 319-11, Japan

Abstract

Number of prompt neutrons as a function of individual fragment mass $\nu(m^*)$ was measured for the thermal neutron-induced fission of ^{233}U and ^{235}U. By measuring the velocities and energies of two fission fragments simultaneously, preneutron-emission fragment mass m^* and postneutron-emission fragment mass m were obtained. Thereafter $\nu(m^*)$ was deduced by subtracting m from m^*. The fragment velocity was measured by the time-of-flight (TOF) method, and the start time was detected by a very thin plastic scintillator film detector. A silicon surface barrier detector was used to measure the fragment energy, which was also used as a stop detector of the TOF. The result of $\nu(m^*)$ for $^{233}U(n,f)$ was in good agreement with other data in the heavy fragment region, but was 20 to 50% larger than those in the light one. $\nu(m^*)$ for $^{235}U(n,f)$ showed a factor of 1.5 to 2 larger in the light fragment region and smaller in the heavy one than the other data.

1. INTRODUCTION

For the study of the shape and excitation energy of a fission fragment, it is important to know the number of prompt neutrons as a function of the individual preneutron-emission (initial) fragment mass $\nu(m^*)$. Average total number of prompt neutrons are reported by many authors[1], but the data of $\nu(m^*)$ are very few.

There are two methods to measure $\nu(m^*)$, namely direct and indirect ones. In the direct method the number of fission neutrons is measured with a large scintillation detector, whereas $\nu(m^*)$ is determined by the mass difference between the initial and postneutron emission (final) fragment mass numbers in the indirect method. These fragment masses are obtained by measuring the velocities and the kinetic energies of both fragments. The velocity of the fragment is usually measured by the time-of-flight (TOF) method[2] and the energy is determined by an ionization chamber or a silicon surface barrier detector (SSB). Andritsopoulos[3] first measured $\nu(m^*)$ for the thermal neutron induced fission of ^{235}U by double-velocity double-energy method. In their experiments, some ingenious devices were used to get the start signal of the TOF. Since then, no similar work has been reported as far as in our knowledge.

We developed the double-velocity double-energy measurement system for fission fragments using thin film detectors (TFD) as the start time detectors of TOF[4]. A newly constructed super mirror neutron guide facility at the Kyoto University Reactor (KUR) was used to get a low background thermal neutron beam.

In this paper, we describe the system briefly and the results of $\nu(m^*)$ for the thermal neutron induced fission of ^{233}U and ^{235}U with this system.

With the energy balance equation, the total kinetic energy is estimated using the $\nu(m^*)$-value and the result is compared with the experimental result. Also using the energy balance equation, the $\nu(m^*)$-values are calculated by assuming the thermal equilibrium at the scission point.

2. EXPERIMENTALS

Principle

Two SSBs were used for measuring the fragment kinetic energy.

The following relationship between the energy and the pulse height is given in [5] as

$$E_i = (a + a'm_i)x_i + b + b'm_i ,$$

where x_i is the pulse height, a, a' and b' are energy calibration constants. The final fragment mass can be obtained as

$$m_i = \frac{ax_i + b}{v_i^2/2 - a'x_i - b'} .$$

The number of emitted neutrons $\nu(m^*)$ is determined by subtracting m_i from the initial fragment mass m_i^*.

Apparatus

The experimental arrangement is shown in Fig.1. We used the super mirror neutron guide tube facility of KUR[6] as a thermal neutron source with extremely low background. The neutrons were guided for 11.7m from the KUR core by this facility . The size of the beam and the neutron flux at the uranium target position are 1cm x 7cm and 5×10^7 n/cm^2/s, respectively.

The fragment velocity was obtained by the TOF method. The flight path was 59cm. To obtain the start time of the fragment, very thin plastic scintillator film detectors (TFDs) were used. On each end of the flight tubes, three SSBs (ORTEC F-series detectors) were mounted to detect the stop time and also to measure the fragment kinetic energy. All of the TFDs, SSBs and a uranium target were installed in a evacuated chamber and two flight tubes. The vacuum was kept at about 10^{-4} Torr.

The uranium target was made from a very thin nitrocellulose film in which organic uranium compound was dissolved. The diameter of the target was 8mm, and its thickness was 7 μgU-233/cm^2 and 9 μgU-235/cm^2, which was determined by counting the number of fission events. The enrichment of ^{233}U was 99.47% and that of ^{235}U 90%.

The TFD illustrated in Fig.2 was placed at 3cm apart from the uranium target on each fragment path. TFD was first studied by Muga et al.[7], and we developed a fabrication technique of thinner (less than $\sim$ 50 μg/cm^2) one. The characteristics of TFD were reported elsewhere [8,9]. The TFD used consists of a thin plastic scintillator film (NE-102), two lucite light guides and photomultipliers. The film 20 μg/cm^2 thick was sandwiched by two hemicylindrical light guides. The thickness of the film was determined by measuring the energy loss of the alpha particles of

^{252}Cf in the film. A hole of 1cm diameter was bored in the light guide for the fission fragments passing through the film.

The electronics of the double-velocity double-energy measurement system is shown in Fig.3. The timing signals, start and stop signals, were fed into a time-to-amplitude converter (TAC) through a timing amplifier and a constant fraction discriminator (CFD). Only coincided four signals, two TAC signals and two energy ones, were taken by the Multi Parameter Data Acquisition System [10]. The data were accumulated in 1024 channels for each parameter and stored on a floppy disk event by event.

Calibration and Correction

In order to obtain a, a′ and b′ in the previous formula for determining the fragment energy, we carried out the energy calibration for SSBs with the spontaneous fission fragments of ^{252}Cf.

The time calibration was achieved by making use of the fission fragments of ^{252}Cf and a delay line (Td = 26.302 ns). The time resolution of the TFD-SSB system was determined with the alpha particles (6.118 MeV) of ^{252}Cf. Finally the time resolution of the system was found to be about 13.3 ns.

The energy loss of the fragment in the uranium target and TFD was calculated by the Beth's formula for each fragment.

We followed the way proposed by Mueller et al.[11] to obtain the plasma delay of SSBs. In this calculation, the average neutron emission number was taken from the work of Apalin et al.[1].

3. RESULTS AND DISCUSSION

The mean values and errors concerning fragment mass, kinetic energy and velocity are listed in Table 1. The values of the present results agree well with other works within the errors.

For $^{233}U(n,f)$, the number of prompt neutrons as a function of initial fragment mass is shown in Fig.4. $\nu(m^*)$ is indicated with a solid circle and the total number of prompt neutrons as a function of heavy fragment mass with an open circle. The results of Apalin et al.[1] and of Milton et al.[12] are also plotted, which were obtained by the direct method. Present result is close to the data of Milton et al. in the heavy fragment region, while in the mass region of 100 -110 the result is close to the data of Apalin et al. In the light fragment region the result disagrees with both data, and is 20 to 50% larger than those of Apalin et al. The $\bar{\nu}(m^*)$-values are listed in Table 1.

For $^{235}U(n,f)$, the present result of $\nu(m^*)$ is shown in Fig.5. In the light fragment region, our result is approximately a factor of 1.5 to 2 larger than the other data. On the contrary, it is a factor of 1.5 to 2 smaller than those in the heavy fragment region. The $\bar{\nu}(m^*)$-values are also listed in Table 1.

Using the $\nu(m^*)$-value, we can estimate the total kinetic energy E_k^* from the energy balance equation. The estimated total kinetic energy and the measured one in this experiment are shown in Fig.6 (a) for ^{233}U and (b) for ^{235}U. The agreement between these values is satisfactory.

We tried to calculate $\nu(m^*)$ with the above energy balance equation by assuming that the thermal equilibrium could be achieved at the scission point between the two fragments. However the calculated $\nu(m^*)$-values for both ^{233}U and ^{235}U cases are considerably different from the experimental results. It may suggests that we have to consider the effect of deformation of the fragments at the scission point.

REFERENCES

1. V.F. Apalin Yu.N. Gritsyuk, I.E. Kutikov, V.I. Lebedev and L.A. Mikaelian, Nucl. Phys., 71, 553 (1965).
2. J.C.D. Milton and J.S. Fraser, Can. J. Phys., 40, 1626 (1960).
3. G. Andritsopoulos, Nucl. Phys., A94, 537 (1967).
4. I. Kanno, Y. Nakagome and I. Kimura, J. Nucl. Sci. Tech., 25, 111 (1988).
5. H.W. Schmitt, W.H. Gibson, J.H. Neiler, F.J. Walter and T.D. Thomas, Proc. Symp. Physics and Chemistry of Fission, Vol.1, p.531 (1965) IAEA.
6. T. Akiyoshi, T. Ebisawa, T. Kawai, F. Yoshida, M. Ono, S. Mitani, T. Kobayashi and S. Okamoto, Proc. 1985 Seminar Nucl. Data, JAERI-M-86-080, p.380 (1986).
7. M.L. Muga, D.J. Burnsed, W.E. Steeger and H.E. Taylor, Nucl. Instr. Methods, 83, 135 (1970).
8. I. Kanno and Y. Nakagome, ibid., A244, 551 (1986).
9. I. Kanno and Y. Nakagome, ibid., A251, 108 (1986).
10. S. Uehara, T. Seo, H. Iimura and S. Yamada, Ann. Rep. Res. Reactor Inst., Kyoto Univ., 18, 178 (1985).
11. R. Mueller, A.A. Naqvi, F. Kaeppeler and F. Dickmann, Phys. Rev., C29, 885 (1984).

12. J.C.D. Milton and J.S. Fraser, Proc. Symp. Physics and Chemistry of Fission, Vol.2, p.39 (1965) IAEA.

Table 1. Mean Values and standard deviations of the initial fragment mass, velocity, energy and average number of prompt neutrons for the thermal neutron-induced fission of ^{233}U and ^{235}U.

	$^{233}U(n,f)$		$^{235}U(n,f)$		
	Present	Milton and Fraser[2]	Present	Milton and Fraser[2]	Andritsopoulos [3]
$\langle m_L^* \rangle$ (amu)	94.36±0.23	94.57±0.10	95.93±0.23	96.08±0.10	95.87±0.07
$\sigma(m_L^*)$ (amu)	6.21	5.85	6.26	5.77	6.3
$\langle m_H^* \rangle$ (amu)	139.64±0.23	139.43±0.10	140.07±0.23	139.92±0.10	139.87±0.07
$\sigma(m_H^*)$ (amu)	6.21	5.85	6.26	5.77	6.3
$\langle V_L^* \rangle$ (cm/ns)	1.44	1.442	1.423	1.409	1.415
$\sigma(V_L^*)$ (cm/ns)	0.072	0.068	0.075	0.062	0.051
$\langle V_H^* \rangle$ (cm/ns)	0.975	0.963	0.977	0.966	0.97
$\sigma(V_H^*)$ (cm/ns)	0.073	0.070	0.073	0.071	0.086
$\langle E_L^* \rangle$ (MeV)	101.38±0.72	99.9±1.0	100.55±0.71	99.8±1.0	99.08±0.07
$\sigma(E_L^*)$ (MeV)	5.95	6.2	6.57	6.0	5.9
$\langle E_H^* \rangle$ (MeV)	68.78±0.34	67.9±0.7	69.25±0.34	68.4±0.7	68.190±0.10
$\sigma(E_H^*)$ (MeV)	7.48	7.3	7.73	7.5	8.7
$\langle E_K^* \rangle$ (MeV)	170.16±0.80	167.8±1.7	169.80±0.79	168.3±1.7	167.45±0.2
$\sigma(E_K^*)$ (MeV)	10.65	11.2	11.59	11.41	14.2
$\bar{\nu}(m_L^*)$	1.68±0.69	1.40	1.94±0.69	1.19	1.16±0.09
$\bar{\nu}(m_H^*)$	0.85±0.72	1.03	0.89±0.72	1.23	1.27±0.09

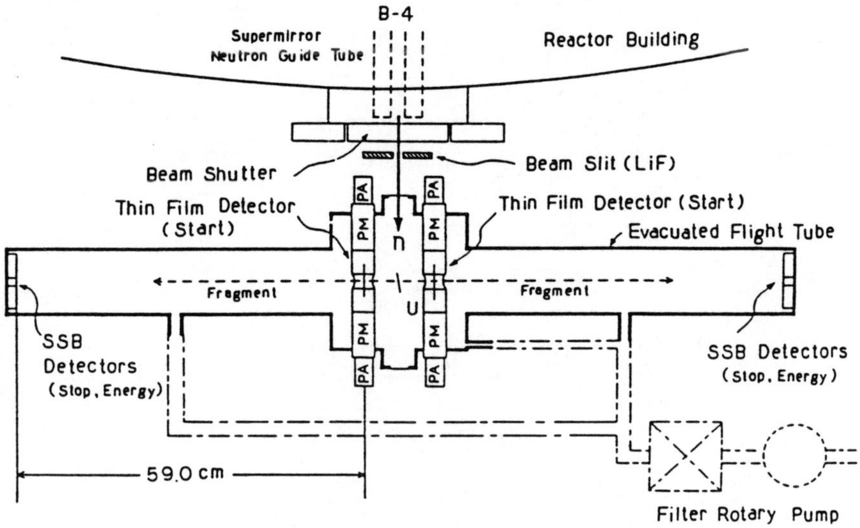

Fig.1 Experimental arrangement for double-velocity double-energy measurement

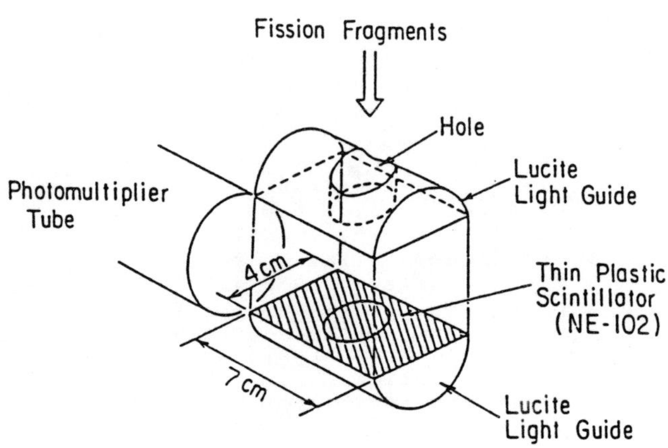

Fig.2 Illustration of thin plastic scintillator film detector (TFD)

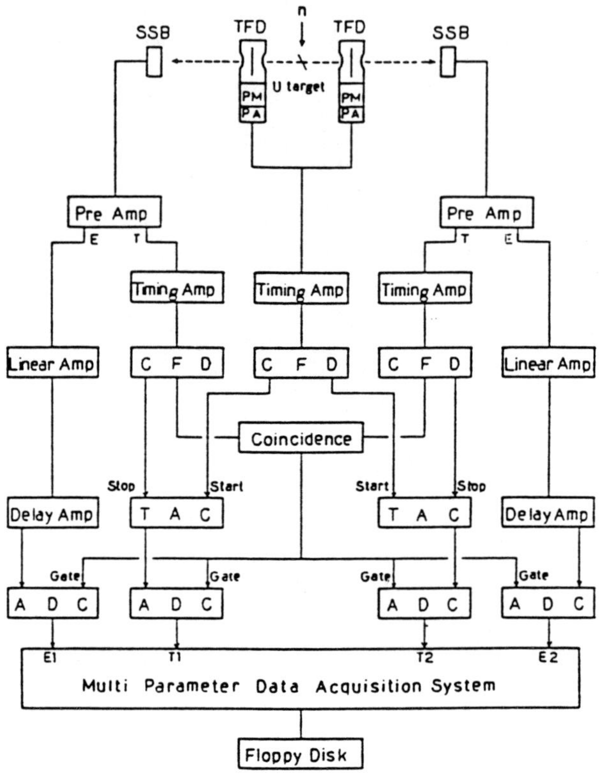

Fig.3 Block diagram of electronic circuit for double-velocity double-energy measurement

TFD : Thin film detector
SSB : Silicon surface barrier detector
PM : Photomultiplier
PA and Pre Amp : Preamplifier
Timing Amp : Timing amplifier
CFD : Constant fraction discriminator
Linear Amp : Linear amplifier
Delay Amp : Delay amplifier
Coincidence : Coincidence circuit
TAC : Time to amplitude converter
ADC : Analog to digital converter

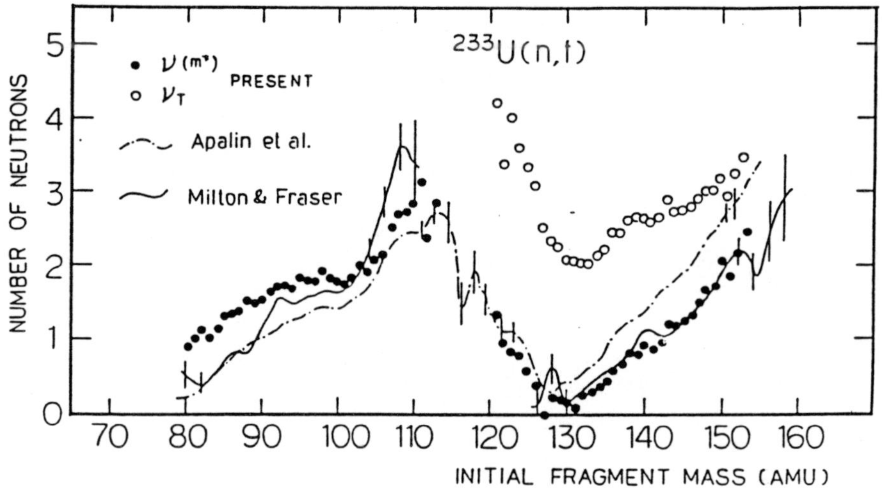

Fig.4 Prompt neutron distribution $\nu(m^*)$ for thermal neutron-induced fission of ^{233}U

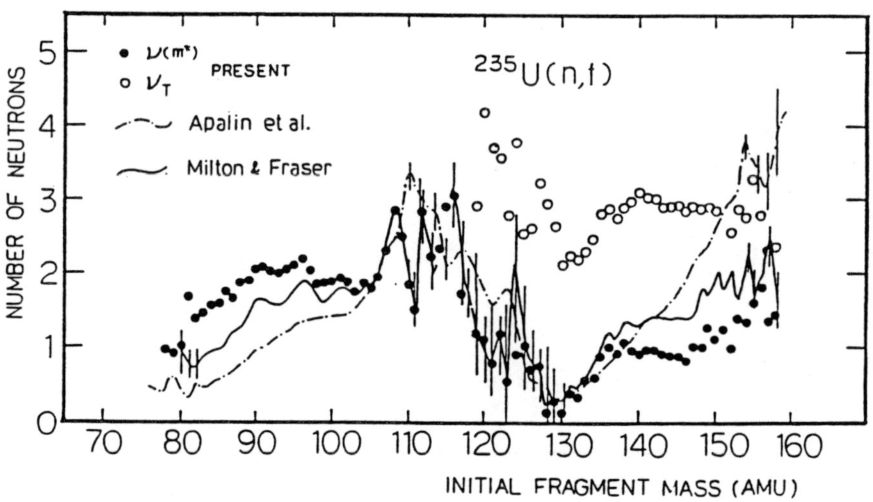

Fig.5 Prompt neutron distribuition for thermal neutron-induced fission of ^{235}U

ODD-EVEN NEUTRON AND PROTON EFFECTS IN LOW ENERGY NUCLEAR FISSION

H.N. Erten[1] and N.K. Aras[2]
[1] Bilkent University, Ankara, Turkey
[2] Middle East Technical University, Ankara, Turkey

INTRODUCTION

Since the discovery of fission fifty years ago(1) the study of mass, charge and kinetic energy distributions of the fission fragments continued at many laboratories around the world. Radiochemical methods have been applied for many years to measure the independent fission yields, IFY, and hence charge distribution of the fission products. However, these measurement are very difficult and only a small fraction of the IFY could be measured. Even for the most heavely studied case of thermal neutron induced fission of ^{235}U, U235 T, there exist many mass chains where fission yield measurements especially at the symetric region and the wings are inadequate.

Recently all available IFY data in U235 T and other low energy nuclear fission systems were complied and published by Wahl(2). These new data include not only radiochemical measurements, but also those obtained from mass spectrometric and fission-product recoil separators, LOHENGRIN and HIAWATA. Ameil and Feldstein(3) studied odd-even systematic in elemental yields in U235 T and U233 T. This effect was further explored by many investigators(4,5). With the availability of more IFY data we have reinvestigated the odd-even effect in low energy fission.

CALCULATIONS

The odd-even effect, simply defined as the enhancement of IFY of even Z, EOZ, and even N, EON, products relative to the odd ones, may be determined from

$$\text{EOZ or EON} = 2Ye/(Ye-Yo) \qquad (1)$$

where Ye and Yo are independent yields of even and odd products respectively. We have calculated EOZ and EON factors using Eq(1) for low energy fission which have enough experimental data. Furthermore we have calculated EOZ and EON values for nine low energy fissioning systems using Zp model of Wahl(6).

RESULTS AND DISCUSSION

A summary of yield data and the results of EOZ and EON calculations in U235 T, U233 T and PU239 T using equation(1) are given in Table 1. It is observed that the yield data is most complete only in the case of U235 T. There is about 10% independent yield yet to be measured in the heavy mass peak region of U235 T. For U233 T and PU239 T the light mass peak region is almost complete whereas about 50% of the yields have not been measured yet in the heavy mass peak region. In the case of Cf252 SF most of the experimental yield data is still missing. This is also the case for all the other low energy fissioning systems. Therefore we presented in Table 1 only the U235 T, U233 T and PU239 T results.

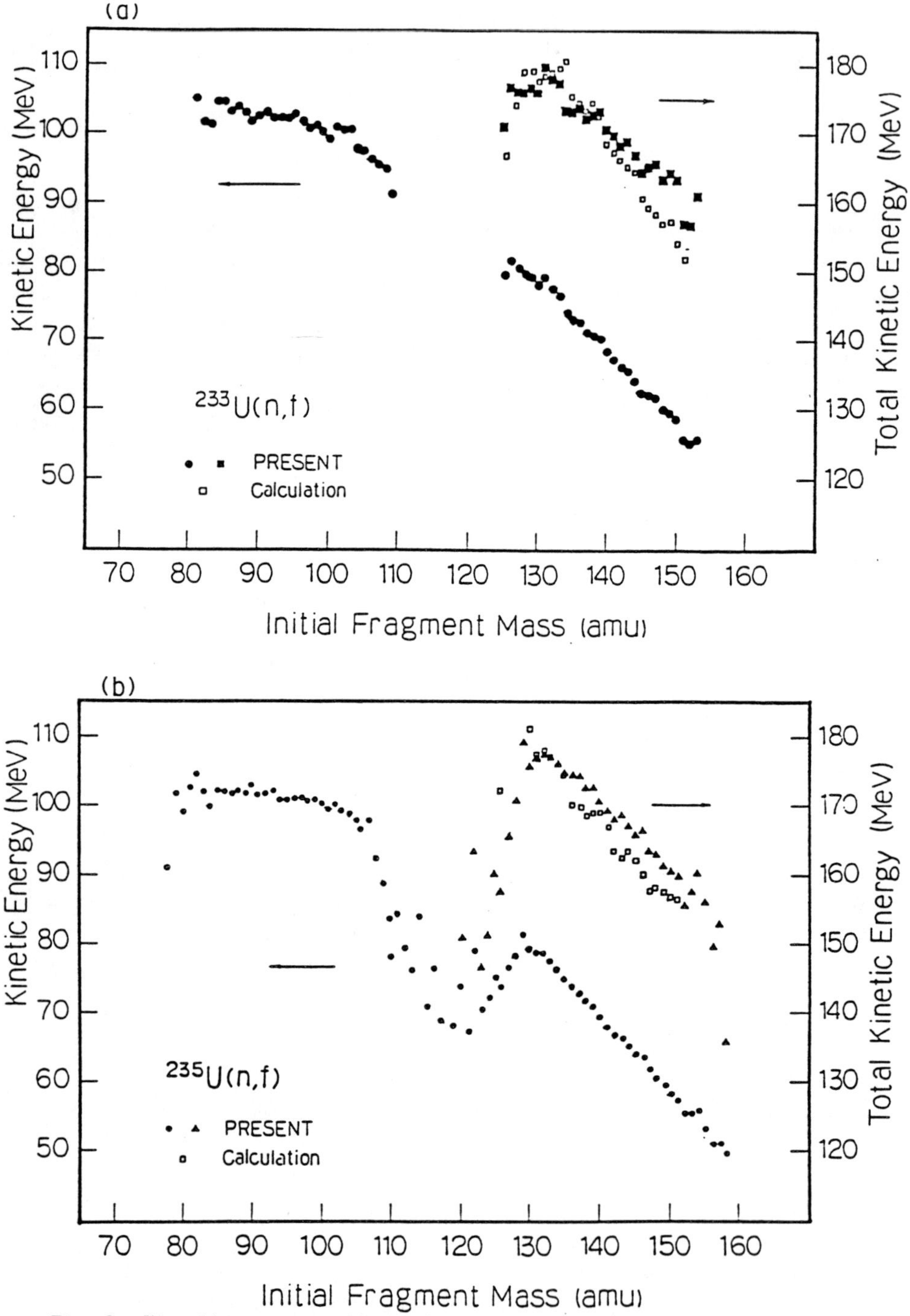

Fig.6 Kinetic energy distribution of fragment for thermal neutron-induced fission of (a): ^{233}U and (b): ^{235}U

It is interesting to observe that the number of even Z or even N products are appreciately less than the corresponding odd ones. On the basis of the pairing effect in nuclear structure, one would expect the opposite to be true, as in the case of stable even and odd isotopes found in nature: 212 versus 54 respectively. It is beleived that this is a result of the nuclear shell effects. On the other hand the total yields of even Z products are significantly greater than the odd Z species as expected. From these considerations we have defined a new term; the Yield Density, YD, which is the independent yield per nuclide. It is seen in Table I that the yield density is above one for even Z products and considerably below one for odd Z products.

The EOZ values calculated using equation(1) as given in Table 1 are pronounced only for U235 T and U233 T whereas the EON values are very close to one indicating no effect. If we repeat the EOZ calculations using YD values in equation (1), the EOZ values obtained are 1.31, 1.23 and 1.15 for U235 T, U233 T and PU239 T respectively. These increased values are probable due to shell effects. We thus estimate yield enhancement due to shell effects to be about 6% in U235 T, U233 T and PU239 T. This estimated magnitude is obtained considering all the fission products in all regions. If one takes into account that the shell effects are most pronounced at the shell regions and their complementaries only, then the magnitude of the shell effects should increase considerably within these regions and would be negligible outside.

These considerations are in line with the findings of Clerc et al.(4) who found Shell-effect modulations in EOZ and EON values for the light peak products separated using LOHENGRIN, and the observations of Wahl(12) in the symmetric region.

Since the available experiment yield data is not sufficient for the above described type of treatment of the even-odd effect in the various low energy fissioning systems, one may use various models of charge distribution to estimate this effect. The Zp model of Wahl(6) is the most widely used one for this purpose. Using the method of least squares and the available yield data for light and heavy peak fission products one obtains value of parameters that describe nuclear charge distribution.

The results of odd-even EOZ and EON effects from the Zp model for several low-energy fissioning system are illustrated in Fig. 1. It is observed that the EOZ effect is well established and pronounced in U235 T and U233 T only, as was found in Table 1. The EON effect appears to be quite small in all cases.

Furthermore recently Wahl(12) using available data in the symmetric region in the U235 T, did not find any odd-even effects in this region. Reisdorf et al. (13) studying nuclear charge distribution in U235 T, U233 T, PU239 T and Cf252 SF by measuring K X-ray yields did not find any significant odd-even effects.

These findings suggest that the odd-even effects in low energy fission must be taken into account with caution and corrections should not be applied to all products in all systems. Apart from U235 T and U233 T, experimental data is lacking to enable an analysis as given in Table 1. More experimental yield measurements should be carried out as suggested by the International Nuclear Data Committee(14) in order to clarify and understand this significant effect.

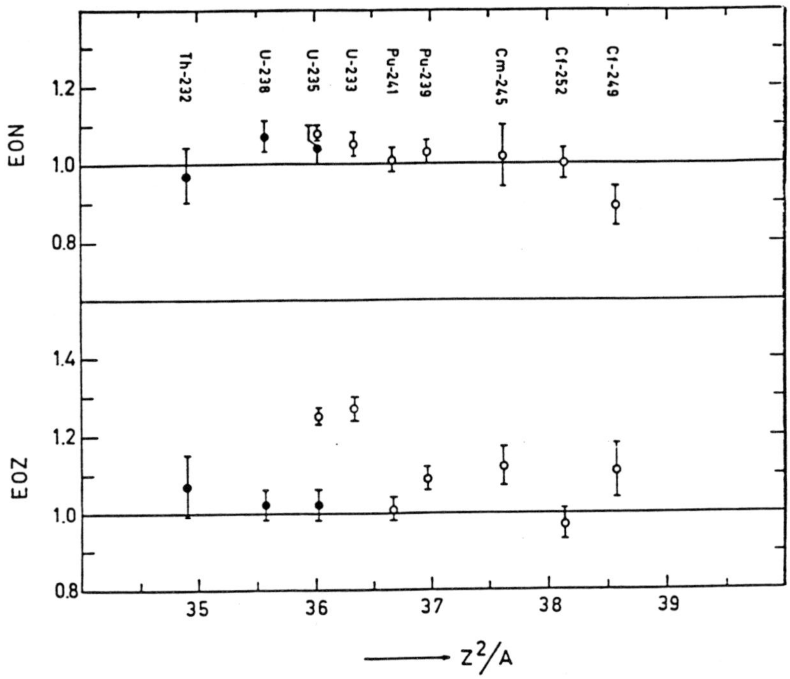

FIGURE 1

Comparison of EOZ and EON values for various Low Energy Fission Processes Calculated According Zp model of Wahl (6). The Experimental Yields are from References (2,5-11)

o Thermal Neutron Fission • Fast Neutron Fission

REFERENCES

1- O.Hahn and F.Strassman, Naturwiss. 27, 11(1939).
2- A.C.Wahl, Atomic Data and Nuclear Data Tables 39, 1(1988).
3- S.Amiel and H.Feldstein, Phys.Rev.C, 11, 845(1975).
4- H.G.Clerc, W.Lang, H.Wohlfarth, K.H.Schmidt, H.Schrader, K.E.Pherdekamper, R.Jungman, Z.Physik A274, 203 (1975).
5- C.Chung, Radiochim. Acta 39, 113(1986).
6- A.C.Wahl J.Radioanal.Chem. 55, 111(1980).
7- R.Brissot, J.Cancon, Ch.Ristore, J.P.Bocquet and A.Moussa, Nucl.Phys. A282, 109(1977)
8- H.R.Von Gunten, K.F.Flynn and L.E.Glendenin Phys.Rev. 161, 1192(1967).
9- J.K.Dickens and J.W.Connell, Phys.Rev. C23, 331(1981).
10- H.H.Maixler, Ph.D.Thesis, Institute für Kernchemie, Universitat Mainz, (1980) unpublished
11- H.N.Erten, A.Grütter, E.Rossler and H.R.von Gunten, Phys.Rev.C 25, 2519(1982).
12- A.C.Wahl, Phys.Rev.C, 32, 184(1985).
13- W.Reisdorf, J.P.Unik, H.C.Griffin and L.E.Glendenin, Nucl.Phys. A177, 337(1971).
14- M.Lammer, IAEA International Nuclear Data Committee, INDC(NDS)-191/G+P(1988).

TABLE 1

Summary of Independent Yield Data and Results of EOZ,EON and Yield Density Calculations in U235 T, U233 T and PU239 T

Parameter Examined	U235 T	U233 T	PU239 T
Number:			
Even Z products	103	84	73
Odd Z products	121	90	84
Even N products	99	88	74
Odd N products	125	86	83
Yield:			
Even Z products	117.74	89.72	82.23
Odd Z products	73.15	60.43	69.30
Even N products	98.92	72.80	76.48
Odd N products	91.97	79.35	75.05
EOZ	1.23	1.20	1.08
EON	1.04	0.96	1.01
Yield Density: (Yield/nuclide)			
Even Z products	1.14	1.07	1.13
Odd Z products	0.60	0.67	0.83
Even N products	1.00	0.83	1.03
Odd N products	0.73	0.92	0.90
Total Yield:			
Light Peak	101.57	98.74	95.39
Heavy Peak	89.32	51.41	56.14
Total	190.89	150.15	151.53

ENERGY BALANCE IN MeV NEUTRON INDUCED FISSION

A. Ruben, H. Märten, and D. Seeliger
Technische Universität, Dresden, GDR

ABSTRACT : General trends of energy balance changes with increasing incidence energy are described in the framework of a simple scission point model including semi-empirical temperature-dependent shell correction energies. In particular, the different behaviour of the total kinetic energy (TKE) dependence for several fissioning nuclei (Th,U,Pu) is explained.

1. INTRODUCTION

In the last years total kinetic energy data of fission fragments were measured by several groups. These data have shown a systematical behaviour of $\overline{TKE}$ (averaged over all fragments) as a function of incidence energy E_i for different fission nuclei. In the case of rather light fission nuclei, e.g. Thorium, $\overline{TKE}$ increases with incidence energy. For the neutron induced fission of ^{235}U all experiments show an increase of $\overline{TKE}$ up to 1 MeV incidence energy and a decrease beyond. Heavier fissioning nuclei, for example Plutonium, are characterized by a descent of the $\overline{TKE}(E_i)$ function. Furthermore, TKE curve as a function of mass asymmetry A_1/A_2 shows a typical structure caused by the fragment shell structure. In this paper, these trends in TKE will be described adopting the Two-Spheroid Model (TSM) [1].

2. THE TWO-SPHEROID MODEL

The TSM [4] as a simple scission point model is based on a general energy balance,

$$\bar{Q}\left(\frac{A_1}{A_2}\right) + B_i + E_i = \underbrace{E_{pre} + \rlap{\overbrace{\phantom{E_{coul} + E^{(1)}_{def} + E^{(2)}_{def}}}^{F}}E_{coul}}_{TKE\left(\frac{A_1}{A_2}\right)} + \underbrace{E^{(1)}_{def} + E^{(2)}_{def} + \overbrace{E_{dis} + E_h}^{E^{(1)}_{int}+E^{(2)}_{int}}}_{E^*\left(\frac{A_1}{A_2}\right)} \qquad (1)$$

According to Terrell [2] we describe the fissioning system with two spheroidically shaped fragments nearly touching at the scission point. The nuclear forces between the fragments cause a small distance at the place of contact.

The deformation energy $E^{(i)}_{def}$ is assumed to be quadratic in radius change with reference to spherical nuclei and linear in a deformability parameter $\alpha^{(i)}$. The main part of TKE is the coulomb interaction E_{coul} of the two charges effectivaly located in the fragment centres. The second part, the pre-scission kinetic energy E_{pre}, corresponds to the relativ motion of the nascent fragments in the fission direction developed in the path between saddle and scission point. Here we use the results of Gönnenwein [3] which also gives the dissipative energy E_{dis} describing the nuclear friction.

As the third term of the fragment excitation energy E^* it's necessary to include for induced fission reactions the heat energy above the fission barrier E_h. In the double humped fission barrier of actinide nuclei there is a relative slow transition of the fissioning system from saddle point A to B. Thus the conclusion can be drawn, that all at the second barrier not redundant collective degrees of freedom have changed into heat energy. Therefore, E_h is determined by the difference between the excitation energy of the fission nucleus, i.e. kinetic energy E_i plus binding energy B_i of the incidence particle, and the second fission barrier $E_{f,B}$ including an energy loss caused by pair breaking. Depending on the value of the pairing gap Δ_p and $E_{f,B}$ exists an incidence energy limit E_i^{lim}. below that the heat energy E_h remains equal zero. Above the limit, E_h increases with E_i.The excitation energy depenence of the pairing gap Δ_p is given by a rough approximation according to Kristiak [4].

The classical principle of minimizing a nuclear potential F, including E_{def} and E_{coul} of complementary fragments results in a set of equations which allow to deduce TKE and the ratio $E_{def}^{(1)}/E_{def}^{(2)}$, if knowing the fragment deformability parameters $\alpha^{(i)}$. According to Kildir and Arras [5], the latter quantities are connected with the fragment shell correction energies $\delta_w^{(i)}$ with reference to the liquid-drop-model deformability.

In the calculation of the semi-empirical shell correction energies $\delta_w^{(i)}$ it is necessary to take into acount the diminution of shell effects due to the intrinsic temperature τ_{int} at the scission point (Bohr and Mottelson [6]). The partition of the intrinsic excitation energy E_{int} containing $E_{h,B}$ and E_{dis} on the fragments is fixed by the condition of equal intrinsic temperatures of complementary fragments $\tau_{int}^{(1)}=\tau_{int}^{(2)}$.

Within this formalism, a set of semi-empirical shell correction energies, reduced to $\tau_{int}=0$, is calculated on the basis of the well known fragment data for $^{252}Cf(sf)$ and $^{235}U(n_{th},f)$. This parameter set will be used in any application.

3. TRENDS OF TKE CHANGES

The total kinetic energy as a function of mass split shows a strong dependence on the fragment shell structure (see Fig.3). The TKE maximum is caused by the extremly negative shell correction energies of the heavy fragments with mass numbers close to 130. Their strong stiffness leads to a nearly spherical shape and thus a smaller distance between the charges. At higher incidence energies shell effects vanish. Thus the negative shell energies increase in conjunction with a descent in the stiffness and, consequently, in the TKE maximum.

In the symmetric mass region charcterized by positive $\delta_w^{(i)}$ the washing-out of shell effects is connected with smaller shell energies. Therefore, stiffness and TKE increase with excitation energy.

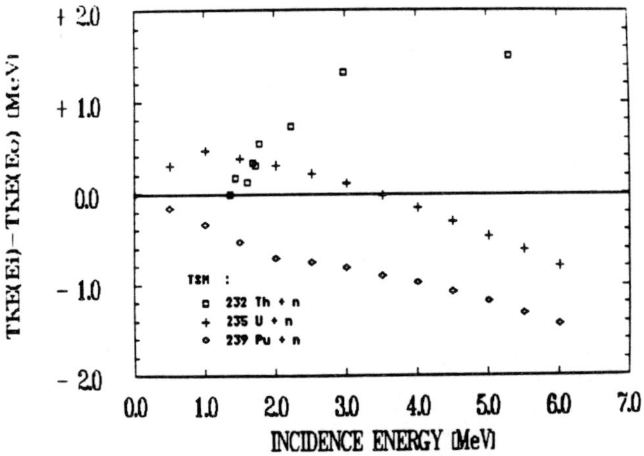

Fig.1 Calculated average total kinetic energy as difference to the value of minimal incidence energy for different incidence energies and fissioning nuclei

Figure 1 shows the dependence of $\overline{TKE}$ (TKE averaged over all fragments) on incidence energy E_i for neutron induced fission of ^{232}Th, ^{235}U and ^{239}Pu. The different behaviour for the several fission nuclei is caused by two reasons :

(i) Up to E_i^{lim} the heat energy E_h is equal zero because of increasing pair breaking. Consequently, TKE increase up to this point in the whole mass region.

(ii) Above E_i^{lim}, the heat energy E_h as a part of the intrinsic excitation energy E_{int} is growing with E_i. The washing out of shell effects leads in this case to a diminution of the TKE maximum and, consequently of the average value $\overline{TKE}$.

The value of E_i^{lim} depends on both the fission barrier hight $E_{f,B}$ and the pairing gap Δ_p. The decrease of the second fission barrier $E_{f,B}$ with mass number for fission actinide nuclei represents one reason of the different TKE curve behaviour.

For the fission of a relative heavy actinide nucleus, e.g. ^{239}Pu(n,f), the value of $E_{f,B}$ is so small that $\overline{TKE}$ decreases within the whole incidence energy range. Figure 2 shows the calculated $\overline{TKE}$ for this fission reaction in comparision with experimental data. The qualitative dependence on incidence energy E_i is well established.The shift between the experimental and calculated data is caused by the mass yield curve.

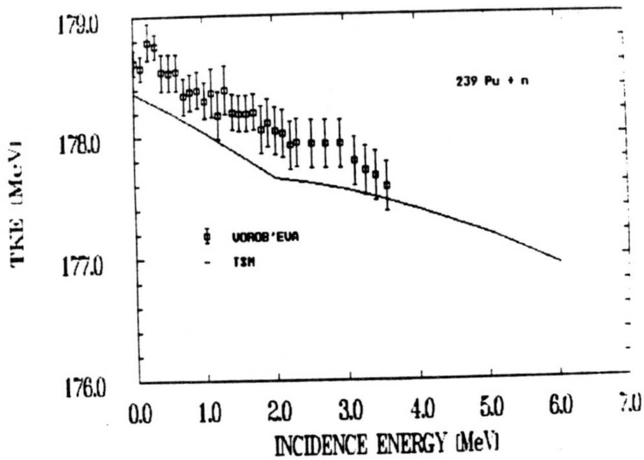

Fig:2 Calculated and experimental $\overline{TKE}$ data as a function of the incidence energy for $^{239}Pu(n,f)$ (exp.data [7])

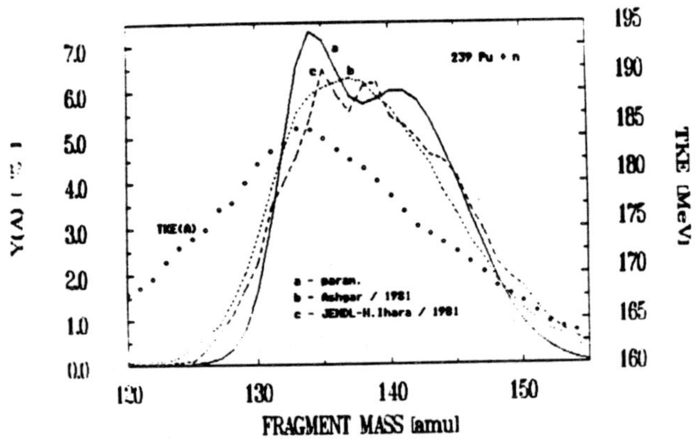

Fig:3 Experimentally received mass distributions (ref.[8],[9]) and TKE as function of fragment mass for thermal induced fission of ^{239}Pu

As depicted in Fig.3 $\overline{TKE}$ is very sensitive to the fragment mass distribution. The plotted different mass distributions cause a difference $\Delta\overline{TKE} \approx 1$ MeV.The neutron induced fission of ^{235}U is characterized by nearly equal barrier hights. The relative high magnitude of the second barrier leads to a disappearing of heat energy E_h, and consequently an increasing $\overline{TKE}$ below 1.2 MeV incidence energy. Above this limit $\overline{TKE}$ decrease. Both the calculated and the experimental data shown in Fig.4 confirm this statement. The shift of the data of Meadows [11] may be caused by different mass distributions.

The neutron induced fission of ^{232}Th as a relative light actinide nucleus is characterized by a large second barrier. Thus the incidence

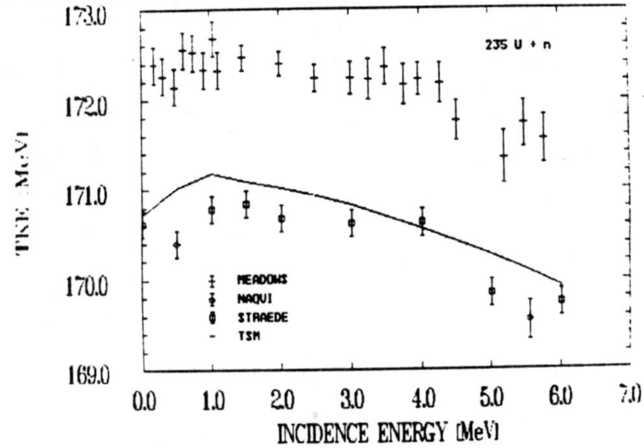

Fig:4 Calculated and experimental $\overline{TKE}$ data as a function of incidence energy for $^{235}U(n,f)$ (exp.data [10],[11],[12])

energy limit E_i^{lim} up to that E_h=0 and $\overline{TKE}$ rise is high (≈ 3 MeV). Additional a removal in the mass distribution toward fragments with higher TKE effects a continuation of $\overline{TKE}$ increase.

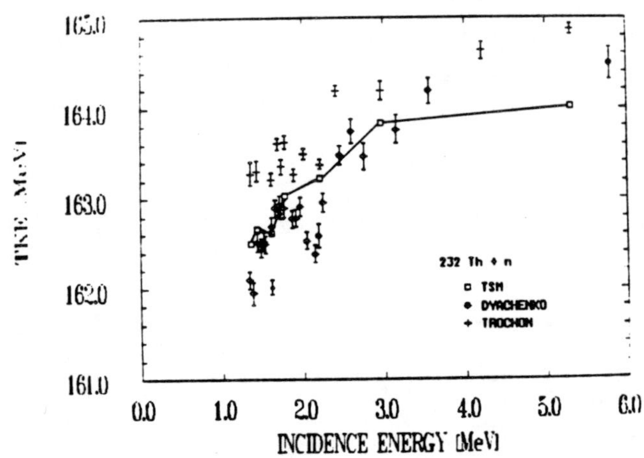

Fig.5 Calculated and experimental $\overline{TKE}$ data as a function of the incidence energy for $^{232}Th(n,f)$(exp.data [13],[14])

Thus, as represented in Fig.5 the $\overline{TKE}$ rise up to an incidence energy of about 5 MeV. Further, the beginning second chance fission reaction (n,n´f) cause a dip in the $\overline{TKE}(E_i)$ curve.
As shown above the fragment mass distribution Y(A) determined by shell effects has a great influence on $\overline{TKE}$. This is one reason of the $\overline{TKE}$-fluctuations in the fission resonance region.
The last figure shows the calculated $\overline{TKE}$ as a function of E_i for some $^{235}U(n,f)$ resonances. The changes in the energy balance due to E_i can be neglected in our model. Thus the alterations of the calculated $\overline{TKE}$ are caused by changes in the mass distribution.

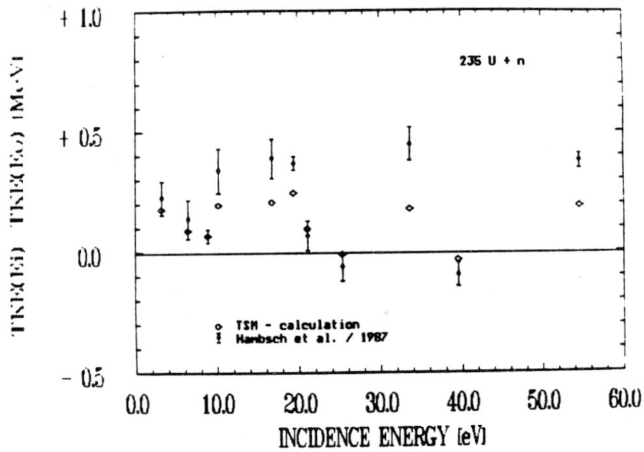

Fig.6 Calculated and experimental $\overline{TKE}$ data as a function of the incidence energy for selected $^{235}U(n,f)$ resonances

Here we have used parameparametrized mass yields according to Hambsch [15]. The striking trends in the fluctuations are reproduced. The differences between calculated and experimentally received values may be caused by a decline of Y(A) changing due to the parametrization4.

4.CONCLUSIONS

The Two-Spheroid model TSM reproduces general trends in the total kinetic energy dependence both on the fragment mass and incidence energy. The $TKE(A_1/A_2)$ function is determined by the fragment shell structure. A realistic description of the different TKE behaviour in the range of small incidence energies requires an accurate energy balance, i.e. pairing energy and barrier shape. A further reason for the $\overline{TKE}$ changes are shifts in the mass distribution. The TSM is used as basis for several applications, e.g. for calculation of fission neutron spectra and multiplicities.

5. REFERNCES

[1] H.Märten,A.Ruben,D.Seeliger, Proc.IAEA Advisory Group Meeting on Properties of Neutron Sources, Leningrad 1986,IAEA-TECDOC-410 (1987) p.153

[2] J.Terrell, Proc. IAEA Symp. on Physics and Chem. of Fission, Salzburg 1965 (IAEA Vienna, 1965), Vol.II, p.3

[3] F.Gönnenwein, 17th Int.Symp.on Nuclear Physics, Gaussig (1987) in print

[4] I.Kristiak, 5th Int.Symp.on Nucl.Ind.Reactions, Smolenice 1988, in print

[5] M.Kildir and N.K.Aras, Phys.Rev.C 25 (1982) p.365

[6] A.Bohr and B.R.Mottelson, Nucleare Structure (Benjamin, New York, 1975), Vol.II, p.607

[7] V.G.Vorobeva′ et al. "Average Kinetic Energies of Fission Fragments" INDC-128, G+Sp. Vienna, IAEA 1979

[8] M.Ashgar et al., Nucl.Phys. A368 (1981) p.319

[9] H.Ihara et al., JENDL Report, JAERI-M-9715 (1981)

[10] Ch.Straede et al.,Nucl.Phys. A462 (1987) p.85

[11] J.W.Meadows and C.Butz-Jorgenson, Report ANL/NDM-64, Jan.1982

[12] A.A.Naqqui et al.,Phys.Rev. C29 (1984) p.885

[13] J.Trochon et al., Nucl.Phys. A318 (1979) p.63

[14] N.P.Diachenko et al., Jad.Konst. 1/40 (1981) p.53

[15] F.-J.Hambsch, Internal Report VG/1796/87

FORMATION OF THE FRAGMENT MASS AND ENERGY DISTRIBUTIONS IN FISSION OF NUCLEI LIGHTER THAN RADIUM

M.G. Itkis, V.N. Okolovich, and G.N. Smirenkin*
Institute of Nuclear Physics, Alma-Ata, USSR
*Institute of Physics and Energetics, Obnisk, USSR

The results of experimental investigations of the mass-kinetic energy distributions of fission fragments for nuclei lighter than Ra are analysed for the excitation energy interval from threshold up to 60 MeV. The influence of the energy surface structure near the saddle point on the formation of general properties of fragments is considered.

Last decade investigations at the Alma-Ata cyclotron were concentrated at the fission process of the cold and heated nuclei lighter than Ra. The measurements covered the mass range from ^{162}Dy to ^{213}At in the fission reactions with protons, ^{3}He and ^{4}He ions. Present report summarizes results, to our view, of the most importance related to the formation of the mass and energy distributions over the broad range of excitation energy /1/.

Investigations of sufficiently cold nuclei yielded two major results. Firstly, we have found and studied in detail the fission asymmetry of the light pre-actinides from ^{204}Pb to ^{213}At and defined physical boundary of existence of this phenomenon at the mass number A=200. Secondly, experimental results appeared to be in excellent agreement with the calculation of Pashkevich /2/, which predicted the existence of valleys in the surface of potential energy

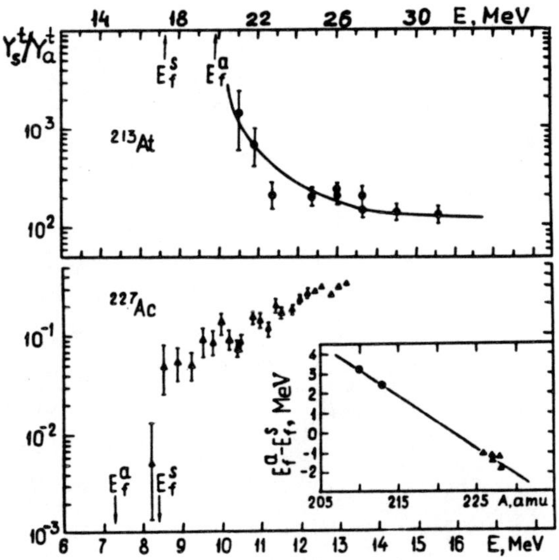

Fig.1 Yield ratio $Y_s/Y_a \simeq \sigma_f/\sigma_f^a$ as a function of the excitation energy E for ^{213}At and ^{227}Ac. The insert shows the difference between the heights of the fission barriers $E_f^a - E_f^s$ for nuclei lighter than Ra.

of deformation. Our investigations made it possible to trace the manifestations of this phenomenon in two prominent stages of evolution of the fissile nucleus. It manifests itself by the difference in the heights of symmetric and asymmetric fission barriers at the saddle point and by difference in the kinetic energies at the scission point.

Figure 1 shows the symmetric to asymmetric yield ratio Y_s/Y_a as a function of energy for ^{213}At. It indicates that the mass-asymmetric valley in the potential energy surface occurs by 2-3 MeV higher than the symmetric one, what is in agreement with theoretical predictions /2/. In contrast, for ^{227}Ac $E_f^a < E_f^s$ and asymmetric fission prevails in the vicinity of threshold. Hence, two modes of fission pass the fission barrier by two distinct paths (valleys). This property is inherent for Ra and Ac as well as for a group of lighter nuclei in the Pb region. Whether these valleys mix during descent (as it was assumed in /3/) or they are completely separated and lead to the distinct scission points /2/? It becomes apparent from fig.2. It shows the cross-sections of the two-dimentional surface of the experimental mass and energy distributions $N(M,E_k)$, which are represented as numbers of fragments integrated over mass and energy intervals of 1 amu and 2 MeV, respectively. It is seen that the single component results from the pure symmetric mode at M=125 (a); the three-humped distribution appears when the three components $Y_{a0}(M)$, $Y_{a1}(M)$ and $Y_s(M)$ contribute with proximate weights (b); the narrow high-energy component Y_{a1} vanishes in the next mass interval (c); in the highly asymmetric mass range $M \geqslant 138$ symmetric component $Y_s(M)$ vanishes and only Y_{a0} remains (d).

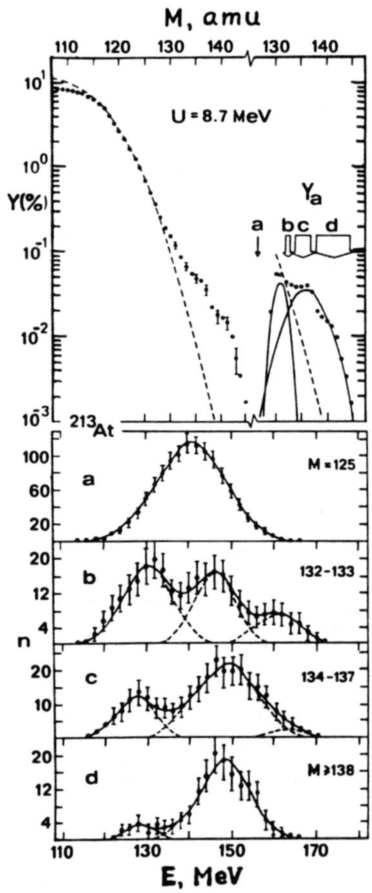

Fig.2 Mass and energy distribution from fission of ^{213}At for various intervals ΔM and ΔE_k.

Thus, distinctions between properties of mass-asymmetric and mass-symmetric valleys are confirmed from saddle to scission. The split of asymmetric component into Y_{a0} and Y_{a1} (anomalously high kinetic energy) is an unique phenomenon together with its analogue in fission of nuclei with $Z \simeq 100$.

In both cases this property is related to the respective valleys arising during descent due to shell effects. The narrow high-energy component corresponds to the valley of spherical fragments and compact scission configurations; the broader component is related to energetically preferable but less compact fragment configurations with closed deformed shells. In contrast, the mass symmetric valley of light nuclei corresponds to mid-shell fragments and positive shell corrections.

As the excitation energy increases, the shell effects vanishes and the liquid-drop valley remains only. Investigations of its properties provided better understanding of the fission process. The data on variances of fragment mass distributions of heated nuclei were helpfull here.

Figure 3 shows experimental data on parameter $q=\Theta/\sigma_M^2$ as a function of the fissility parameter together with the calculations of nuclear stability against asymmetric deformation by Strutinsky Q(X) /4/ and Cohen-Swiatecki K_3(X) /5/. This comparison suggests that saddle point of fissile system affects crucially upon formation of the mass distribution in fission of nuclei up to Th at least. In addition, this conclusion is confirmed by analysis of dependence of σ_M^2 on nuclear temperature over broad mass interval from ^{201}Tl to ^{227}Ac. The time of mass asymmetric oscillations τ_M was estimated from the experimental value of $\hbar\omega_M=(0.5\pm0.2)$MeV and appeared to be equal to $5\cdot10^{-21}$s, what is in agreement with the data from studies of quasi-fission. The dissipation time τ_E was obtain-

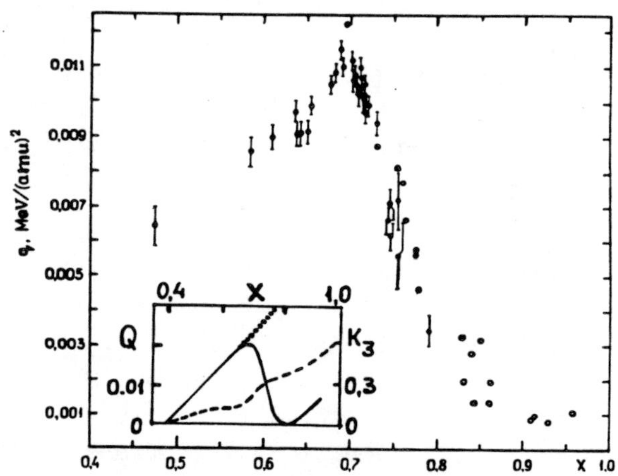

Fig.3 Nuclear stiffness q as a function of fissility X. The insert shows the parameter of stability against mass asymmetric deformation. For saddle point: solid curve - Q(X) /3/, dashed curve - K_3 /4/; for scission point: dotted curve - Q(X) /3/.

ed from the temperature dependence of kinetic energy variance and appeared to be in an order of magnitude lower ($\hbar\omega_E$=3.5 MeV).

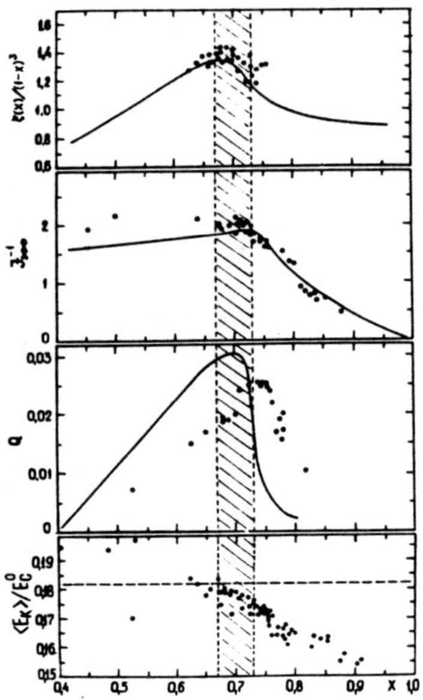

Fig.4 Comparison between experimental characteristics of the fission process and the LDM calculations (see text).

Experimental characteristics of the fission process of the heated nuclei are summarized in fig.4:fission barriers represented as $\xi(X)/(1-X)^3$, inverse effective moments of inertia $J^{-1}_{ef}(X)$, parameters of stability against mass-asymmetric deformation $Q(X)=A^2 \cdot q(X)/32E^0_s$, fragment kinetic energy (in units of E^0_c). Experimental results are compared with respective static LDM calculations by Strutinsky with Γ=-0.1. Whereas general agreement is easily seen, the divergence between experiment and calculations increases from up to down as the dynamic effects increase. It is therefore clear fromthis comparison that carefull description of the properties of symmetric fission of sufficiently heated nuclei in the framework of LDM (static and dynamic especially) might provide new step towards quantitative theory of asymmetric fission, which would be more complicated due to shell effects.

REFERENCES

1. Itkis M.G., Okolovich V.N., Smirenkin G.N. Particles and Nuclei. Joint Inst.Nucl.Res., Dubna, Moskva, Atomizdat. 1988, V.19, P.701.
2. Pashkevich V.V. Nucl.Phys. 1971, V.A169, P.275.
3. Mustafa M.G., Mosel U., Schmitt H.W. Phys.Rev. 1973, V.C7, P.1519.
4. Strutinsky V.M. Zh.Teor.Eksp.Fiz. 1963, V.45, P.1900.
5. Cohen S., Swiatecki W.J. Ann.Phys. 1963, V.22, P.42.

A NEW APPROACH TO DETERMINE ELEMENTAL YIELD, CHARGE POLARIZATION AND ODD-EVEN EFFECTS IN FISSION

T. Datta, P.K. Pujari, S.K. Das, B.S. Tomar, S.B. Manohar, A. Goswami, H. Naik, and Satya Prakash
Radiochemistry Division, Bhabha Atomic Research Centre, Trombay, Bombay-400 085, India

Abstract: A new approach is formulated to determine fission fragment elemental yield and charge dependent kinetic energy distributions along with charge polarization parameter. The experimental technique involves measurement of coincident fragments kinetic energies using a pair of Si-surface barrier detectors and mass dependent energy calibration. Subsequently the coincident kinetic energy spectrum is deconvoluted in terms of fragment charge (Z) using model-independent mass-charge correlation followed by empirical corrective method for detector response function.

INTRODUCTION

Experimental investigations on fission fragment charge, mass and kinetic energy distributions provide insight into the mechanism of the nuclear fission process involving large-amplitude nuclear motions together with single particle effects[1)]. The present work describes a new approach for deconvolution of the coincident fragment kinetic energy spectra as a function of fragment charge to obtain the elemental yield distributions in ^{252}Cf(sf). The assumed mass-charge correlation leads to evaluation of charge polarization parameter which is treated as a free parameter, as well as odd-even effect showing also the influence of fragment shell effects.

EXPERIMENTAL

A ^{252}Cf source of strength 1.7 ng (1055 fission/sec) was electrodeposited on a nickel foil of thickness 50 $\mu g/cm^2$. The coincident fragment kinetic energies were measured using a pair of Si-surface barrier detectors of resolution ~ 15 keV at 5.49 MeV α peak of ^{241}Am and leakage current < 2 μA in a vacuum chamber. Singles kinetic energy spectrum in each of the detectors was seen to have the shape-characteristics within the prescribed limits of ideal spectrum[2)] on similar detectors. The coincidence kinetic energy spectra of the fission fragments from ^{252}Cf(sf) were acquired on a 4x4K channel multi parameter system in dual parameter mode using 128x128 channel array with coincidence time width of 0.5 μsec. The difference in the pulse rise time of the two detectors was < 5 nsec.

The well established[2)] mass dependent energy calibration procedure was used to deduce the kinetic energy (E_i) of any fragment of mass (M) at a channel X_i

$$E_i = (a+a'M)X_i + (b+b'M) \tag{1}$$

where a, a', b, and b'are the usual calibration constants deduced as prescribed by Schmitt et al.[2)]. In the fission process the mass-to-charge (M/Z) ratio of the fission fragments is expected to be the same as that of the fissioning nucleus on the basis of UCD hypothesis[3)]. In low energy fission, however, charge distribution studies[3)] show that the most probable charge (Z_p) for a given isobaric mass (M) chain often differs from the UCD value by $\pm$ ΔZ referred to as charge polarization.

In view of these observations, in the present work a model independent mass-to-charge correlation is proposed as

$$M = (Z \pm C)*(A/Z)_{FN}. \tag{2}$$

C is a free parameter, equivalent to the charge polarization parameter the (+)ve or (-)ve sign refers to the Z in the heavier or lighter mass region, respectively. On this basis and the usual mass-energy correlation we have,

$$E_L/E_H = M_H/M_L = (Z_H+C)/(Z_L-C);\ Z_L+Z_H = Z;\ E = E_L+E_H. \tag{3}$$

In the same frame-work the mass dependent energy calibration (equation 1) is transformable into the form

$$E_{ij} = [a+a'(Z \pm C)_j*f]X_i + b + b'(Z \pm C)_j*f, \tag{4}$$

where $f = (A/Z)_{FN}$ and the calibration constants a, a', b, and b' maintaining their usual meaning/values. These transformations were, therefore, made essentially maintaining the mass dependence of energy calibration on one hand and conservation of momentum on the other.

The coincident kinetic energy spectrum in channel by channel array, $N(X_1,X_2)$, is also transformed into fragment charge/kinetic

energy array, $N(Z_1,E)$, using appropriate Jacobian J_Z as

$$N(Z_1,E) = N(X_1,X_2)\ J_Z \begin{bmatrix} X_1 & X_2 \\ Z_1 & E \end{bmatrix}. \tag{5}$$

For a given fragment of charge Z_1, that C-value at which the sum $N(Z_1)$ of the $N(Z_1,E)$ over the whole range of total energy (E) compared with the similar sum, $N(Z_C)$, of its complementary fragment of charge Z_C within 2 %, was taken as the appropriate C-value for that pair of fragments (Z_1 and Z_C) and the corresponding average of $N(Z_1)$ and $N(Z_C)$ values was taken as the yield for that pair of fragments. Thus, the C-value was determined for each fragment pair only from the condition of charge-complementarity. The corrections for energy resolution and neutron evaporation were applied in the same manner as in mass yield distribution as given by the equation (6).

$$N_C(Z_1) = N(Z) - \sigma^2/2*d^2N(Z)/dZ^2, \tag{6}$$

where $N_C(Z_1)$ is the resolution corrected $N(Z_1)$. The second derivative was obtained numerically and σ is the rms width parameter of the charge resolution functions given by $\sigma^2 = \sigma_1^2 + \sigma_2^2 + \sigma_3^2$, where $\sigma_1^2 = (4/3)\nu_T E_n Z_1 Z_C/ZE$, $\sigma_2^2 = \sigma_E^2(Z_1^2+Z_C^2)/E^2$, $\sigma_3^2 = (1/3)(Z*\Delta E/E)^2$, ν_T and E_n are the total number of neutrons and their average energies, respectively. E is the energy loss of the fragment in the backing (1.2 MeV), σ_E is the energy resolution of the detectors (0.41 MeV corresponding to 1.5 MeV FWHM) and E is the average total kinetic energy for the complementary pair Z_1-Z_C. The elemental yields $Y(Z_i)$ were obtained after iterative correction[2] using equation (18) for two/three steps to obtain $N_C(Z_1)$ followed by normalization of the total yield to 200 %. Charge dependent total, $E(Z_1)$, and individual, $E_k(Z_1)$, kinetic energy distributions were obtained from $N(Z_1,E)$ as

$$E(Z_1) = \sum_E E*N(Z_1,E) \Big/ \sum_E N(Z_1,E);\quad E_k(Z_1) = (Z_C/Z)*E(Z_1). \tag{7}$$

Comparison of the average quantities pertaining to the mass yield, charge yield, and kinetic energy distribution as obtained in the present work with the same from literature[2,4,5] is shown in table-1, indicate goodness of the experimental data and choice of the resolution function.

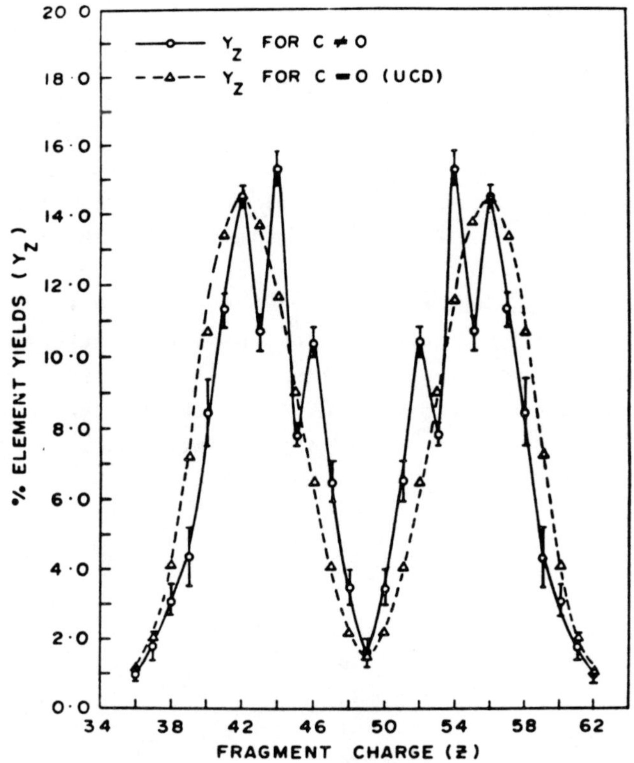

Figure-1 Elemental yield distribution in $^{252}Cf(sf)$

RESULTS AND DISCUSSION

Figure-1 shows the resolution corrected elemental yield distribution profiles based on UCD hypothesis (C=0 for all Z) and involving the free parameter (C≠0). It is seen that for C=0 at all Z, there is no structure in the elemental yield distribution in contrast to the expected[3,6] odd-even fluctuations. This observation indicates (a) need for invoking the free parameter as a physical reality for charge polarization ($|\Delta Z|$) and (b) that the appearance of odd-even fluctuations is not merely due to the charge resolution correction. Odd-even effect and charge polarization (C) or ΔZ as a function of various charge-splits along with the average fragment mass for each Z were shown in fig.-2. It is seen that the ΔZ parameter tends to zero for symmetric split as is expected and is maximum around Z=52 with an average value of ΔZ≈0.46 showing also an odd-even fluctuation. High value of charge polarization at Z=52 is expected from relative gain in

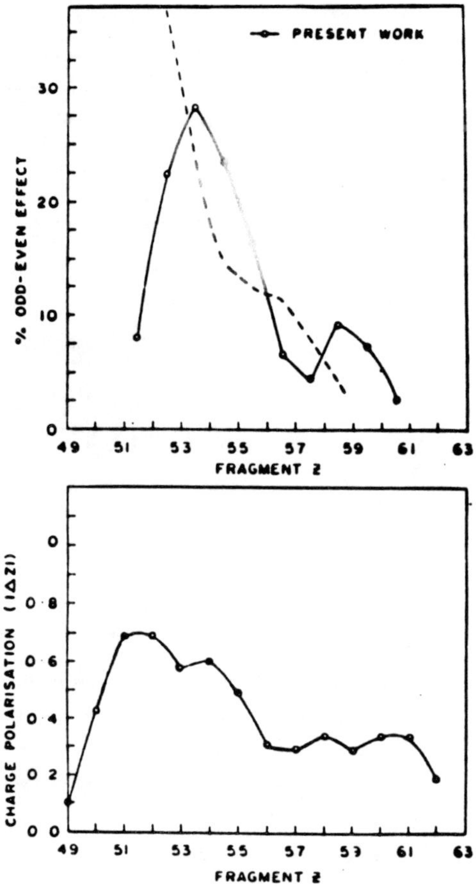

Fig.-2 Percentage odd-even effect and charge polarization in $^{252}Cf(sf)$

potential energy for fragments with 82n shell configuration and 52 protons resulting in N/Z ratio close to that of the fissioning nucleus[1]. Odd-even fluctuation in ΔZ is expected due to preferential formation of even Z fragments in low energy fission[1]. The average odd-even fluctuation in ΔZ (0.46) is in close agreement with the radiochemical observations[3] in various low energy fissioning nuclei. The average odd-even effect in $^{252}Cf(sf)$ is 12 % as expected. In figure-2 it is seen that magnitude of the odd-even effect gradually increases with increasing fragment Z according to present work as well as ΔE-E measurement[6]. The present work, however, shows a decrease in the Z region 41(57) to 43(55) and nearer to the symmetric region unlike near-constant increase shown by Mariolopoulos et al.[6]. Low (reduced nuclear pairing effect) is indeed expected in the region of

Table-1 Average quantities of mass, elemental yields, and kinetic energy distributions in 252-Cf(sf).

Quantity	Present Work	Literature Work
$\langle M_L \rangle$	107.95 ± 0.50	108.55[a], 107.8[b], 108.5[c]
$\langle M_H \rangle$	142.97 ± 0.48	143.45[a], 144.2[b], 143.5[c]
$\langle \sigma_A \rangle$	7.02 ± 0.40	6.72[a], 7.27[b], 7.1[c]
$\langle Z_L \rangle$	43.06 ± 0.15	42.46[d]
$\langle Z_H \rangle$	55.04 ± 0.15	55.34[d]
$\langle \sigma_Z \rangle$	2.77 ± 0.05	3.22[d]
$\langle E_T \rangle$	186.65 ± 1.28	186.5±1.2[a], 182.1±1.7[b], 185.9[c]
$\langle E_L \rangle$	105.88 ± 1.12	106.2±0.7[a], 104.4±1.0[b], 105.7[c]
$\langle E_H \rangle$	80.76 ± 0.2	80.3±0.5[a], 78.3±0.7[b], 80.2[c]
σ_{EK}	12.65 ± 3.67	12.0[a], 15.2[b], 11.6[c]
ΔZ	0.46 ± 0.20	0.3[d]
%δ	12 ± 1	12±2

(a - Ref. 2, b - Ref. 4, c - Ref. 5, d - Ref. 6)

neutron-shell (e.g. 66n along with 88n and 82n) configuration[1)] and symmetric region where UCD hypothesis is valid as seen from this work. Fragment kinetic energy profiles were seen to be in good agreement with the literature data[2)] in terms of the first and second moments (table-1). Fragment kinetic energy as function of mass and charge show the typical features, e.g. peak at mass 132 for doubly magic shell configuration (50p,82n) and kinetic energy deficit of 7 MeV[2)] at the symmetric split of mass 126 and charge 49. Significant odd-even fluctuation, however, was not observed in the kinetic energy distribution as a function of fragment charge in contrast to the observation of Mariolopoulos et al.[6)]. Absence of odd-even fluctuation on observed kinetic energy might be due to larger experimental uncertainty compared to the expected odd-even effect (E_k) on kinetic energy in ^{252}Cf(sf).

REFERENCES

1. B.D. Wilkins, E.P. Steinberg, and R.R. Chasman, Phys. Rev. **C14**, 1832 (1976).
2. H.W. Schmitt, J.H. Neiler, and F.J. Walter, Phys. Rev. **141**, 1146 (1966). H.W. Schmitt, W.E. Kiker, and C.W. Williams, Phys. Rev. **137**, B837 (1965).

3. A.C. Wahl, A.E. Norris, R.A. Rouse, and J.C. Williams, Proc. IAEA Symp. on Physics and Chemistry of Fission, Vienna (1969) p. 813. A.C. Wahl, Proc. 2nd Adv. Group Meeting on Fission Product Nuclear Data, Report INDC(NDS)-87, 215 (1977). T. Izak-Biran and S. Amiel, Phys. Rev. **C16**, 266 (1977).
4. J.S. Fraser, J.C.D. Milton, H.R. Bowman, and S.G. Thomson, Can. J. Phys. **41**,2080 (1963).
5. J.P. Unik, J.E. Gindler, L.E. Glendenin, F.F. Flynn, A. Gorski, and R.D. Sjoblom, Proc. IAEA Symp. on Physics and Chemistry of Fission, Rochester (1973), Vol. II, 19.
6. G. Mariolopoulos, C. Hamelin, J. Blachot, J.P. Bocquet, R. Brissot, J. Crancon, H. Nifenecker, and C. Ristori, Nucl. Phys. **A361**, 213 (1981).

RADIOCHEMICAL INVESTIGATION ON MASS-RESOLVED FRAGMENT ANGULAR DISTRIBUTION IN MEDIUM ENERGY FISSION

S.B. Manohar, T. Datta, A. Goswami, and Satya Prakash
Radiochemistry Division, Bhabha Atomic Research Centre, Trombay, Bombay-400 085, India

Abstract: Studies on angular distribution of fission fragments as a function of their mass are likely to provide independent understanding of several aspects of recent interest in fission and heavy-ion reactions. Theses aspects include the extent of coupling of the various collective degrees of freedom, e.g. rotational and mass-asymmetry/shape degrees[1], the relative extent of validity of the static scission (SSM) and the transition state model (TSM)[2] and, in particular, the question regarding whether mass distribution is frozen at the second saddle[3]. The present work describes experiments on determination of mass-resolved angular distribution in helium ion induced fission of ^{232}Th. The results have been analyzed.

EXPERIMENT AND ANALYSIS

The angular distribution of fission products as a function of their mass was determined radiochemically in the 28.5 MeV α-particle induced fission of ^{232}Th. Attempt was made to deduce the mass-resolved angular distribution of the fragments theoretically on the basis of liquid-drop shape-parameterization of the rotating fissioning nucleus for deformed fragments at various degrees of mass-asymmetry in the framework of the TSM and the two-mode hypothesis.

The measurements on the angular distribution of fission products were carried out at the 88" (inch) variable energy cyclotron at Calcutta using 30±0.3 MeV external α-particle beam. Targets of ^{232}Th of thickness 150 $\mu g/cm^2$ were prepared by electrodeposition on 25 μm thick superpure aluminum backings. The measurements were based upon recoil-catcher technique of collection followed by γ-spectrometric assay of the fission product activities. An irradiation chamber of cylindrical geometry was used to house the target-catcher assembly. The target was placed at one end of the chamber at 45 deg angle with respect to the collimated (dia. 5 mm) beam axis. Superpure aluminum catcher foil of thickness 25 μm was fixed on the inner surface of the chamber covering azimuthal angles from 0 deg to 180 deg to collect the fission products at various pre-decided solid angles. The irradiations were carried out with typical integral current measured by a Faraday cup in the range of 8 to 20 μAh. The mean angles of collection (Θ) varied from 8.2 deg to 84.4 deg with respect to the beam direction while the deviation ($\Delta\Theta$) due to finite widths of these strips varied from 4.5 deg to 11.2 deg. At the end of the irradiations, the catcher strips were counted under identical geometry for specific fission product activities on a 60 cc. HPGe detector coupled to a 4K MCA having resolution of 1.8 keV at the 1332 keV γ-line of ^{60}Co. The dead time was less than 5 %. In all 13 different fission products pertaining to both asymmetric and symmetric modes of mass-division were studied. In each experiment the peak area (A) of the γ-line for individual fission product from catcher strips were least-square fitted to the equation

$$A_i/\sin\Theta_i \; \Delta\Theta_i = a + b \cos^2\Theta_i, \tag{1}$$

where Θ_i and $\Delta\Theta_i$ are the mean angle of emission and its deviation, respectively, at a given strip (i). On the basis of the fitting parameters a and b the angular anisotropy[3)] for any particular fission product (j) was obtained as

$$[W(0)/W(90)]_j = 1 + (b/a)_j . \tag{2}$$

Table-1 shows the obtained anisotropy, W(0)/W(90), for 13 individual fission products averaged from our four measurements. The error limits quoted are precision of the measurements that varied within 10 to 15 % for all the 13 fission products. Figure-1 shows the angular distribution profiles of typical asymmetric and near-symmetric fission products. It is seen that anisotropy is higher for the asymmetric products compared to symmetric products.

On the basis of the transition state model angular distribution of fission fragments is described by the K-distribution assumed to be frozen from the saddle onwards. In medium energy fission where nucleon pairing effect is much less significant, angular distribution of fragments for spin zero target and projectile is described as

$$W(\Theta) = \sum_{l=0}^{\alpha} (2l+1)^2 T_l \exp[-(l+0.5)^2 \sin^2\Theta/4K_0^2] *$$

$$J_0[i(l+0.5)^2 \sin^2\Theta/4K_0^2]/\mathrm{erf}[(l+0.5)/\sqrt{2}K_0^2], \tag{3}$$

where T_l are the transmission coefficients for partial l waves, J_o is the zero-order Bessel function and $K_0{}^2$ is variance of the

K-distribution to be determined. on the basis of the experimentally observed angular anisotropy of the individual fission product and equation (3), K_0^2 for various symmetric and asymmetric mass splits were deduced and are given in table-1.

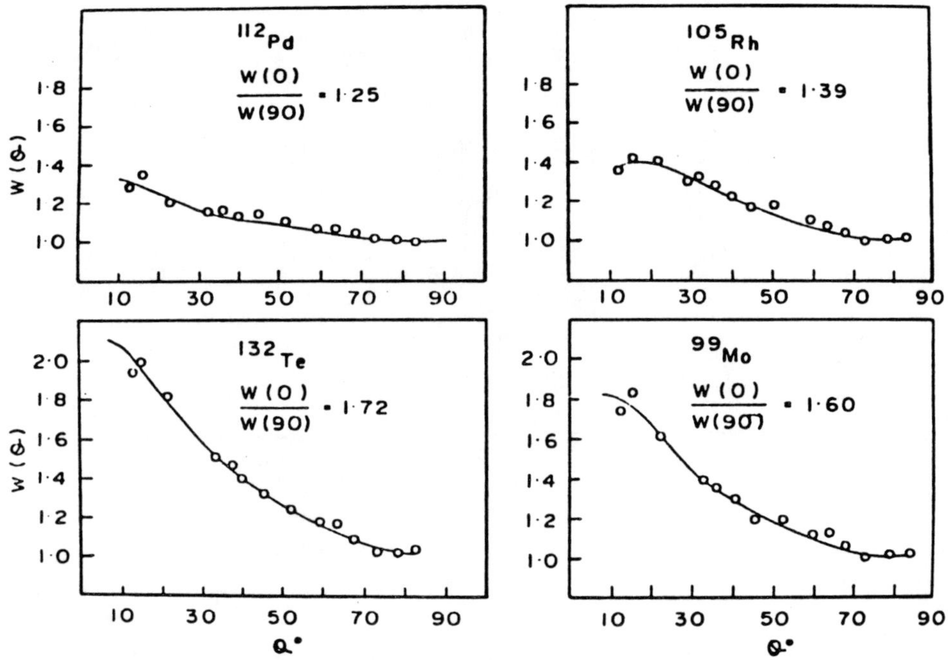

Figure-1 Experimental angular distribution for symmetric and asymmetric products in helium ion induced fission of ^{232}Th

RESULTS AND DISCUSSION

It is seen from table-1 that in spite of the error limits there is a distinct variation of the angular anisotropy W(0)/W(90) of the fission products as a function of their mass. In order to examine the observed dependence of W(0)/W(90) on mass asymmetry it is necessary to consider the effect of multi chance fission and dependence of K_0^2 on the shape of the fission nucleus for specific mass asymmetric configuration. Evaluation of the extent of various-chance fissions was carried out on the basis of ref. 8 of the ratio Γ_f/Γ_n of the decay-widths of the compound nucleus for fission and neutron emission using appropriate fission barrier and neutron separation energies for ^{236}U, ^{235}U, and ^{234}U (ref. 9) and assuming that the level density parameter for fission is 1.1 times that for neutron emission. It was seen that

Table-1: Fission product angular anisotropy and the corresponding K_0^2-values in $^{232}Th(\alpha_{28.5MeV},f)$

Fission Product	W(0)/W(90)	$K_0^2[\hbar^2]$
^{91}Sr	1.64 ± 0.03	45.0
^{92}Sr	1.66 ± 0.03	44.0
^{99}Mo	1.50 ± 0.02	55.0
^{105}Ru	1.41 ± 0.02	67.0
^{112}Pd	1.25 ± 0.03	105.0
^{115}Cd	1.28 ± 0.03	92.0
^{127}Sb	1.59 ± 0.04	48.0
^{129}Sb	1.57 ± 0.05	50.0
^{132}Te	1.69 ± 0.02	42.0
^{133}I	1.67 ± 0.03	43.0
^{139}Ba	1.80 ± 0.03	37.0
^{142}La	1.72 ± 0.04	40.0
^{143}Ce	1.82 ± 0.02	36.0

in $^{232}Th(\alpha_{28.5MeV},f)$ the extent of first, second, and third chance fission events are 89, 7, and 4 %, respectively. As a result, only the first chance fission events need to be considered in the present work. The excitation energy (E^*) and the average angular momentum of the net fissioning nucleus were 23.4 MeV and 10.2 ħ, respectively.

THEORETICAL EVALUATION OF K_0^2: SHAPE PARAMETERIZATION

Evaluation of K_0^2 requires[7)] moments of inertia of the fissioning nucleus for rotations parallel ($I_\parallel$) and perpendicular ($I_\perp$) to the fission axis and temperature (T) above the barrier.

$$K_0^2 = I_{eff}\, T/\hbar^2, \text{ where } I_{eff}^{-1} = I_\parallel^{-1} - I_\perp^{-1}. \qquad (4)$$

These moments of inertia were deduced on the basis of ref. 10 of shape parameterization for a rotating charged liquid-drop fissioning nucleus for axially symmetric but reflection asymmetric and symmetric shapes. The shape of the fissioning nucleus is given by

$$R = R_0/\lambda*[1 + \alpha_2 P_2(\cos\theta) + \alpha_3 P_3(\cos\theta)], \qquad (5)$$

where $R_0 = 1.2249\ A^{1/3}$ and λ is the volume conservation constant deduced in usual way[10] for specific α_2- and α_3-values. The conditions to determine the α_2- and α_3-parameters for asymmetric saddle shapes were as given in ref. 12.

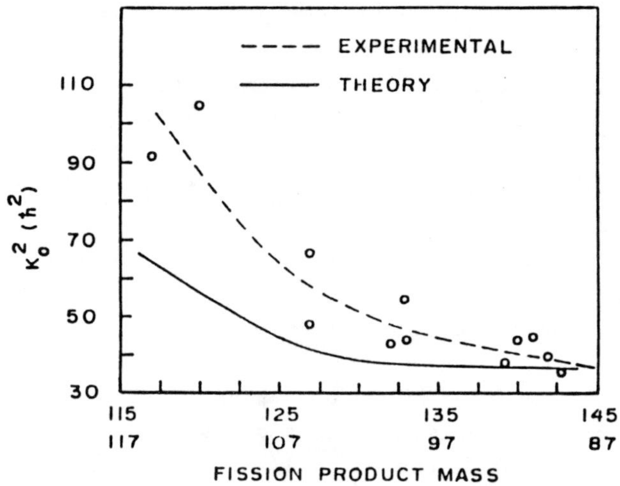

Figure-2: Comparison of experimental and theoretical K_0^2 values

It must be considered at this point that more accurate evaluation of α_2- and α_3-values necessitates microscopic calculations. However, in the absence of any proven/accurate microscopic approach (particularly at medium energy) we consider only RLDM-shapes since the single-particle effects become less significant with increasing excitation energy. From the deduced α_2- and α_3-values for the various mass-asymmetric configurations from either of the modes, the moments of inertia ($I_\parallel$ and $I_\perp$) of the fissioning nucleus were deduced[12]. The K_0^2 values for specific mass division configurations were obtained subsequently using equation (4).

Since the observed angular distribution of the fission products are resultant from both the modes, the calculated K_0^2 values for various mass divisions from the two modes were compounded considering average %-contribution of the two modes at various mass divisions for comparison with the experimental observations. This was done on the basis of statistical model[11] using the fission barrier heights and curvatures [11] for the symmetric and asymmetric modes in ^{236}U. The resultant K_0^2 values for various mass-asymmetric configurations or fission product masses were seen to be in reasonable agreement with the experimental observations as shown in figure-2. Although the dependence of K_0^2 on mass asymmetry was interpreted earlier[3] on the basis of the difference in the effective excitation energy or fission-barrier heights for the two modes through the temperature (t), variation in T alone is inadequate to account for such dependence due to small difference in these barrier heights. On the other hand, variation in I_{eff} for different mass-asymmetric configurations as well as the extent of occurrence of the two modes of mass division appear to be more significant in the observed variation of K_0^2 with fragment mass. To conclude, present work shows that (a) fragment angular anisotropy is strongly dependent on their mass or the extent of mass-asymmetry; (b) in medium energy fission the collective rotational (tilting) and the mass-asymmetry degrees are strongly coupled following TSM and the two-mode hypothesis.

The authors acknowledge the encouragement provided by Dr. P.R. Natarajan, Head of Radiochemistry Division with keen interest. They also thankfully acknowledge the cordial assistance by the cyclotron crew, VECC, Calcutta.

REFERENCES

1. I.G. Moretto and G.J. Wojniak, Ann. Rev. Nucl. & Part. Sci. **34**, 189 (1984).
2. H. Rossner, J.R. Huizenga, and W.O. Schroder, Phys. Rev. **C33**, 560 (1986).
3. H. Kudo, Y. Nagame, H. Nakahara, K. Miyano, and J. Kohno, Phys. Rev. **C25**, 909 (1982).
4. P. Moeller and S.G. Nilsson, Phys. Lett. **31B**, 283 (1970).
5. B.I. Cohen, B.L. Ferrel-Bryan, D.J. Coombe, and M. Hullinga, Phys. Rev. **98**, 685 (1955).
6. J. Blachot and C. Fiche, Ann. Phys. 6(1981).
7. J. Halpern and V.M. Strutinski, Proc. 2nd United Nations Conf. on Peaceful Uses of Atomic Energy **15**, 1513 (1958).
8. R. Vandenbosch and J.R. Huizenga: Nuclear Fission (Academic Press, New York, 1974).
9. S. Bjornholm and J.E. Lynn, Rev. Mod. Phys. **52**, 725 (1980).
10. W.J. Swiatecki, Proc. 2nd UN Conf. on Peaceful Uses of Atomic Energy **15**, 248 (1958), J.R. Hisker, URCL Report URCL-9275 (1958).
11. S. Baba, H. Umezawa, and H. Baba, Nucl. Phys. **A175**, 177 (1971).
12. T. Datta, S.P. Dange, S.M.S.K. Sahakundu, Satya Prakash, and M.V. Ramaniah, Z. Phys. **A330**, 103 (1988).
13. S. Cohen and W.J. Swiatecki, Ann. Phys. (N.Y.) **22**, 406 (1963).
14. R. P. Schmitt and M. Tirion, Phys. Rev. **C31**,701 (1985).

STUDY OF PROMPT NEUTRON AND GAMMA EMISSION IN THERMAL NEUTRON FISSION OF ^{239}Pu AND ^{241}Pu BY USING THE MONTE-CARLO TECHNIQUE

G. Yener
Ege University, Institute of Nuclear Sciences, Bornova-IZMIR, Turkey

Abstract: Fragment de-excitation in thermal neutron fission of 239-Pu and 241-Pu was studied using two different models for scission point configuration. The charge distribution of the secondary fragments as well as the neutron and the gamma emission were investigated using statistical and adiabatic models. Deviation of most probable charge from unchanged charge distribution for secondary fragments were also studied and compared with the experimental results. Adiabatic model was found to give better agreement with the experimental results. Monte-Carlo method was used for the calculations of neutron evaporation from the fragments.

INTRODUCTION

The scission point configuration and the fragment de-excitation in low energy fission have been studied extensively using different models in numerous theoretical works. Among these, Kluge and Lajtai[1)] used a statistical approach for neutron emission in thermal fission of ^{235}U, ^{239}Pu and spontaneous fission of ^{252}Cf. Nörenberg[2)] used single particle molecular model to study scission configuration of ^{240}Pu and ^{242}Pu. Terrell[3)] investigated prompt neutron emission in thermal neutron fission of ^{239}Pu on the basis of fragment-deformation theory using shell effect together with adiabatic model assumption of Wheatstone[4)]. A similar approach was used by Kildir and Aras[5)] for spontaneous fission of ^{252}Cf. Although majority of the theoretical works have supported the adiabatic models for low energy fission, in the present work both adiabatic and statistical models were applied to thermal neutron fission of ^{239}Pu and ^{241}Pu using Monte Carlo technique in the calculations for the de-excitation of the fragments and the results were compared with the experimental data.

CALCULATIONS

In statistical models[1,6,7)] of fission, complementary fragments are accepted to have equal nuclear temperatures at scission point and according to Fermi-gas model[8)] excitation energy of the light and the heavy fragments are given in terms of the nuclear temperature T and the energy surface density $\bar{a}$ by the equations

$$E_{xl} = \bar{a}_l T_l^2, \quad E_{xh} = \bar{a}_h T_h^2,$$

where l and h stand for the light and the heavy fragments, respectively. Average total excitation energies E_x in general is given by

$$E_x = E_{xl} + E_{xh} = E_T - E_k,$$

where E_T is the mass difference and E_k is the average total kinetic energy of the fragment pair and is equal to the Coulomb energy between the fragments. Using the assumption of $T_h = T_l$ in statistical model and the above relations together with Ignatyuk's[7] formulation for energy surface density, the following equation was obtained for the calculation of the excitation energies of the individual fragments:

$$\frac{E_{xh}(\alpha' A_l+\beta' A_l^2)\ \{1-[1-\exp(-\gamma'(E_x-E_{xh}))]\ \dfrac{\delta W_l}{E_x-E_{xh}}\}}{(E_x-E_{xl})(\alpha' A_h+\beta' A_h^2)\ \{1-[1-\exp(-\gamma' E_{xh})]\ \dfrac{\delta W_h}{E_{xh}}\}} = 1,$$

where $\alpha' = 0.154$, $\beta' = -6.3 \times 10^{-5}$, and $\gamma' = 0.054\ \mathrm{Mev}^{-1}$ were used in the calculations and δW is the shell energy. Since this relation between the excitation energies is not linear, E_{xl} and E_{xh} were evaluated numerically using a computer.

In adiabatic models[2,9] the primary fragments are supposed to have zero temperature at scission and the deformation energy E_D before the scission is converted into excitation energy of the fragments just after the scission having the same sort of the division of excitation energy between the complementary fragments as

$$E_{Dl}/E_{Dh} = E_{xl}/E_{xh}.$$

Using the deformation energy given by Terrell[3] in terms of stiffness parameter C_2 with the Coulomb energy, the ratio of the excitation energies in adiabatic model was obtained[5] as

$$\frac{E_{xl}}{E_{xh}} = \frac{(7.985-\delta W_h)(7.985+\delta W_l)\{4.103-7.112[(N_h-Z_h)/A_h]^2-0.079Z_h^2/A_h}{(7.985+\delta W_h)(7.985-\delta W_l)\{4.103-7.112[(N_l-Z_l)/A_l]^2-0.079Z_l^2/A_l}$$

Stiffness parameters of the fragments were obtained using the correlation deduced by Kildir[5] between the experimental C_2 values of Wong[10] and C_2 calculated from shell energies of Myers and Swiatecki[11]. Myers formula was used also for mass calculations. Total kinetic energies were taken from Schmitt[12] and the most probable charges or primary fragments were taken from Wolfsberg[13]. The charge distribution of the secondary fragments were also investigated and the deviation of the most probable charge Z_P from the unchanged charge distribution Z_{UCD} were studied and the results were compared with the experimental values. Treating the neutron evaporation as a statistical process, Monte Carlo method was used for the calculations as it was applied for other low energy fission studies[5,14].

RESULTS AND DISCUSSIONS

Average number of prompt neutrons $\nu(A)$ emitted from individual fragments and average total number of neutrons $\nu_T(A)$ released per fragment pair were calculated by statistical and adiabatic models. Fig. 1 shows the results obtained by two models and by experiment of Apalin[15]. Although the deviation $\nu(A)$ with A calculated by statistical model is completely different from the results of adiabatic model and also from the experimental values, the variation of total neutron number $\nu_T(A)$ with A does not show any discrepancy. It should be concluded that then, the fraction of the total energy used for the excitation does not depend on the scission model, but neutron evaporation from individu.l fragments does strictly depend on the structure of the fragment. It could be also said that scission point configuration is best explained by an adiabatic model for low energy fission. Fig. 2 gives the results for ^{241}Pu using the adiabatic model again showing the good agreement with literature values.

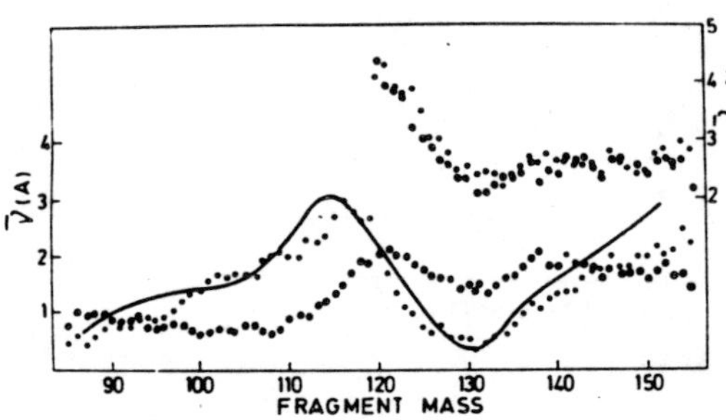

Fig. 1 Average neutron number against mass number for ^{239}Pu fission (• - statistical model, o adiabatic model, line - exp. data, ref. 15)

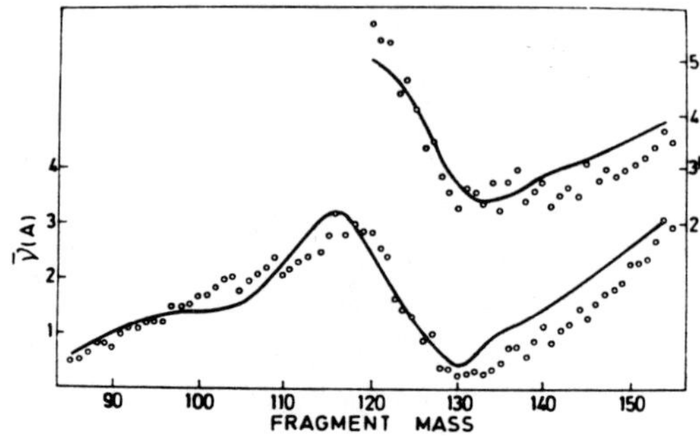

Fig. 2 As for fig. 1, but for ^{241}Pu (o - this work, line - exp. data, ref. 17)

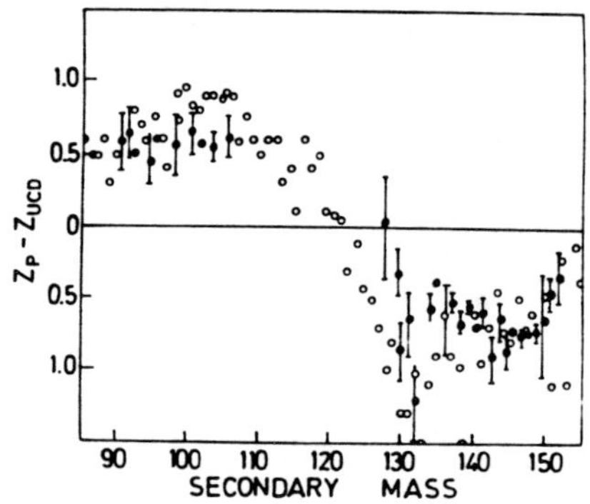

Fig. 3 Deviation of most probable charge from UCD for 239-Pu (o - this work, ɸ - ref. 9)

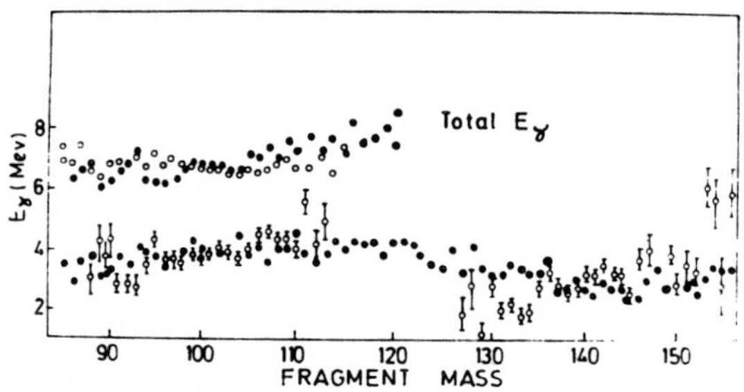

Fig. 4 Variation of $E\gamma$ with fragment mass in 239-Pu fission (o - this work, ɸ - ref. 16)

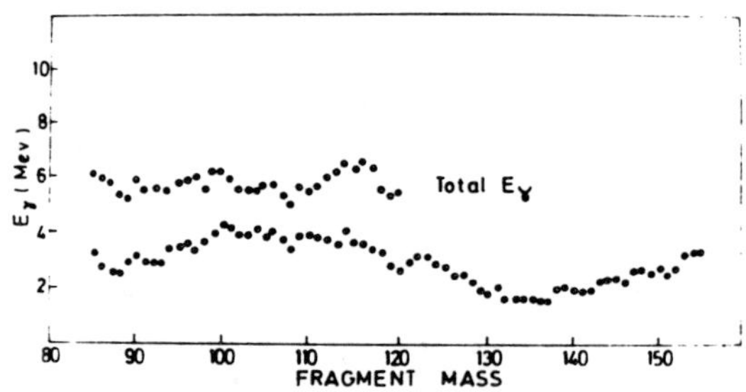

Fig. 5 Variation of $E\gamma$ with fragment mass in ^{241}Pu fission

The deviation of Z_P from the unchanged charge distribution Z_{UCD} calculated with adiabatic model for ^{239}Pu was plotted against mass number in fig. 3 together with Denschlag's[9)] experimental results supporting again the model and the technique used in the work.

Finally the variation of gamma ray energies with mass number were given in figs. 4 and 5 for ^{239}Pu and ^{241}Pu. The results show that the gamma energies are independent of the fragments masses as it was experimentally observed[16]. The charge dependence of gamma ray energies were also examined and prompt gamma rays of even Z isobars in a given mass chain were found to carry higher energy than those of odd Z fragments.

REFERENCES

1. Gy. Kluge and A. Lajtai, Phys. Lett. **30B**, 311 (1969).
2. W. Nörenberg, Phys. Rev. **C5**, 2020 (1972).
3. J. Terrell, Proc. IAEA Symp. Phys. Chem. Fission, Salzburg, 1965, **Vol. II**, 3 (1965).
4. S.L. Wheatstone, Phys. Rev. **114**, 581 (1959).
5. M. Kildir and N.K. Aras, Phys. Rev. **C25**, 365 (1982).
6. P. Fong, Phys. Rev. **102**, 434 (1956).
7. A.V. Ignatyuk, Sov. J. of Phys. **9**, 208 (1969).
8. J.R. Huizenga and L.G. Moretto, Ann. Rev. Nucl. Sci. **22**, 427 (1972).
9. H.O. Denschlag, Fission Product Nuclear Data, **Vol. II**, 421 (1977).
10. C.Y. Wong, Nucl. Data Sect. **A4**, 271 (1968).
11. W.D. Myers and W.S. Swiatecki, Nucl. Phys. **81**, 1 (1966).
12. H.W. Schmitt, J.H. Neiler, and F.J. Walter, Phys. Rev. **141**, 146 (1966).
13. K. Wolfsberg, LA-5553 (UC-34C, 1974).
14. G.E. Gordon and N.K. Aras, Proc. IAEA Symp. Phys. Chem. Fission, Salzburg, 1965, **Vol. II**, 73 (1965).
15. V.F. Apalin et al., Nucl. Phys. **71**, 553 (1965).
16. F.P. Pleasonton, Nucl. Phys. **A213**, 413 (1973).
17. J.H. Neiler, F.S. Walter, and H.W. Schmitt, Phys. Rev. **149**, 894 (1966).

MASS-ENERGY SPECTRUM OF THERMAL NEUTRON-INDUCED ^{242M}Am FISSION FRAGMENTS

A.A. Alexandrov, A.A. Goverdovskij, N.N. Demidovich, Yu.A. Korzhuk, S.I. Sitnikov, A.I. Slyusarenko, Yu.F. Pevchev, S.L. Podshybyakin, Yu.V. Pyatkov, S.L. Sharov, A.N. Shemetov, R.A. Shehmametiev, and I.A. Shlyapina

The measurements of mass-energy spectrum of thermal neutron - induced ^{242m}Am fission fragments have been made at the time-of-flight spectrometer on a vertical channel of the Moscow Engineering Physical Institute reactor. The density of thermal neutron flux at the target was $4*10^{10}$ n/cm^2*s. An alectrostatic particle guide was used to augment spectrometer's aperthee [1,2]. The velocity and energy of fission fragments were measured by means of two channel-plate detectors [3] (overall resolution 120 ps) and a surface-barrier detector, respectively. About $3.4*10^6$ events were recorded. The calibration procedure for detectors was performed on ^{235}U(n_{th},f) fission fragment measurements.

The total mass spectrum of ^{242m}Am(n_{th},f) fission fragments fits reasonably with the radiochemical data [4]. It is interesting to note, that the relation of the most probable yield (140/103) and the mass-symmetric fission fragments yield (121/122) runs into an anomalously high value P/V=290 ± 30 (350 according to [4]) as compared to the similar characteristics for target nuclei ^{241}Am and ^{243}Am [5]. The following table contains, in addition to P/V, the neutron binding energy B_n values and the fission external hump barrier B_f of americium isotopes.

Fissile nucleus	B_n, MeV	B_f, Mev	P/V
Am-242	5.54	5.70	117 ± 10
Am-243	6.43	5.60	290 ± 30
Am-244	5.37	5.60	82 ± 5

It follows from the table that for the ^{241}Am and ^{243}Am targets the thermal neutron-induced fission is a subbarrier one and the ^{242m}Am a upbarrier one, that is why the exitation energy in the scission point of the final nucleus seems to exceed the first two. However, an americium isomer mass-symmetric fission fragments yield is essentially lower, that is beyond the scope of the statistical approach (e.g. [6]) in the description of mass distributions. In addition, a high value of P/V for ^{242m}Am is beyond the scope of the global dependence of symmetric group fission fragments yield on the Z^2/A nucleus fissility parameter. According to Gonnenwein's estimates [7] of the intensities of energy exchange between collective and single-particle degrees of freedom on way from saddle to scission, the exitation energy E^*– 10 MeV of fissile ^{243}Am in the scission point is twice as large as E^* for uranium or neptunium, and nevertheless a mass-symmetric fragments yield is roughly the same. The neutron-induced mass-symmetric nuclear fission fragments yield being formed may be suggested to be dependent on individual properties of the fissile system. However it is still problematic to identify these specific systems.

A figure shows mass spectra of ^{242m}Am cold fragmentation (CF) for various kinetic energies of light fragments E_L and energy window width 1 MeV. A fine structure with a period ~ 2 amu, which can be related to discreteness and even-odd difference of light fragment group charges yield at low energies of fissile nucleus exitation [8] is plainly displayed in the spectra. An availability of two components with the centres in the regions M_L= 105 amu and M_L= 85 amu is another salient feature of the CF mass spectrum. It is worth focussing attention on the CF group with M_L= 105 amu predominating over the whole range E_L mentioned. A pair of nuclei ^{158}Sm and ^{85}As meeting the requirements of light fragment charge excess over an equilibrium charge Z_{UCD} may well be in agreement with mass spliting (158/85). The reaction energy in this case is Q=192 MeV, and the total kinetic energy of fragments E_K= 182.5 MeV (for E_L= 119 MeV). The difference $D=Q-E_K$ is rather high (9.5 MeV), however high is the total quadrupol deformation of two fragments: $\varepsilon_2(158)$=0.85 and $\varepsilon_2(85)$=0–0.1, ε_2^n=0.95 [6].

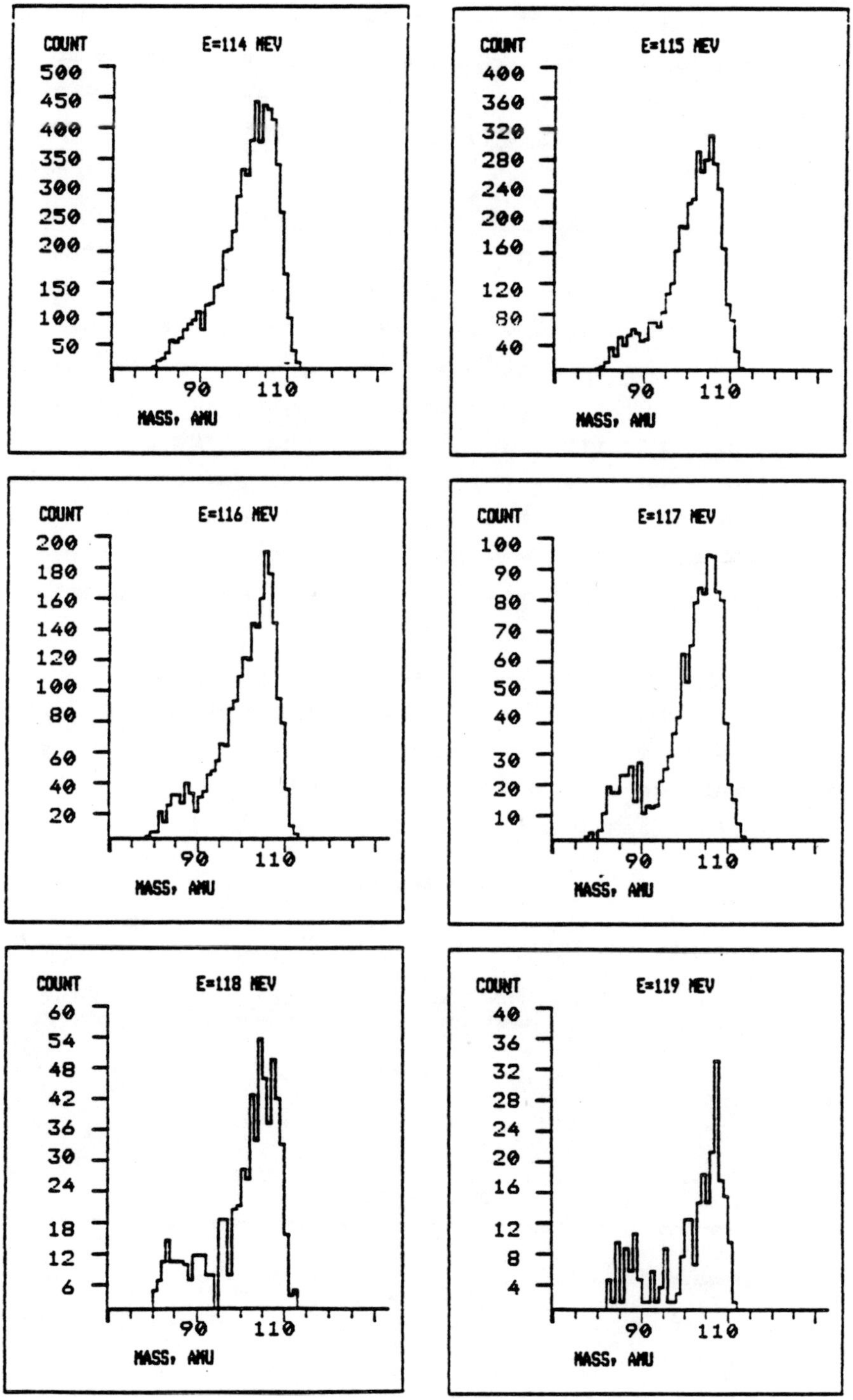

COUNT
E=114 MEV
500
450
400
350
300
250
200
150
100
50
90
110
MASS, AMU
COUNT
E=115 MEV
400
360
320
280
240
200
160
120
80
40
90
110
MASS, AMU
COUNT
E=116 MEV
200
180
160
140
120
100
80
60
40
20
90
110
MASS, AMU
COUNT
E=117 MEV
100
90
80
70
60
50
40
30
20
10
90
110
MASS, AMU
COUNT
E=118 MEV
60
54
48
42
36
30
24
18
12
6
90
110
MASS, AMU
COUNT
E=119 MEV
40
36
32
28
24
20
16
12
8
4
90
110
MASS, AMU

A pair of fragments ^{138}I and ^{105}Mo is the most probable one for the main component of the CF mass spectrum. The corresponding energy responses in this case are: Q=212.7 MeV, E_K= 208.8 MeV and D= 3.9 MeV. The total deformation of fragments is as high as ε_2^n= 0.5. It seems to be a relatively low total deformation (138/105) heat resolues cold fragmentation of this pair in a wide range of investigated values E_L. The energy balance fission analisis makes it possible to understand, why americium CF with the formation of a double-magic tin - 132 does not occur, though it is this nucleus whose nucleonic composition is most advantageous for mass splitting (132/111). In this case the value D is about 5.5 MeV, and the total deformation of the pair exceeds ε_2^n= 1.1. That is why an energy prohibition for americium fragmentation with the formation of pair (132/111) occurs over the whole region of variation E_L> 114 MeV. The 132 amu fragments can be observed for lower values E_L.

The CF line M_H= 132 - 134 amu is alternatively emphasized for a fissile system ^{234}U with a light fragment energy growth and it is due to essentially alternative (than in the case of americium) balance Q and E_K resolving CF M_H= 132 amu in a wide range of variations E_L.

References

1. Oakey N.S., McFarlayne R.D. // Nucl. Instr. Methods. 1967. v. 49., N 2. p. 220 - 227.
2. Pyatkov Yu.V., Romanov I.V., Slyusarenko A.I., //Book./Experimental methods in Nuclear Phisics.M. Energoatomizdat.1985.p. 3 - 8.
3. Alexandrov A.A., Pyatkov Yu.V., Volkov N.G.// PTE.1981,N6,p. 21.
4. Wolfsberg K., Ford G. P. // Phys. Rev. 1971. v. C3. p. 1333 - 1340.
5. Ashgar M., Gaiticoly F., Perrin P. et.al. // Nucl. Phisics. 1980. v. A334. p. 327 - 335.
6. Wilkins B. D., Steinberg E. P., Chasman R. R. // Phys. Rev. 1979. v. 14. N 5. p. 1832 - 1863.
7. Gonnenwein F. // Proc. Int. Conf. on Phys. of Fission, Obninsk, 1987. / Vopr. At. 1988. N 1. p. 14 - 23.
8. Knitter H.-H., et. al. // Proc. Int. Conf. on Nuclear Data for Fund. and Appl. Physics, Santa-Fe, 1985. NBS. 1986. v. 1. p. 409.

II. FUNDAMENTAL FISSION PROBLEMS

DISSIPATION AND FRICTION IN NUCLEAR FISSION*†

R.W. Hasse
Gesellschaft für Schwerionenforschung GSI, D-6100 Darmstadt, Germany

Abstract

We give a short review on experimental evidence for dissipation in nuclear fission as well as on the development of theories of nuclear dissipation and friction applied to the fission process. Theories span from two-body viscosity and one-body dissipation via the time dependent Schrödinger equation to linear response theory and the Fokker-Planck equation.

1. Introduction

50 years of nuclear fission research means also 50 years of theoretical investigations in the fission process. Already in 1939 Bohr and Wheeler [1] set the paths and Hill and Wheeler [2] in 1953 substantiated the lines to persue followed by Swiatecki with his contributions to the liquid drop model (LDM) and Wilets summarized the open problems in his still timeless booklet [3]. It was soon realized that statics is not enough to explain many fission phenomena, in particular fragment mass and kinetic energy distributions and their dependence on excitation energy. Hence Nix studied the dynamics of the LDM from 1965 on.

Bjørnholm [4,5] then noted that as going e.g. from ^{240}Pu spontaneous fission via ^{239}Pu neutron induced isomer fission to induced 4.6 MeV resonance fission the additional excitation energy mainly shows up as increased average fragment kinetic energy ($\bar{E}_{\rm kin}^{\rm total}$), for newer data [6] see Fig.1. These processes therefore seem to be superfluid and leave no room for dissipation (the straight line in Fig. 1). Increasing again the excitation energy, for instance for thermal neutron induced fission (6.4 MeV above ground state) and higher there is a drop in the $\bar{E}_{\rm kin}^{\rm total}$ followed by a gentle decrease. Therefore he introduced the concept of damping during the descent from saddle to scission in order to account for internal excitation and concluded that the gain in energy from saddle to scission very likely is smaller than the theoretical static value ΔE_{ss}. According to Fig.1, for thermal neutron induced fission the dissipated energy should amount to at most 7.6 MeV. In 1974, Nifenecker *et al.* [7] deduced from fission fragment neutron and γ-spectra of ^{252}Cf spontaneous fission that the internal excitation energy was not more than $5 \cdots 10$ MeV. The most complete analysis of the dissipated energies from odd-even effects in the charge yields and the total fragment kinetic energies in slow neutron fission is provided by Gönnenwein *et al.* [8]. They found that about one third of ΔE_{ss} is converted to internal heat, i.e. from 4 to 12 MeV out of ΔE_{ss} = 12 to 30 MeV as going from Thorium to Californium, see Fig.2. At high excitation energy, from heavy ion induced fission experiments with compound nuclear temperatures of $1 \cdots 5$ MeV Hilscher *et al.* [9] deduce that most of the excitation energy

*Review prepared for the XVIIIth International Symposium on Nuclear Physics - Physics and Chemistry of Fission, Gaussig (Dresden), GDR, Nov. 21-25, 1988

†Dedicated to my colleagues and friends Rainer Beck, Friedrich Dickmann and Baron Ernest Stückelberg who deceased recently

is lost *before* scission by neutron emission. In this region, hence, damping is very strong. At the scission point the system is essentially cold because the neutron evaporation time becomes much shorter than the time of fission.

These observations stimulated a large literature on theoretical investigations on the damping mechanism of nuclear fission. In this review we outline the major trails followed. However, by virtue of its complexity and by the lack of manpower this problem is still far from being settled. For an old review on nuclear friction and dissipation see ref. [10].

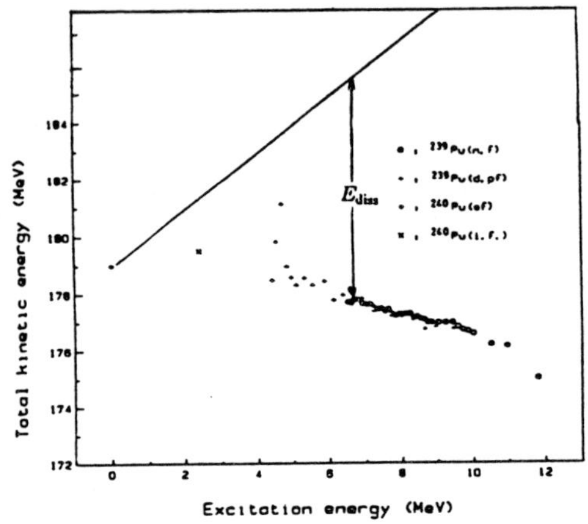

Figure 1:
Average total fragment kinetic energies against excitation energy of the fissioning system ^{240}Pu (after [6]).

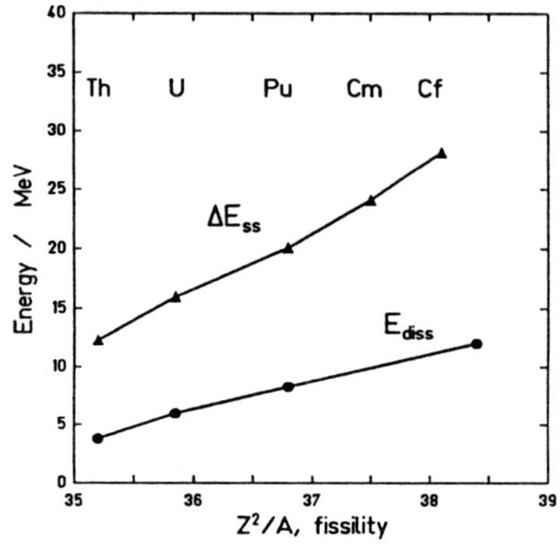

Figure 2:
E_{diss}, dissipated energy vs. fissility for thermal neutron fission of various nuclei and ΔE_{ss}, the gain in potential energy from saddle to scission (from [8]).

2. Dissipation and Friction

There are no isolated systems in nature which do not interact with others, neither classically nor quantum mechanically. Classically, because an isolated system must be confined in a container and then interacts with its walls and quantum mechanically, because the tails of the wave functions have a long range. If all or at least the most important interactions, H_{int}, of a system H_{coll}, described by an ensemble of *collective* coordinates q with other (*intrinsic*) degrees of freedom ξ as well as the intrinsic degrees of freedom themselves, H_{intr}, could be treated,

$$H(q,\xi) = H_{coll}(q) + H_{intr}(\xi) + H_{int}(q,\xi), \tag{1}$$

there would be no need of introducing the concept of damping. However, this is only possible in ideal cases like a system of coupled harmonic oscillators [11] or very low temperature physics. Hence any theoretical treatment of only a few collective coordinates without interaction with the surrounding is an idealization. The next step beyond this idealization is to neglect the motion of the irrelevant *fast* intrinsic coordinates and to mock up the effect of the interaction on the *slow* collective coordinates by modifying the equations of motion or the Hamiltonian of the collective coordinates. Since in many cases the intrinsic system initially is in its ground state and the collective system is in a metastable state with large potential energy, the collective system looses energy and, hence, undergoes a damped motion. This has been noted already 35 years ago by Hill and Wheeler [2] in saying that *"Damping is one way to speak of the exchange of energy between vibration and nucleonic excitation"*. The term *dissipation* usually refers to an exchange of energy (or angular or linear momentum) by all kinds of damping mechanisms from collective motion to intrinsic heat, whereas *friction* is a special damping process associated to (not necessarily linear) frictional forces.

When talking about *collective* degrees of freedom we tacitly have assumed their existence, which is nontrivial. In principle they have to be defined via expectation values of operators with respect to the total many-body wave function. As this is not practicable, they are usually defined *ad hoc* as classical parameters or quantum mechanical generalized coordinates. Consequently, the concept of friction hinges upon the chosen collective coordinates as well as on the subspace of the intrinsic coordinates. For the same collective coordinate, the dissipated energy, hence, will depend for instance on whether it is calculated with respect to the adiabatic basis or the co-moving diabatic basis of the single-particle Hamiltonian. Experimentalists need not care about such details but theoreticians often arrived at different conclusions because they did not draw the dissipated energy from the same heat reservoir.

Nuclear fission is a typical example of damped collective motion. Here the collective coordinates are a set of prescribed shape parameters and the intrinsic system corresponds to the nucleonic degrees of freedom. Thus any coupling of the shape degrees of freedom to the nucleonic motion results in a damping of the collective motion and an intrinsic excitation. At least three shape parameters are necessary to describe adequately elongation, necking-in, and asymmetry; for a large choice of families of shapes see ref. [12].

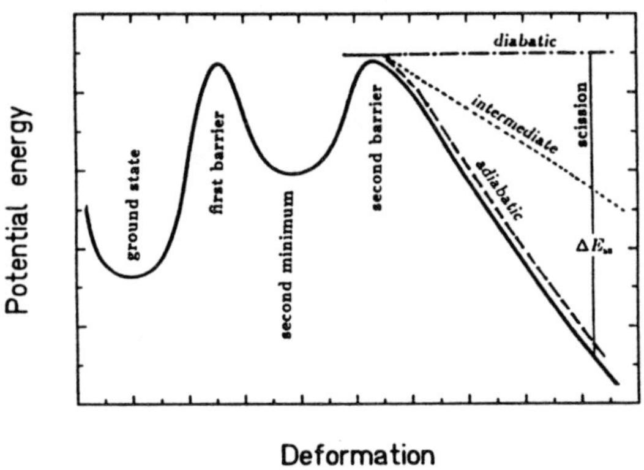

Figure 3:
Schematic view of adiabatic, intermediate and diabatic descent from saddle to scission.

3. History of Theory

Since little was known about the average excitation energy accumulated before scission nor on ΔE_{ss}, everything could and has be assumed in the early fission calculations. The first *dynamic* models of fission [13,14] were *macroscopic* models and were based on solutions of the classical equations of motion for a few shape degrees of freedom and the Hamiltonian consisting of the liquid drop potential energy surface $V(q)$ (Coulomb plus surface energy and, sometimes also the curvature energy and shell corrections) and the kinetic energy involving hydrodynamic inertias $M_{ij}(q)$. This model is completely *adiabatic* since the trajectory follows the potential energy without any coupling to the intrinsic degrees of freedom, cf. Fig.3. The other extreme has been persued in the *statistical* model of Fong [15] by assuming complete thermal equilibrium in the total system, thus dissipating the total ΔE_{ss} into internal excitation energy. In the Wilkins model [16], on the other hand, intermediate coupling between the collective and intrinsic degrees of freedom at the scission point is assumed. Typically, 10 MeV is in the collective motion with a collective temperature of about 1 MeV and about 20 MeV in the intrinsic system with an intrinsic temperature of about 0.75 MeV. Hereby, excitation energy is converted into *intrinsic temperature* according to the formula of Lang and LeCouteur [17] obtained from neutron evaporation spectra,

$$E^* = aT^2 - T \ . \tag{2}$$

The common use of $a = A/8\text{MeV}$ for the single particle level density parameter a at the Fermi energy, however, is only justified at very low temperature because the existing correlations are washed out at higher temperatures yielding $a \approx A/15\text{MeV}$ at $T \approx 5$ MeV, see [18]. The *collective temperature* is obtained from charge dispersion data.

The Wilkins model is closely related to Nörenberg's [20] quasistatic scission point model of intermediate coupling. He evaluates the some 10 collective modes at the scission point like bending, wriggling, twisting, e.t.c. and deduces from them the collective temperature which came out to be close to the one of Wilkins. Recently, a major deficiency of the Wilkins model has been removed by Koczon [21], namely by replacing the Boltzmann distribution by a microcanonical one, assuring that no part of the potential energy surface beyond the Q-value is explored. This is an important feature for *cold* fission events which come close to the Q-value.

In order to refine the dynamic LDM, without much worries about the validity it was natural to include the Navier-Stokes viscosity with an adjusted viscosity constant η in order to account for a damping of the descent from saddle to scission [22,23]. In solving the Navier-Stokes equation one gets the rate of energy loss

$$-\dot{W} = \sum_{ij} Z_{ij}\dot{q}_i\dot{q}_j \ , \tag{3}$$

with the viscosity coefficients $Z_{ij}(q) \propto \eta$ which are calculated in a similar way as the hydrodynamic inertias. The equations of motion then contain generalized linear frictional forces (the second term in the first sum),

$$\sum_j \left(M_{ij}\ddot{q}_j + Z_{ij}\dot{q}_j\right) + \sum_{jk}\left(\frac{\partial M_{ij}}{\partial q_k} - \frac{1}{2}\frac{\partial M_{jk}}{\partial q_i}\right)\dot{q}_j\dot{q}_k + \frac{\partial V}{\partial q_i} = 0 \ . \tag{4}$$

For e.g. ^{235}U thermal neutron fission the static gain in potential energy from saddle to scission was about $\Delta E_{ss} \approx 25$ MeV. Assuming that out of this about 10 MeV is dissipated, Wieczorek *et al.* [22] arrived at an estimate of $\eta \approx 10^{-23}\,\text{MeV}\,\text{sec}\,\text{fm}^{-3}$. Sierk and Nix [24] lateron used about the same value. Refinements of the potential energy due to curvature effects [25] and shell effects and also constrained Hartree-Fock calculations [27] now yield less steep potentials and give a much smaller value of ΔE_{ss}. Furthermore, the scission point probably does not coincide with the point of geometrically touching fragments but is situated before at the *exit point* [26] where the neck ruptures rapidly. Here the *collective* velocity becomes comparable to the *intrinsic* velocity so that the hydrodynamical concept breaks down. Accordingly, η should be appreciably smaller. From the spectra of particles emitted in high energy heavy ion reactions, on the other hand [28], one needs a value of η about 15 times larger which indicates a strong temperature dependence of η, unfortunately in the wrong direction because one expects that hot matter is less viscous than cold one. Attempts to deduce η from the trend of the average fragment kinetic energies vs. fissility did not yield decisive values because only the heavy actinides have a long way from saddle to scission and because their prescission contributions are rather small which could be accounted for as well by slight changes of the LDM parameters.

However, one can also conclude from this that the viscous mechanism is not applicable to low energy nuclear fission. Navier-Stokes viscosity is a typical two-body effect stemming from many collisions of particles with *short mean free paths* in a system at thermal equilibrium. This assumption, however, does not hold in nuclear matter nor in finite nuclei. As is well known to condensed matter physicists, the short mean free path (λ) limit of the Boltzmann equation for a Fermi gas corresponds to the *hydrodynamic* limit, where the viscosity constant is proportional to the mean free collision time $\tau = \lambda/v_F$, viz. $\eta = p_0\tau$ with the static pressure $p_0 = \rho_0 v_F^2/5$ and the average density ρ_0.

In the *long mean free path* or *zero sound* limit damping vanishes up to the order $1/\tau$, thus being in the collisionless regime, cf. e.g. [29]. This was noted by Wegmann in the context of nuclei [30] in 1974 who realized that another damping mechanism is more appropriate for nucleons with a long mean free path in nuclei, namely the nowadays called *one-body* damping of a *Knudsen gas*. This was originally studied by Landau, see [31], and results purely from the interaction of the particles in a container with its flexible walls. Particles in the interior with average density ρ_0 and average velocity $\bar{v}$ heading towards the wall are assumed to be completely randomized. If the wall moves locally with the normal velocity $\dot{\mathbf{n}}$, they are reflected in a specular manner and experience a momentum loss close to the surface. Then they are assumed to be randomized again when heading towards the interior. This manifests itself in an overall energy loss which reads in its simplest form (the *wall formula*)

$$-\dot{W} = \bar{v}\rho_0 \int \dot{\mathbf{n}}^2 \mathrm{d}S, \tag{5}$$

where the integral extends over the surface of the wall.

The classical background of this formula has been worked out by Swiatecki and collaborators [32], for a derivation via the Boltzmann equation see ref. [33], and it has been applied enthusiasticially to giant resonance vibrations, fission and heavy ion scattering with several modifications by Nix and others (wall-and window-, surface-plus-window one-body dissipation) to correct for translational invariance [24], and for surface enhancement [34]. However, with the normal parameters of the nuclear density and $\bar{v} = 3v_F/4$, where v_F is the Fermi velocity, any type of collective motion, in particular the descent from saddle to scission, turned out to be overdamped. In these calculations, hence, the total ΔE_{ss} is converted into internal excitation energy at the scission point.

Moreover, the formula cannot be valid for low-multipole vibrations because by symmetry of the shapes classical trajectories are closed orbits which do not allow for immediate randomization of the reflected particles. This is intimately related to the nowadays fashionable theories of *chaos* where integrable systems do not exhibit chaotic motion and, thus, are unlikely to exhibit damping phenomena. The entirely classical nature of the wall formula has several drawbacks. Firstly, it does not account for the Pauli principle which is least effective in the Fermi surface because of the large phase space available for collisions, see [19]. The mean free path of a nucleon in a nucleus is inversely related to the imaginary part of the optical potential, the latter vanishing at zero temperature at the Fermi energy. As a rule of the thumb, Bertsch [35] derives from the formula of Morel and Nozières [36] for the imaginary part of the optical potential at finite temperature T,

$$W \propto \epsilon^2 + \pi^2 T^2 \tag{6}$$

the relaxation time τ of the Fermi surface

$$\tau \approx 1 \mathrm{GeV\,fm}\, c^{-1} \,/\, \epsilon\,, \tag{7}$$

where ϵ is the energy per particle of a participating nucleon above or below the Fermi energy, viz. $\lambda \approx 250$ MeV fm/ϵ. For higher energies, $\epsilon > 50$ MeV, Blin *et al.* [19] calculate a rather constant mean free path of $\lambda \approx 5$ fm. Furthermore, the quadratic dependence on temperature is only valid in a small range of $|\epsilon|$ around zero which even breaks down in the nuclear surface. In using a finite range effective interaction, for nuclear matter the dependence becomes linear for $T > 5$ MeV. At low temperature the mean free path, thus, becomes extremely long on the Fermi surface. Calculated with a realistic interaction, the imaginary part of the optical potential is strongly surface peaked (surface absorption). In the nuclear surface, hence, the mean free path becomes short which favours a *surface peaked* damping mechanism. Secondly, the nuclear *wall* does not correspond to an infinitely high potential but rather to a Woods-Saxon shape. Thus particles scattered to high lying states can escape rather than being forced to rescatter back to the bottom of the potential. The effective phase space available, hence, is concentrated around the Fermi surface up to zero energy, see [37]. This can reduce the one-body damping mechanism by up to 90%. Thirdly, pairing is not included which prevents from excitations below the pairing gap and, thus, giving no damping at all.

The *microscopic* origin of the damping mechanism in fission has been considered from the beginning by Hill and Wheeler [2] and Wilets [3], mainly on the basis of level crossings. If the residual interaction (H') is sufficiently small to be treated as a perturbation and the collective velocity ($\dot{q}$) is large and constant, the probability of jumping from the lower level (before the crossing $|a\rangle$ and after the crossing $|b\rangle$) to the upper level (before the crossing $|b\rangle$ and after the crossing $|a\rangle$) is high according to the Landau-Zener-Stückelberg (LZS) formula

$$P_{\mathrm{low-up}} = \exp\left[-\frac{2\pi|\langle a|H'|b\rangle|^2}{\hbar\dot{q}\frac{\partial}{\partial q}(E_a - E_b)}\right] . \tag{8}$$

The adiabatic lowest collective potential energy, however, follows the level coming down, thus resulting in an occupation of higher levels and, accordingly, in an excitation energy with respect to the adiabatic surface. The time rate of change of the this excitation energy, however, by virtue of the nonanalytic behaviour of the LZS formula in $\dot{q}$, cannot be written as a biquadratic form in $\dot{q}$ and, hence, one cannot associate to it a linear frictional force.

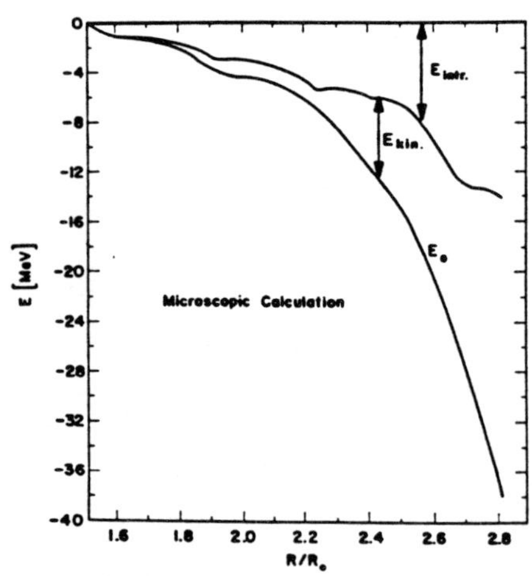

Figure 4:
Excitation energy of a fissioning ^{236}U as a function of elongation in the model of Schütte [38].

The LZS formula only crudely describes the real situation, i.e. level crossings are not isolated, the collective velocity is not constant and the residual interaction often is too large to be treated in first order perturbation theory. Therefore more realistic calculations were performed [38,39,40] including large configuration spaces of many-quasiparticle excitations $|\nu\rangle$, thus allowing for pairing effects and for pair breaking. By solving the time dependent Schrödinger equation, the occupation numbers $c_\nu(t)$ are obtained which define the excitation energy

$$E^*(q,\dot{q}) = \sum_\nu 2\epsilon_\nu |c_\nu|^2 \, . \tag{9}$$

The velocity at each instant of time is computed from the constancy of the total energy

$$E = V(q) + \tfrac{1}{2} M_q \dot{q}^2 + E^* \, . \tag{10}$$

This procedure works only in one dimension because only the *energy* is known and not the individual *forces*. For ellipsoidal deformations beyond the saddle point of a ^{236}U compound nucleus about half of the gain in potential energy in Schütte's model is converted into excitation energy, see Fig.4. Walliser *et al.* with a similar model [40] arrive at only 10 ⋯ 15%.

As an extension of this work, Wong and Tang [41] turned from the single particle basis to the time dependent Hartree Fock picture and derived a master equation for the occupation numbers. Work along these lines is still under way, in particular the inclusion of the correct formulation of the Boltzmann collision term.

Another *microscopic* approach has been applied to nuclear frictional problems since 1973, namely the *linear response theory* of Bohm and Pines [42], first by Beck and Gross [43] and lateron by Hofmann and Siemens [44]. By expanding the interaction Hamiltonian in first order in the collective variable,

$$H_{\text{int}}(q,\xi) = (q - q_0) F(\xi) \, , \tag{11}$$

or, more general, $F(\xi, q) = \partial H_{\text{int}} / \partial q$, one obtains the formal expression for the friction coefficient at zero temperature by expanding the Fourier transform of the response function around $\omega = 0$,

$$Z = 2\pi \langle 0 | F \delta'(E_0 - H_{\text{intr}}) | F | 0 \rangle \equiv 2\pi \sum_{n>0} |\langle n | F | 0 \rangle|^2 \delta'(E_0 - E_n) \, . \tag{12}$$

In choosing $q_0(t)$ with reference energy $E_0(t)$ as the collective path containing already the influence of the coupling in the past this procedure can be applied to small time steps in the future, thus applying perturbation theory properly. The derivative of the delta function in eq.(12) has to be replaced by the derivative of a Lorentzian via the imaginary part of the optical potential due to the finite width of the single particle states or a Strutinsky like temperature smoothing has to be performed. At finite temperature it is replaced by the derivative of a Gaussian. Yamaji *et al.* [45] in this way have obtained $Z = (1 \cdots 3) 10^{-21}$ MeV sec fm^{-2} at $T = 2$ MeV for symmetric fission of ^{212}Po as going from the ground state via the saddle point to the scission point. This compares roughly to 25% of the corresponding value of the one-body dissipation wall formula and yields underdamped motion.

It should be noted that according to eq.(6) the zero-frequency limit restricts the range of applicability of this procedure to finite temperatures $T \gg \omega/\pi \to 0$, large as compared to the intrinsic frequencies, i.e. to induced nuclear fission. The same holds for the theory of Grangé *et al.* [46] in which, viewing fission as a diffusion process, the Fokker-Planck equation

$$\frac{\partial P}{\partial t} + \frac{p}{\mu} \frac{\partial P}{\partial q} + \frac{\partial V}{\partial q} \frac{\partial P}{\partial p} = \beta \frac{\partial p P}{\partial p} + \beta T \mu \frac{\partial^2 P}{\partial p^2} \tag{13}$$

for the probability density $P(q, p; t)$ of the shape variable q and its conjugate momentum p is solved. Here $V(q)$ is the potential and μ is the reduced mass assumed to be constant. β is a friction constant estimated from experiments in the range $(0.1 \cdots 1) \cdot 10^{21}$ sec^{-1}. In this model, two regimes can be distinguished depending on the local curvature $|\omega_q|$ of the potential energy. In the first potential well (close to the ground state) one gets underdamped motion with $\beta < 2\omega_q$ whereas beyond the saddle $\beta > 2\omega_q$ and the descent towards scission becoming overdamped.

4. Summary and Outlook

We have outlined in short the theories and models which have been proposed and applied to the problem of dissipation in nuclear fission since about 25 years. They are all rather disconnected and often based on contrary assumptions, like short or long mean free path, strong and weak coupling between collective and intrinsic degrees of freedom (fast and slow intrinsic relaxation), classical and quantum mechanics e.t.c. yielding overdamped or only weakly damped motion.

This is partly due to the lack of experimental information on the internal excitation energy at the scission point. This situation improved only recently by the now available analyses of even-odd effects in thermal neutron fission. However, still missing are complete measurements of the fragment's excitation energies, i.e. of the neutrons and γ's emitted before and after scission for given mass split and given kinetic energy. Here one could learn what parts of the potential energy landscape are explored from the transition from cold compact fission with highest kinetic energies ($E_{\rm kin}^{\rm total}$) via the relatively hot events with medium $E_{\rm kin}^{\rm total}$ to the probably also cold deformed events with lowest kinetic energies [21,47]. On the other hand, from the $\bar{E}_{\rm kin}^{\rm total}$ of fission induced by slow neutrons one can learn about the washing out of shell effects and the onset of pair breaking, i.e. the onset of dissipation.

A unified theory of nuclear fission including damping, on the other hand, by the proper inclusion of pairing should distinguish between superfluid fission events below the pairing gap and those above. It should also account for realistic coupling strengths and mean free paths, thereby utilising all the experience we have already gained from the imaginary part of the optical potential. Here one can think of an RPA model with a realistic two body interaction and with a Landau-Boltzmann type collision term, see [48], or of its semiclassical analogue, the Vlasov-Uehling-Uhlenbeck equation.

References

[1] N. Bohr, J.A. Wheeler, *Phys. Rev.* **56** (1939) 426

[2] D.L. Hill, J.A. Wheeler, *Phys. Rev.* **89** (1953) 1102

[3] L. Wilets, *Theories of Nuclear Fission*, Clarendon, Oxford (1964)

[4] W.J. Swiatecki, S. Bjornholm, *Phys. Rep.* **4C** (1972) 326

[5] S. Bjornholm, Proc. 27th *Nobel Symp. on Super-heavy Elements – Theoretical Predictions and Experimental Generation*, Ronneby 1974, *Phys. Scripta* **10A** (1974) 110

[6] C.Wagemans, E.Allaert, A. Deruytter, R. Barthélémy, P. Schillebeeckx, *Phys. Rev.* **C30** (1984) 218

[7] H. Nifenecker, J. Lachkar, Y. Patin, J. Sigaud, Proc. *Third IAEA Symp. on the Physics and Chemistry of Fission*, Rochester (1974) vol.II, p.117

[8] F. Gönnenwein, J.P. Bocquet, R. Brissot, Proc. *XVII Int. Symp. on Nuclear Physics*, Gaussig, GDR, 1987

[9] D. Hilscher, D.J. Hinde, H. Rossner, Proc. *Texas A&M Symp. on Hot Nuclei*, College Station 1987, to be published

[10] R.W. Hasse, *Rep. Prog. Phys.* **41** (1978) 1027

[11] G.W. Ford, M. Kac, P. Mazur, *J. Math. Phys.* **6** (1965) 504

[12] R.W. Hasse, W.D. Myers, *Geometrical Relationships in Macroscopic Nuclear Physics*, Springer (1988)

[13] J.R. Nix, W.J. Swiatecki, *Nucl. Phys.* **71** (1965) 1

[14] R.W. Hasse, *Nucl. Phys.* **A118** (1969) 609

[15] P. Fong, *Phys. Rev.* **102** (1956) 434; *Statistical Theory of Nuclear Fission*, Gordon and Breach (1969)

[16] B.D. Wilkins, E.P. Steinberg, R.R. Chasman, *Phys. Rev.* **C14** (1976) 1832

[17] J.M.B. Lang, K.J.LeCouteur, *Proc. Phys. Soc.* **A67** (1954) 586; K.J.LeCouteur, D.W. Lang, *Nucl. Phys.* **13** (1959) 32
[18] R.W. Hasse, P. Schuck, *Phys. Lett.* **179** (1986) 313
[19] A. Blin, R.W. Hasse, B. Hiller, P. Schuck, *Phys. Lett.* **161B** (1985) 211; R.W. Hasse, P. Schuck, *Nucl. Phys.* **A438** (1985) 157; R.W. Hasse, J. Jaenicke, P. Schuck, C. Grégoire, B. Remaud, *Prog. Part. Nucl. Phys.* (1988), in print
[20] W. Nörenberg, Proc. *Second IAEA Symp. on the Physics and Chemistry of Fission*, Vienna (1969) p.51
[21] P. Koczon, *Thesis, TH Darmstadt* (1988)
[22] R. Wieczorek, R.W. Hasse, G. Süssmann, Proc. *Int. Conf. on Nuclear Physics*, München (1973) vol.1, p.585; Proc. *Third IAEA Symp. on the Physics and Chemistry of Fission*, Rochester (1974) vol.I, p.523
[23] K.T.R. Davies, A.J. Sierk, J.R. Nix, *Phys. Rev.* **C28** (1983) 679
[24] A.J. Sierk, J.R. Nix, *Phys. Rev.* **C3** (1980) 982
[25] J. Treiner, R.W. Hasse, P. Schuck, *J. Physique Lettres* **44** (1983) L-733
[26] M. Asghar, R.W. Hasse, Proc. *Workshop on Semiclassical Methods in Nuclear Physics*, Grenoble 1984, *J. Physique Coll.* **45** (1984) C6-455
[27] J.F. Berger, *these Proceedings*
[28] B. Schürmann, preprint (1988) subm. to *Mod. Phys. Lett.* A
[29] R.W. Hasse, Proc. *Workshop on Fluid Dynamical Approaches to the Many-body Problem: Fundamental and Mathematical Aspects*, Catania 1984, *Nuovo Cim.* **87A** (1985) 109
[30] G. Wegmann, *Phys. Lett.* **50B** (1974) 327
[31] I.L. Bekarevich, I.M. Khalatnikov, *JETP (USSR)* **39** (1960) 1699; Engl. transl. *Sov. Phys. JETP* **12** (1961) 1187
[32] J. Blocki, Y. Boneh, J.R. Nix, J. Randrup, M. Robel, A.J. Sierk, W.J. Swiatecki, *Ann. Physics* **113** (1978) 330
[33] R.W. Hasse, *J. Phys. Lett.* **G5** (1979) L101
[34] J.R. Nix, A.J. Sierk, Proc. *Int. Symp. on Perspectives in Nuclear Physics*, Madras 1987
[35] G. Bertsch, *Z. Phys.* **A289** (1978) 103
[36] P. Morel, P. Nozières, *Phys. Rev.* **126** (1962) 1909
[37] C. Yannouleas, *Nucl. Phys.* **A439** (1985) 336; J.J. Griffin, M. Dworzecka, *Nucl. Phys.* **A455** (1986) 61
[38] G. Schütte, L. Wilets, *Nucl. Phys.* **A252** (1975) 21; G. Schütte, *Phys. Rep.* **80** (1981) 113; G. Schütte, Proc. *Int. Symp. on Nuclear Fission and Related Collective Phenomena and Properties of Heavy Nuclei*, Bad Honnef 1981, *Lecture Notes in Physics, Springer* (1982) p.176
[39] D. Glas, U. Mosel, *Nucl. Phys.* **A264** (1976) 268
[40] H. Walliser, K. Wildermuth, F. Gönnenwein, *Z. Phys.* **A329** (1988) 209
[41] C.Y. Wong, H.H.K. Tang, *Phys. Rev.* **C20** (1979) 1419
[42] D. Bohm, D. Pines, *Phys. Rev.* **32** (1953) 609
[43] R. Beck, D.H.E. Gross, *Phys. Lett.* **47B** (1973) 143
[44] H. Hofmann, P. Siemens, *Nucl. Phys.* **A257** (1976) 165; *Nucl. Phys.* **A275** (1977) 464; A.S. Jensen, K. Reese, H. Hofmann, P.J. Siemens, Proc. *Fourth IAEA Symp. on the Physics and Chemistry of Fission*, Jülich 1979, IAEA (1980) vol.II, p.423
[45] S. Yamaji, A. Iwamoto, *Z. Phys.* **A313** (1983) 161; S. Yamaji, H. Hofmann, R. Samhammer, *Nucl. Phys.* **A475** (1987) 487
[46] P. Grangé, H.A. Weidenmüller, *Phys. Lett.* **96B** (1980) 26; P. Grangé, Li Jun-Qing, H.A. Weidenmüller, *Phys. Rev.* **C27** (1983) 2063; P. Grangé, *Nucl. Phys.* **A428** (1984) 37c; H.A. Weidenmüller, *Nucl. Phys.* **A471** (1987) 17c; S. Hassani, P. Grangé, *Z. Phys.* **A325** (1986) 95
[47] R.W. Hasse, Proc. *All-Union Symp. on Physics and Chemistry of Fission*, Obninsk 1987, *Vop. At. Nauki Techn. Ser. Yad. Konst.* **1** (1988) 3
[48] A.A. Abrikosov and I.M. Khalatnikov, *Rep. Prog. Phys.* **22** (1959) 329

INFLUENCE OF DIABATICITY ON FISSION FRAGMENT MASS ASYMMETRY

M.T. Matev and B.D. Slavov
Sofia, Bulgaria

The theoretical description of a heavy nucleus, undergoing spontaneous or induced fission leads to analysis of solutions of equations of motion in a properly chosen configuration space of collective (deformation) coordinates $\{q_i, i = 1, ..., n\}$. Presuming that the wave function satisfies the collective Schrödinger equation with a collective potential in this configuration space, essential information about the process can be extracted from its potential energy surface (PES). The static approaches result in static PES, which, one can expect, provides information about the process in the adiabatic approximation. However in certain regions of the PES the increase of the collective kinetic energy due to a fall of the potential energy of the system makes the adiabatic approximation unapplicable and even contradictory. With the increase of the collective velocities the nucleons experience time dependent interaction potential, this causes couplings of single-particle to collective degrees of freedom to appear, leading to diabatic excitations of single-particle states which result in "dynamical heating" of the nucleus. Evidence for considerable diabatic effects has been obtained for the fission process even at low energies [1], during the descent of the nucleus from saddle to scission point, thus their influence on the PES cannot be neglected and is of great interest. The authors of that work studied the evolution of the occupation probabilities by applying the Landau-Zener asymptotic formula in a basis of quasiparticle excitations.

The microscopic dissipation mechanisms have been later studied in details for nucleus-nucleus collisions in ref.2, where in terms of time-dependent shell model a dynamical basis of single particle states has been introduced, which incorporated main parts of the interaction. This led to a Liouville equation for the density matrix $\rho(t)$ of the form (with $\hbar = 1$)

$$i\frac{\partial \rho(t)}{\partial t} = L(t)\rho(t), \tag{1}$$

represented in single-particle basis by

$$i\frac{\partial \rho_{\alpha\beta}(t)}{\partial t} = \sum_{\gamma\delta} L_{\alpha\beta,\gamma\delta}(t)\rho_{\gamma\delta}(t) = \sum_{\gamma\delta} \left(u_{\alpha\gamma}\delta_{\delta\beta} - u_{\delta\beta}\delta_{\alpha\gamma}\right)\rho_{\gamma\delta}(t) \tag{2}$$

where the coupling matrix elements in the Liouville operator $L(t)$ were defined as

$$u_{\beta\alpha}(t) = \exp\left(-i\int^{t} dt'(\epsilon_\alpha - \epsilon_\beta)\right)\langle\beta|\, H'(q(t)) - i\sum_k \dot{q}_k \frac{\partial}{\partial q_k}\,|\alpha\rangle \tag{3}$$

($H'(q(t))$ stays for the nondiagonal part of the hamiltonian in the basis). In the weak coupling limit after projection of $\rho(t)$ on its diagonal part, Liouville equation was transformed to a system of differential equations for the occupation probabilities $n_\alpha = \rho_{\alpha\alpha}(t)$ of the single particle states

$$\frac{\partial n_\alpha}{\partial t} = \sum_\beta W_{\alpha\beta}\left(n_\beta - n_\alpha\right) \tag{4}$$

The authors of this outstanding work obtained expressions for the dynamic dependence of various terms which appeared in the equation of motion for the collective variables q_i, like mass parameters, friction forces etc. and suggested that it must be solved simultaneously with the equation for the density matrix (2).

In ref.1 an evolution of the process was traced along only one deformation coordinate (namely the elongation), which we consider was not realistic enough for the case of ^{236}U. Asymmetric degree of freedom should have been used as a collective coordinate as well, which appears to be extremely essential beyond the second (asymmetric!) saddle point of the PES.

For that in the present work we utilized a three-parameter asymmetric two-center shell model (ATCSM), introduced in [4] with shell correction renormalization for the total static potential energy. The model itself had been specially designed for the reproduction of the fragment mass asymmetry and had described the system at large elongations (beyond the saddle point) quite satisfactory [5]. PES was taken for fixed value of the fragment elongation parameter $\alpha = 1.0$ and the evolution of the system was studied on an equidistant mesh for the other two (d, κ), where d is the distance between the centers of the fragments and $\kappa = A_H/A_L$ is the asymmetry coordinate.

We tried to solve the system of differential equations for the occupation probabilities precisely and simultaneously with equation of motion:

$$\frac{1}{2}\sum_{ij} B_{ij}\dot{q}_i\dot{q}_j + V(q(t)) = E = const \tag{5}$$

, in this two-dimensional configuration space. The term $V(q(t)) = E_{stat}(q) + E_{dyn}(q,t)$ now represents the local "dynamic" potential energy surface, which consists of a static part E_{stat}

(from the static PES) and a dynamic contribution E_{dyn} (equivalent to intrinsic excitation energy E^*) depending on the history of the process (i.e. on the trajectory of the nucleus).

The starting deformation (initial conditions for the dynamic model) was chosen to be the saddle point of the static landscape, assuming that it is realistic for the energy interval being studied (6-8 MeV above the ground state).

We applied consistent iterative procedure for determination of the collective velocity from the solution of the equation of motion (unlike [1], where authors used interpolation formula for the collective velocity $\dot{d}$ – free fall between *a priori* chosen points along the elongation axis).

Collective motion of the system was traced on the mesh with iterative determination of $T_{coll} = \frac{1}{2}\sum_{ij} B_{ij}\dot{q}_i\dot{q}_j$. Up to now only diagonal components of the mass tensor were considered, presuming that $B^2_{d\kappa} \ll B_{dd}B_{\kappa\kappa}$. Hidrodynamic estimate [3] for B_{dd} was used in the concrete calculations and $B_{\kappa\kappa}$ was omitted. Hence, the collective dynamics with respect to d. as mentioned above, was estimated consistently by the equation of motion (5) and the equation for the occupation probabilities (4), while the motion along the asymmetry coordinate was treated by steepest descent minimization.

By now, in $W_{\alpha\beta}$ we took into account only 1-body friction term in the form of the Landau-Zener transition rate value

$$W_{\alpha\beta} = \frac{2\pi\left|H'_{\alpha\beta}\right|^2}{\dot{q}\left|\partial(\epsilon_\beta - \epsilon_\alpha)/\partial q\right|}$$

where $H'_{\alpha\beta} = \langle\beta\,|H'(q)|\,\alpha\rangle$.

In the hamiltonian we utilized schematic nondiagonal interaction between single particle states of the form

$$H'_{\alpha\beta} = C\,(j_z(\alpha)/j_c)^2\,\delta_{j_z(\alpha),j_z(\beta)} \tag{6}$$

where $j_z(\alpha)$ stays for the z projection (on the elongation axis d) of the total spin of nucleon state α and $j_c = 10$. The shell model single particle basis was so far restricted for the sake of simplicity to 100 states respectively for protons and for neutrons.

In order to obtain reasonable dynamics, we varied so far the constant C of this ansatz with the aim to reproduce roughly the time scale of $10^{-20}s$ for the saddle-to-scission descent time interval [1]. In our case this led to C-values between 0.006 and 0.01 MeV.

More elaborate calculations are now in progress and the following items are paid special attention:

1. The Landau-Zener approximation for $W_{\alpha\beta}$ is far from the physical situation. It will

further on be replaced by terms, which are nonlocal in time and represent more correctly the dissipation mechanisms [2].

2. Dynamical coupling term $\dot{d}\frac{\partial}{\partial d} + \dot{\kappa}\frac{\partial}{\partial \kappa}$ in $u_{\alpha\beta}$ of eq.(3) was omitted for the sake of computational simplicity. It must be taken into account and is expected to reduce considerably some of the elements in the transition rate matrix $W_{\alpha\beta}$ in (4).

3. The basis, especially for neutrons, needs to be enlarged.

4. An additional term, accounting for the pairing correlations correctly in $H'_{\alpha\beta}$ may be of major importance.

References

[1] G.Schütte,P.Moller,J.R.Nix,A.J.Sierk,*Z.Phys.*A297 (1980) 289.

[2] S.Ayik, W.Nörenberg,*Z.Phys.*A309,(1982) 121.

[3] G.Schütte,*Phys.Rep.*80 (1981) 113.

[4] B.D.Slavov,J.E.Galonska,A.Faessler,*Phys.Lett.*31B (1971) 483.

[5] B.D.Slavov,A.Faessler,*Z.Phys.*A271 (1974) 161.

SPACE PARITY VIOLATION IN NUCLEAR FISSION

G.A. Petrov
LNPI, Gatchina, Leningrad District, USSR

Space parity violation in low energy fission was observed by the Soviet physicists in 1977 in the course of the angular distribution investigations of the light and heavy fragments in the polarized thermal neutron fission of $^{233,235}U$ and ^{239}Pu [1]. Unexpected large values of the P-odd asymmetry coefficients ($\bar{\alpha}_{nf} \sim 10^{-4}$) were obtained in the angular distribution

$$W(\theta) = 1 + \bar{\alpha}_{nf} (\vec{\sigma}_n \cdot \vec{p}_f) . \quad (1)$$

So large values of $\bar{\alpha}_{nf}$ looked very strange because of the existance of numerous different exit channels in fission process ($\sim 10^8 - 10^{10}$) and a small relative value of nonconserving space parity potential of the weak NN-interaction ($\sim 10^{-7}$).

The discovery of this P-odd asymmetry was interesting not only as an example of the weak interaction revealing in a complex nuclear reaction but because it promised a new way do recieve information about fission process itself.

In addition to the P-violating asymmetry $\bar{\alpha}_{nf}$ P-conserving right-left asymmetry of the fission fragment angular distribution

$$W(\theta) = 1 + \bar{\alpha}^{RL}_{nf} \vec{p}_f \cdot [\vec{\sigma}_n \times \vec{p}_f] \quad (2)$$

was observed in 1979 [2]. This type of asymmetry arises as a result of interference of the s- and p-resonances in the reaction differential cross sections. In the formulas (1) and (2) $\vec{\sigma}_n$ is a unit pseudovector of a neutron transversal polarization, $\vec{p}_f$ and $\vec{p}_n$ are the unit vectors of the light fission fragments and neutron momenta.

The main goal of the new experimental investigations of P-odd and P-even effects in fission was a search of the possible relationships between the asymmetry coefficients and the characteristics of the entrance and exit channels in slow neutron fission. The information of this type is quite necessary to construct an adequate theory the first versions of which were introduced shortly after this experimental programme began to be realized [3,4].

In this paper the brief review of the experimental results obtained by different groups is given. The main part of these results have been obtained at the WWR-M reactor of Leningrad Nuclear Physics Institute (LNPI).

1. Integral coefficients of P-odd and P-even asymmetries of the light fission fragment emission

All known results of the integral asymmetry coefficient measurements for the light fission fragments are presented in Table I.

Table I

	^{233}U	^{235}U	^{237}Np	^{239}Pu	^{241}Pu	^{245}Cm	**
$\bar{\alpha}_{nf}\cdot 10^4$	2.8(3)	1.5(4)		-4.8(8)			A
	4.5(4)	1.2(2)		-7.0(7)			
	3.8(6)			7.8(I.3)			B
	3.6(3)	0.8(1)					C
		0.84(6)					
	4.8(4)	1.0(4)	7(26)	6.2(4)	0.3(3)	0.1(2)	D
	5.3(3)						
	4.4(1)						
	4.9(5)						E
$\bar{\alpha}^{RL}_{nf}\cdot 10^4$	-3.2(3)*	1.6(1)					C
	-6.4(5)*			1.3(3)	-0.4(8)		D

* These results were obtained with different energies of polarized neutrons.

** Exp.groups: A - Danilyan et al., B - Andreev et al., C - Lobashov et al., D - Petrov et al., E - Bondarenko et al.

The experiments were carried out with thermal neutrons ($\bar{\lambda} \simeq 1.5 - 4$ Å) reflected on the magnetized mirror. One can see that P-odd asymmetry obviously exists in the cases of $^{233,235}U$ and ^{239}Pu fission alongside with the P-even right-left asymmetry of fission fragment emission.

It is very interesting to note that the first attempts to search P-odd effect in ternary fission gaves following results

^{233}U: $\bar{\alpha}^{tern}_{n\alpha} = -(0.4 \pm 0.6)\cdot 10^{-4}$ (Petrov et al.)

^{239}Pu: $\bar{\alpha}^{tern}_{nf} / \bar{\alpha}^{bin}_{nf} = (0.67 \pm 0.20)$ (Danilyan et al.)

The observation of approximally the same results for the coefficient $\bar{\alpha}_{nf}$ in ternary and binary fission and the absence of asymmetry of this type for light particle emission favour the quite different mechanisms of the fragment and light particle creation in ternary fission.

It is rather strange that the both types of asymmetry were not observed at the level $\sim 10^{-4}$ in the case of ^{241}Pu fission in spite of a similarity of the ^{239}Pu and ^{241}Pu fission properties.

P-odd fission fragment emission asymmetry was not detected also in the case of ^{245}Cm and ^{237}Np. The last case is especially interesting because the great enhancement of P-odd effect can be expected as a result of a possible difference in barrier heights for s- and p-resonance fission [5].

So up to now only three cases of fissioning heavy nuclei are known where P-odd effect $\sim 10^{-4}$ was observed. This situation may be a result of the different fission properties influence or it is a question of experimental accuracy achieved.

2. Dependence of $\bar{\alpha}_{nf}$ and $\bar{\alpha}^{RL}_{nf}$ coefficient values on the fragment masses and total kinetic energies

Masses and total kinetic energies of fragments in ^{233}U fission were estimated by the measurement of two fragment energies with rather poor resolution ($\Delta M \simeq 6.8$ a.m.u., $\Delta E \simeq 2$ MeV). As a result of the experiments two-dimentional matrices of the experimental asymmetry coefficient values were obtained. These matrices were fitted by the expansion

$$\bar{\alpha}_{nf} \simeq \bar{\alpha}_0 + \bar{\alpha}_1(M - M_0) + \bar{\alpha}_2(E - E_0) . \quad (3)$$

Here M, E are the fragment masses and total kinetic energies and M_0, E_0 are their most probable values. The expansion coefficients α_0, α_1 and α_2 after including all corrections are presented in Table II. One can see that within the accuracy of the data there is no visible dependence of the asymmetry coefficients on the masses and total kinetic energies of fission fragments. The same results were obtained also for the right-left asymmetry coefficients.

Table II

Polarized neutron beam			Depolarized neutron beam		
$\bar{\alpha}_0 \cdot 10^4$	$\bar{\alpha}_1 \cdot 10^4$	$\bar{\alpha}_2 \cdot 10^4$	$\bar{\alpha}_0 \cdot 10^4$	$\bar{\alpha}_1 \cdot 10^4$	$\bar{\alpha}_2 \cdot 10^4$
4.4(1)	0.03(2)	0.01(1)	0.2(2)	0.002(28)	-0.03(2)

3. Dependence of $\bar{\alpha}_{nf}$ and $\bar{\alpha}^{RL}_{nf}$ coefficient values on neutron energy

The polarized monochromatic neutrons with the energies up to ~ 2 eV were obtained by the neutron reflection method on magnetized monocrystal of Geisler alloy. Because of relatively low neutron intensities the experimental accuracy $\sim 10^{-5}$ could be obtained only for rather low neutron energies. All results of the experiments are presented on Figs. 1, 2 and 3. Except the ^{235}U where the statistical accuracy of the data is not enough to do definite conclusion about asymmetry coefficient behaviour the asymmetry coefficients $\bar{\alpha}_{nf}$ and $\bar{\alpha}^{RL}_{nf}$ have irregular energy dependences near well known strong s-resonance 0.29 eV (^{239}Pu) and in the vicinity of a weak 0.17 eV resonance of ^{233}U. As to the case of ^{239}Pu fission the dependence of $\bar{\alpha}_{nf}$ coefficient on neutron energies can be explained if one supposes the 0.29 eV resonance spin to be 1^+ and the spin of so called negative neutron resonances giving visible contribution in neutron fission cross section to be 0^+. Then for the asymmetry coefficient value observed in experiment one can write

$$\bar{\alpha}^{obs}_{nf} \simeq \bar{\alpha}_{nf}(1^+) \frac{\sigma_{nf}(1^+)}{\sigma_{nf}(1^+) + \sigma_{nf}(0^+)} \quad . \tag{4}$$

Here $\bar{\alpha}_{nf}(1^+)$ is the asymmetry coefficient for pure 0.29 eV resonance, $\sigma_{nf}(1^+)$ and $\sigma^{tot}_{nf} = \sigma_{nf}(1^+) + \sigma_{nf}(0^+)$ are proper fission cross sections as a function on neutron energy. To understand the energy behaviour of $\bar{\alpha}_{nf}$ and $\bar{\alpha}^{RL}_{nf}$ coefficient for ^{233}U it is necessary to use the following approximate theoretical equations for the vicinity of p-resonance [3,4]

$$\begin{aligned} \bar{\alpha}_{nf} &\simeq \text{const}\cdot(E - E_p) \ , \\ \bar{\alpha}^{RL}_{nf} &\simeq \text{const}\cdot\sqrt{E}\ \left[(E - E_p)\,\Gamma_p - (E - E_s)\,\Gamma_s\right]. \end{aligned} \tag{5}$$

4. Main conclusion for the fission physics

Recent theory of space parity violation in fission [3,4] proposes that P-odd mixing of nuclear states with opposite parities is appeared at the compound stage of fissioning nucleus exited by slow neutron capture. Then the exited compound state with mixted parity can be fissioned through the transition states near the barrier top with the definite quantum numbers J and K which can be splitted on parity if the pearshape deformation exists there. Each member

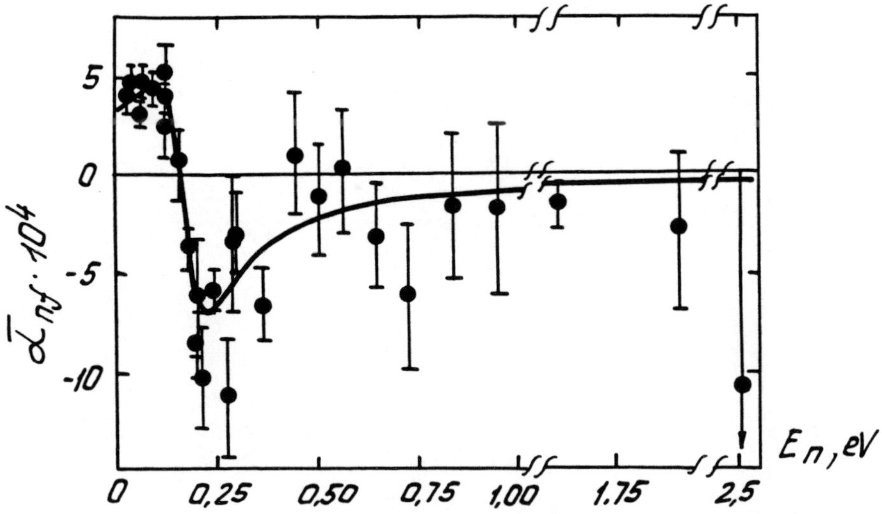

Fig.1. Dependence of the P-odd asymmetry coefficient on neutron energy for ^{233}U fission and theoretical description of the data for E_p = 0.16 eV and Γ_p = 0.14 eV.

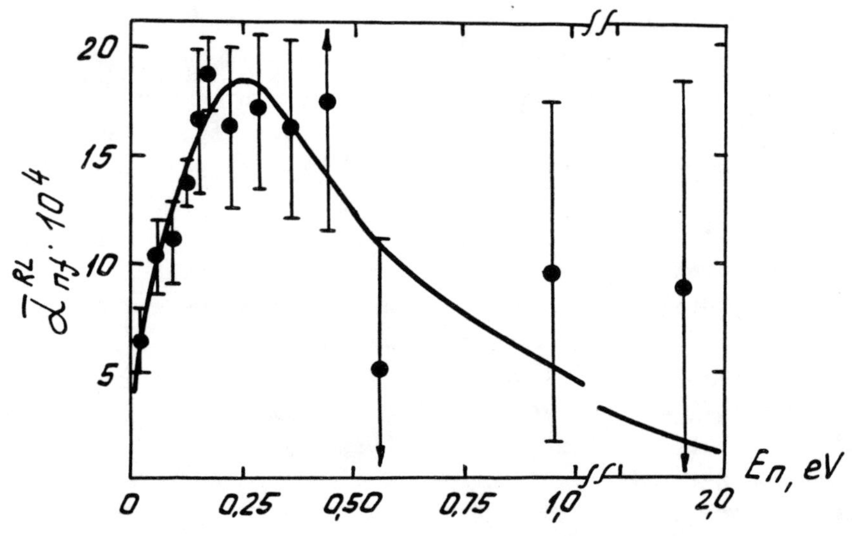

Fig.2. Dependence of the right-left asymmetry coefficient on neutron energy for ^{233}U fission and theoretical description of the data for E_p = 0.18 eV and Γ_p = 0.29 eV.

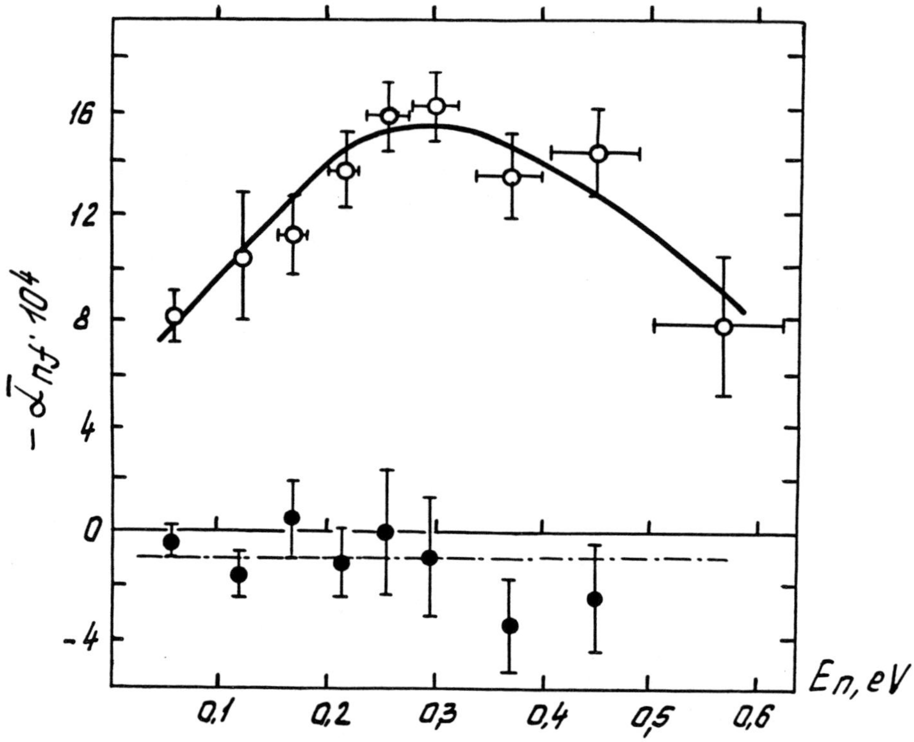

Fig.3. Dependence of the P-odd asymmetry coefficients for ^{239}Pu (⊢ο⊣) and ^{235}U fission on neutron energies. The solid curve is theoretical description by the equation (4), dashed line is average value of coefficient.

of such parity dublet is characterized by definite orientation of pear shape drop axis relative to the spin polarized partly after neutron capture. If such relative orientation and the pear shape deformation of charged drop are conserving up to the scission point the P-odd asymmetry of light (and heavy) fission fragments emission will appeare. One can see that in the frame of such theory many fission characteristics of a great importance are involved in the P-odd (and P-even) effect forming and the proper investigations of this effects can give an interesting information. Even now when only the first experiments have been carried out it is clear that the typical quantum mechanical characteristic as a space parity can play important role in such quasi-classical process as nuclear fission.

The existance of the both types of asymmetries and independence of the asymmetry coefficients on the masses give the possibility to conclude that the general characters of mass distribution in low energy fission of heavy nuclei is forming already near the top of the external barrier and the mutual orientation of spin and symmetry axis of deformed nucleus is conserving along the way to the scission point.

The other possible consequences of the first experiments about the ternary fission mechanism and parameters of neutron resonances have been mentioned above.

The progress in modern accelerator and nuclear reactor technology gives a real chance to use the P-odd and P-even effect in fission to recieve the information of a great importance about this very complex nuclear process.

References

1. Danilyan G.V. Usp.Fiz.Nauk, 131 (1980) 329.
2. Borovikova N.V., Lobashev V.M. et al. Pis'ma Zh.Eksp.Teor.Fiz., 31 (1980) 704.
3. Flambaum V.V., Sushkov O.P. Usp.Fiz.Nauk, 136 (1982) 3.
4. Bunakov V.E., Gudkov V.P. Nucl.Phys., A401 (1983) 93.
5. Vladimirski V.V., Andreev V.N. Zh.Eksp.Teor.Fiz., 41 (1961) 663.

THEORY OF PARITY VIOLATION IN NUCLEAR FISSION

V.V. Flambaum and O.P. Sushkov
Institute of Nuclear Physics, 630090 Novosibirsk 90, USSR

Abstract: The parity violation in nuclear fission induced by slow neutrons is considered. The effect is caused by the mixing of opposite parity rotational states of a cold, strongly deformed nucleus. This mixing is due to weak interaction in the compound nucleus.

Spatial parity violation induced by polarized thermal neutrons has been observed in the fission of ^{233}U, ^{235}U and ^{239}Pu[1-6]. The effect consists of the correlation of the momentum of the light fission fragment with the spin of the initial neutron. The asymmetry is the order of 10^{-4}. When one attempts to give a theoretical treatment of this effect the following problems arise:

(1) The relative strength of the parity violating weak interaction in nuclear forces is known to be of the order $Gm_{\pi}^2 \approx 10^{-7}$. Why is the experimental asymmetry much larger?

(2) How does the two-particle weak interaction affect the collective, really macroscopic motion of the system?

(3) The number of final states of the fission fragments is very large. Thus, the effect seems to be eliminated because of statistical averaging.

We suppose that the P-odd angular distribution of the fission fragments is formed at the cold stage similar to the usual P-even one. At this stage the nucleus is a pear-shaped top. The P-odd asymmetry is due to the mixing of opposite parity rotational states of the top. This mixing is induced by the weak interaction in the compound nucleus*). The natural magnitude of the effect is the same as for reaction (n, γ), i.e. ~ 10^{-4}.

Nuclear fission at not too high energy is known to go through the following stages:

(1) Neutron capture and formation of a hot compound nucleus.

(2) Cold pear-shaped nucleus.

(3) Scission of the neck.

All the variety of final states are formed at the last stage. Let ψ be the wave function (WF) of the nucleus after the capture of a neutron. Expand ψ in products of one-quasi-particle WFs (quasi-particles are nucleons, holes and core vibrations):

*)The supposition that the compound state plays an important role was made firstly by G.V. Danilyan and E.S. Rzhevsky

$$\psi_\eta = \sum_i^N a_i \varphi_i^\eta + A|a>^\eta \tag{1}$$

Here η is the parity of ψ. The states φ_i correspond to excitation of one, two ..., etc. quasi-particles. The number of terms in ψ is of the order of 10^6. On the other hand, it is known that the fission goes through a small number of fission channels. For simplicity we consider the case of one channel. This means that in ψ only one term, which corresponds to this fission channel, has a transition into the continuous spectrum. We isolate this term $|a>$ in eq. (1). The fission probability is proportional to $|A|^2$. The state $|a>$ is obviously the one with maximal deformation corresponding to the fission stage. Indeed, all the energy here is concentrated in the deformation of the core (core vibration degree of freedom), and there are no other excitations. In the state $|a>$, the nucleus is a pear-shaped top. The corresponding WF can be represented in the form[7]:

$$|a> \equiv |a,K>_{JM}^{\varphi} = \sqrt{(2J+1)/8\pi}\,\{D_{MK}^{J}(\varphi,\theta,\Omega)\,|a,K> +$$

$$\eta(-1)^{J+K}\, D_{M,-K}^{J}(\varphi,\theta,\Omega)\,|a,-K>\} \tag{2}$$

where η is the parity, J is the total angular momentum, K is projection of J on the axis of the top, $|a,K>$ is an internal state of nucleus. Here we suppose that $K \neq 0$ since, if $K = 0$, there are no rotational states of opposite parity at fixed J. The form of the WF of opposite parity is similar to that of ψ_η:

$$\psi_{\bar\eta} = \sum_k b_k \varphi_k^{\bar\eta} + B\,|a,K>_{JM}^{\bar\eta} \quad , \quad \bar\eta \equiv -\eta \tag{3}$$

The level density in the compound nucleus is very large (at $A \approx 240$, the distance between the levels is of the order of 1 eV). In refs. 8-11 it is shown that a high level density leads to enhancement of the mixing of nearest opposite parity levels. The enhancement factor is ~ $\sqrt{N} \sim 10^3$, where N is the typical number of terms in the wave functions (1,3), $N \sim \Delta E/D \sim 10^6$, $\Delta E \sim$ MeV is the interval between single-particle levels, $D \sim$ eV is the interval between compound-state levels. The relative strength of the parity violating weak interaction in nuclear forces is of the order of $Gm_\pi^2 \sim 10^{-7}$.

Thus, from perturbation theory

$$\psi = \psi_{\eta} + \beta\psi_{\bar{\eta}} = \Sigma\ a_i\varphi_i^{\eta} + \beta\Sigma\ b_k\varphi_k^{\bar{\eta}} + A\ [\ |a,K\rangle_{JM}^{\eta} + \beta\ \frac{B}{A}\ |a,K\rangle_{JM}^{\bar{\eta}}\ ],$$

$$\beta = \frac{\langle\ \psi^{\bar{\eta}}\ |\ H_W\ |\ \psi^{\eta}\ \rangle}{E - E_{\bar{\eta}} + \frac{1}{2}i\ \Gamma_{\bar{\eta}}} \sim \sqrt{N}\ Gm_{\pi}^2 \tag{4}$$

Here E is the initial energy, $E_{\bar{\eta}}$ and $\Gamma_{\bar{\eta}}$ are the energy and the width of the admixed level. The part of the WF corresponding to the cold state can be written as follows:

$$|a,K\rangle_{JM}^{(cs)} = |a,K\rangle_{JM}^{\eta} + \alpha\ |a,K\rangle_{JM}^{\bar{\eta}},$$

$$\alpha = \sum_{\nu} \beta_{\nu} \frac{B_{\nu}}{A_{\nu}} = \sum_{\nu} \frac{\langle\ \bar{\eta},\nu |\ H_W\ |\eta\ \rangle}{E - E_{\nu} + \frac{i}{2}\ \Gamma_{\eta}}\ (\Gamma_{\nu}/\Gamma)^{1/2}\ e^{i(\varphi_{\nu}-\varphi)} \tag{5}$$

Here Γ and φ are the fission width and the phase of the transition into the cold stage. We denote these parameters for admixed states by Γ_{ν} and φ_{ν}. The phases φ and φ_n are due to the instability of the nucleus. Both the basic and admixed WF in (5) correspond to the same internal state of the nucleus and differ only in the microscopic rotation: their expansions include orbital angular momenta of opposite parity. Therefore the amplitudes of fission from $|a,K\rangle^{\eta}$ and $|a,K\rangle^{\bar{\eta}}$ into any internal state of fragments coincide, and the WF for $\tau \to \infty$ will be also of the form (5). The angular distribution of the fragments can be using $|a,K\rangle_{JM}^{\eta}$ from eq. (2):

$$W(\theta) = 1 + a\ \cos\theta, \qquad a = \frac{2K}{I+1/2}\ (-1)^{J-I-1/2}\ \mathrm{Re}(\alpha) \tag{6}$$

Here θ is the angle between the neutron spin and the momentum of the light fragment, I is the spin of the unpolarized target nucleus. If the fission goes through several channels with different K, then $a = \Sigma\ W_K\ a_K$, where W_K is the corresponding fission probability.

Stress that the effect has resonance energy dependence near p-wave compound-resonance ($a \sim \mathrm{Re}(e^{i\beta}/(E-E_{\nu}))$, E is the energy of p-wave resonance (state $\psi_{\bar{\eta}}$) which has the parity opposite to that of s-wave resonance (state ψ_η). We suppose that fission by s-wave neutron dominates at low energies.

We emphasize that the effect is due to the orientation of the nucleus before scission. Indeed, there is a mean orientation of the nucleus along $\vec{J}$ in the state (5). The correlation $\vec{J}\,\vec{n}$ is both p- and T-odd ($\vec{n}$ is the axis of the nucleus). Therefore it will arise because of the finite life-time of the nucleus only, when Γ_ν or $\varphi-\varphi_\nu \neq 0$. In other words, one can say that it is the usual T-even correlation $\vec{J}\,\langle\vec{p}\rangle$ where $\langle\vec{p}\rangle \sim \Gamma\,\vec{n}$ is the averaged fragment momentum before scission. In real situation interferences of several capture resonances can occur. An interesting effect arises when resonances with different J overlap: the asymmetry is not equal to zero even for K = 0, when there are no rotational levels of opposite parity at fixed J.

Now we formulate our answers to the questions posed at the beginning of this paper:

(1) The effect is induced by the weak interaction at the stage of the compound nucleus. Due to a high level density the magnitude of the mixing of opposite parity states of the compound nucleus is ~ 10^{-4};

(2) The cold stage of the fission, where the nucleus has a pear-shaped deformation, is important. In the cold stage, parity is connected with the rotational state of a nucleus. Therefore P-odd mixing in the compound nucleus turns into mixing of rotational states. The value of the mixing retains the same order of magnitude.

(3) At the cold stage the process goes through a small number of channels (~1). Due to parity non-conservation and the finite life-time of the nucleus the direction of the fission axis is correlated with the neutron spin even before scission. Therefore the large number of final states of the fragments does not lead to statistical suppression of the effect.

There is also an interesting question why the nucleus does not "forget" the weak mixing of initial compound states during the long complicated process of fission. Experimentally, the excitation energy of a nucleus is specified very accurately, within $\delta E \sim 0.01$ eV, which is the spread of the neutron energy. In this case it is meaningless to speak in terms of a temporal resolution of the fission process into sequential steps, since the uncertainty relation $\delta t\ \delta E \sim \hbar$ tells us that under the condition $\delta E < \Gamma$ the uncertainty in time $\delta t > T$, where T is the life time of the nucleus. Of course, there is no "forgetting" in this case. In the opposite case $\delta E \gg D \geq \Gamma$ a temporal description is possible. In that case the initial hot step is "forgotten" upon the transition to the cold state, and the effect caused by the weak interaction in the hot step is suppressed by the factor $(D/\delta E)^{1/2}$ (number of resonances involved to the process is $N_\delta \sim \delta E/D$. The signs of the amplitudes A, B, β are random. Therefore, interference term $a \cos\theta$ in equ. (6) is suppressed by a factor $N_\delta^{-1/2}$).

A detailed description of our work can be found in ref. 11 and in the review[12].

CONCLUSION

The observation of parity non-conservation in nuclear fission shows us that the fission channels with different parity or angular momentum interfere in the angular distribution of fragments. This means that they are related to a rotational band constructed on the same intrinsic state of a cold pear-shaped nucleus. The angular distribution is formed before scission of nucleus into fragments. The existence of dynamical enhancement of small interaction (e.g. of weak interaction) is confirmed.

We have considered parity violating correlation $\vec{\sigma}\ \vec{p}_f$ ($\vec{\sigma}$ is neutron spin, $\vec{p}_f$ light fragment momentum). There are also odd harmonics in the angular distribution of the fragments which are due to the interference of s- and p-wave neutron capture and don't violate parity: $\vec{p}_n\ \vec{p}_f$ and $\sigma(\vec{p}_n \times \vec{p}_f)$, where $\vec{p}_n$ is neutron momentum (see e.g. ref. 12). The magnitude of these correlation is $\sim 10^{-4}$- 10^{-3} for thermal neutrons.

After averaging over energy, all odd harmonics in the fragment angular distribution disappear in standard fission mechanism due to their interference nature. This fact can be used for the search for the non-standard mechanism of fission which have no compound nucleus state. Therefore, the experimental investigation of odd harmonics in fission by the particles (n, γ, e, p, ...) with large energy spread seems to be very interesting (see details in ref. 13).

REFERENCES

1. G.V. Danilyan et al., Pis'ma Zh. Eksp. Teor. Fiz. **26**, 198 (1977).
2. B.D. Vodennikov et al., Pis'ma Zh. Eksp. Teor. Fiz. **27**, 68 (1978).
3. G.V. Danilyan et al., Yad. Fiz. **27**, 42 (1978).
4. V.N. Andreev et al., Pis'ma Zh. Eksp. Teor. Fiz. **28**, 53 (1978).
5. A.K. Petukhov et al., Pis'ma Zh. Eksp. Teor. Fiz. **30**, 470 (1979).
6. N.V. Borovikova et al., Pis'ma Eksp. Teor. Fiz., **30**, 527 (1979).
7. A. Bohr and B.R. Mottelson, Nuclear Structure, Vol. II (Benjamin, New York, 1974).
8. R. Haas, L.B. Leipuner, and R.K. Adair, Phys. Rev. **116**, 1221 (1959).
9. R.J. Blin-stoyle, Phys. Rev. **120**, 181 (1960).
10. I.S. Shapiro, Usp. Fiz. Nauk **95**, 647 (1968).
11. V.V. Flambaum and O.P. Sushkov, Phys. Lett. **B94**, 277 (1980); Yad. Fiz. **33**, 59 (1981); Sov. J. Nucl. Phys. **33**, 31 (1981).
12. O.P. Sushkov and V.V. Flambaum, Usp. Fiz. Nauk. **136**, 3 (1982); Sov. Phys. Usp. **25**, 1 (1982).
13. V.V. Flambaum, Yad. Fiz. **42**, 578 (1985); Sov. J. Nucl. Phys. **42** (1985).

PARITY VIOLATING WEAK NN-POTENTIALS WITHIN THE STANDARD MODEL

M. Kirchbach
Central Institute for Nuclear Research, Rossendorf, Germany

Abstract: The applicability of the standard model of electroweak interactions for describing parity-mixing effects in nuclei is proofed.

The asymmetry of coordinate space relative to mirror reflexion (violation of spatial parity) becomes apparent, for example, through the weak interaction between the nucleons as observed in the $(\vec{n},\gamma)$-reaction with heavy nuclei. The effect is signaled by the circular polarization of the γ-rays (P_γ) and is caused by the interference of compound-resonance states with different response to mirror reflexion (i.g. opposite parity), predominantly S- and P-wave resonances /1/. A rather similar effect of parity violation (PV) is observed in light nuclei in which parity doublets (closely lying J^+-J^- -states) of low-excitation energies are found. In such cases the interference of E1- and M1-multipoles produces a circular polarization of the γ-rays emitted during the deexcitation to the ground state. Both PV-effects are proportional to the matrix element of the PV-part of the weak NN-potential between the respective doublet-states,

$$P_\gamma \sim \langle J^- | V_{PV} | J^+ \rangle \quad . \tag{1}$$

Whereas in the $(\vec{n},\gamma)$-reaction the matrix element of the PV-part of the weak NN-potential between the compound-resonance states is usually replaced by a semiempirical value of $\langle S|V_{PV}|P\rangle \sim 10^{-4}$ eV, the P-odd effect in the light nuclei is calculated microscopically by exploiting detailed knowledge on the structure of the nuclear states involved. In this way PV-observables are related to parameters of the fundamental weak hadronic interaction. The form of V_{PV} is well understood in terms of an effective one-boson-exchange model in which a meson is coupled to a nucleon via the weak interaction at the one vertex (dashed circle in the fig.1) , and the strong interaction on the other. The weak interaction conserves the number of baryons (B), but it does not conserve their electric charge as seen from the neutron-β-decay

$$n \rightarrow p + e^- + \bar{\nu}_e \quad , \tag{2}$$

where the electric charge of the nucleon changes by one unit (ΔQ^N=1). As a consequence, nuclear β-decays take place between the members of an isobar triad

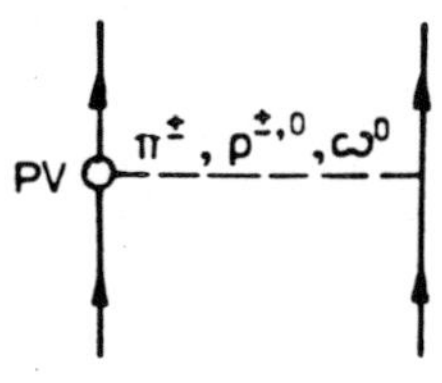

Fig.1

$$A(Z,N) \xrightarrow{\beta^\pm} A(Z \mp 1, N \pm 1) + e^\pm + \nu_e(\bar{\nu}_e) \quad . \tag{3}$$

From the correspondence between the electric charge, the baryonic number, and the quantum number of isospin ($T^B_{m=\pm 1,0}$),

$$Q^B = T^B_0 + \frac{B}{2} , \qquad (4)$$

one reads off that PV-forces connect nuclear states with isospin differing by one unit (ΔT=1-forces). In other words, V_{PV} should mix in a given A(Z,N)-nucleus the isobar partners of the neighbouring A(Z$\pm$1, N$\mp$1)-excitations to the remaining nuclear states. For example, the ΔT=1-(isovector) PV NN-potential is responsible for the mixing between the 0^+;1.042MeV-level (the isobar analogue of the ground state of ^{18}Ne) to the 0^-;1.081 MeV -state in ^{18}F (see fig.2). This mixing leads to an asymmetry (A_γ) between the numbers of γ-quants transmitted through a magnetic polarimeter when the electron polarization is parallel (N_+), or antiparallel (N_-) to the photon-direction,

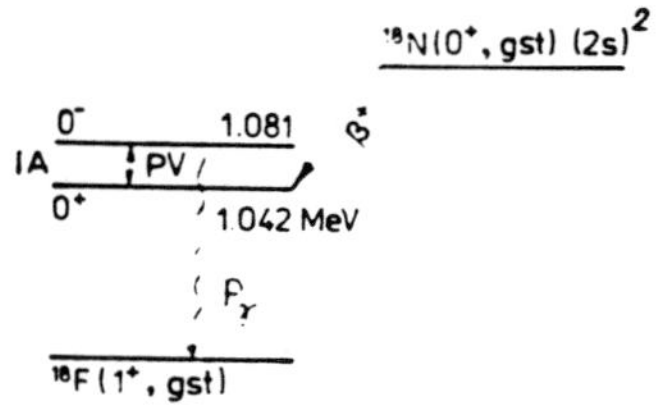

Fig.2

$$A_\gamma = \frac{N_+ - N_-}{N_+ + N_-} . \qquad (5)$$

Therefore, the γ-rays emitted in the E1-(M1)-transitions from the respective $0^-(0^+)$-states to the ground state are circular polarized. Due to structural effects the circular polarization of the 0^-;1.081MeV-γ-rays appears to be enhanced by a factor of$\sim 10^4$ relative to that of the γ-rays in the M1-transition. The value of P_γ(1.081MeV) has been measured in the last years with an increasing accuracy by several groups, and has been found to be $(1.7 \pm 5.8)\times 10^{-4}$ by Page et al. /2/, and $(2.7 \pm 5.5)\times 10^{-4}$ by Bini et al. /3/. This corresponds to an averaged experimental value of the nuclear matrix element of

$$\langle 0^+ | V^{\Delta T=1}_{PV} | 0^- \rangle^{exper} = (0.048 \; ^{+0.12}_{-0.09}) \text{ eV} . \qquad (6)$$

The weak PV NN-potential is due to spin-momentum ($\vec\sigma\cdot\vec p$)-correlations of the interacting nucleons

$$V^{\Delta T=1}_{PV} = V^1_\pi + V^1_\rho + V^1_\omega , \qquad (7)$$

$$V^1_\pi(i,j) = \frac{h^1_\pi g_r}{4\sqrt{2}} \left[\tau(i)\times\tau(j)\right]_0 (\vec\sigma(i)+\vec\sigma(j))\cdot\left[\vec p(i)-\vec p(j), \exp(-m_\pi r)/r\right] .$$

Here, $V^1_{(M=\pi,\rho,\omega)}$ denote the contributions to $V^{\Delta T=1}_{PV}$ due to the respective M-meson exchange, h^1_M stands for the (isovector) coupling strength in the weak MNN-vertex. As seen from the fig. 1, $V^{\Delta T=1}_{PV}$ is in its nature a predominantly charge-exchange force. This NN-force is rather similar to the mesonic exchange correction (MEC) to the β-decay transitions of the type 0^+-0^- .

In order to get an idea how a charged pion is created by the nucleon in the weak NN-interaction, one has to look on the level of the nucleon constituents (the quark level). Consider, for concreteness, the $np\pi^-$ -vertex (see fig.3). The neutron is built up by one u-quark with an electric charge of $Q_u = 2/3$, and two d-quarks with $Q_d = -1/3$. The three quarks are confined by the exchange of electric neutral bosons (the so called gluons (G)). This means that the u-(d-) quarks are once for all u-(d-) quarks ("flavor" conservation through gluon exchange). The quark-flavor is further conserved by the electromagnetic interaction mediated via the γ-quants. In weak interaction-processes, however, the quarks can change their flavor through the exchange of the charged $W^\pm$ -bosons (they mediate the $\beta^\pm$ -decays), or they can conserve the flavor if the neutral partner W_0 should be exchanged. However, in rather short-ranged flavor-conserving processes quarks do not distinguish between electromagnetic and weak interactions but interact "electroweak" by the exchange of a new kind of weak boson, the so called Z_0 -boson, to which both the photon and the W_0-boson are joined, $Z_0 = f_1(\theta_W)W_0 + f_2(\theta_W)\gamma$. The coefficient functions $f_{1,2}(\theta_W)$ depend on a parameter known as the "Weinberg-angle".

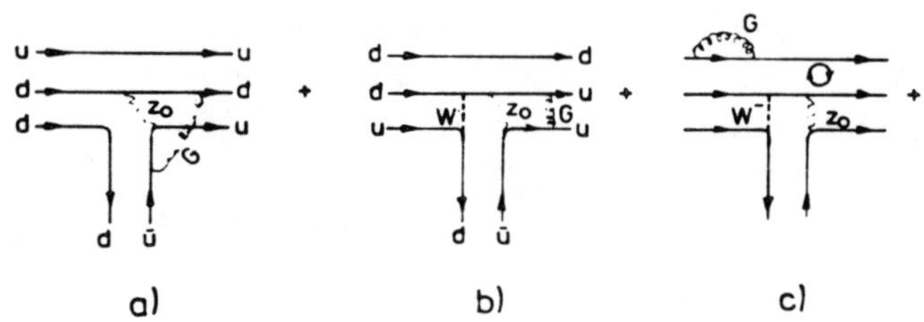

Fig.3

Now, one of the three nucleon-constituent (valence-) quarks can virtually emit a Z_0-boson which then decays into a quark-antiquark ($u\bar{u}$-) pair. A pertinent recombination between the constituent quarks and the $u\bar{u}$-pair leads to a virtual emission of a charged pion by the neutron within the nucleus (fig.3). This way, isovector parity mixing effects in light nuclei provide an insight into the dynamics of weak neutral current-current interactions. In studying such phenomena, the applicability of fundamental theory of weak interactions to nuclear physics can be proofed. The theory underlying these ideas has been developed by Glashow, Weinberg and Salam (GWS) who showed that at the energy scale of $E \sim M_{Z_0} c^2 \sim 100$ GeV, electromagnetic and weak interactions "amalgamate" to a more general interaction, the unified electroweak interaction. In taking into account the influence of the gluon exchange on the weak qq-interaction, the so called $SU_c(3) \otimes SU_L(2) \otimes U(1)$-standard model is obtained. Now, the weak meson-nucleon couplings are calculated microscopically as

$$h^1_M = \langle MN|\mathcal{H}^w|N\rangle \quad . \qquad (8)$$

Here, $\mathcal{H}^w$ is the weak Hamiltonian of the standard model. The calculation of h^1_M becomes rather difficult because of the influence of the vacuum-state of matter (the quark-sea) which introduces quark-loops into the respective graphs (see, for example, graph c) in fig.3).Applying some renormalization-group techniques, the contribution of such graphs can be kept small. So, the weak meson-nucleon coupling constants h^1_π, h^1_ρ, and h^1_ω can be calculated consistently by exploiting the MIT-bag model for the wave functions of the valence quarks, the PCAC-relation, as well as the SU(6)- symmetry behaviour of the spatial overlap of the hadronic wave functions in the MNN-vertex (the overlap of the wave functions of all spin-0(1) -mesons with all baryons is almost a constant). Such calculations have been performed at the JINR Dubna by Dubovik and Zenkin (DZ). These authors concluded that the PV part of the weak NN-potential is mainly controlled by the $\pi^\pm$ -exchange ($\approx$80%), and partly by the ω-meson exchange ($\approx$20%) /4/. With the help of the DZ-weak meson-nucleon coupling constants and by the use of shell-model wave functions for the nuclear states involved, the parity-mixing matrix element, determining the circular polarization of the γ-rays in the E1-transition from the 0^-;1.081 MeV-state to the ground state in ^{18}F has been calculated to be /5/

$$\langle 0^- |\, V^{\Delta T=1}_{PV}\, | 0^+\rangle = 0.13 \text{ eV} \quad . \qquad (9)$$

This result is consistent with the upper bound of the empirical value of eq.(6).Consequently, the standard model of electroweak interaction leads to results which are compatible with nuclear-physics measurements.

References:

/1/ O.P.Sushkov, V.V.Flambaum, Sov.Phys.Usp. 25 (1982) 1
/2/ S.A.Page et al. , Phys.Rev. 35 (1987) 1119
/3/ M.Bini et al., Phys.Rev.Lett. 58 (1987) 1616
/4/ V.M.Dubovik, S.V.Zenkin, Ann.Phys. 172 (1986) 100
/5/ M.Kirchbach, J.Phys.G:Nucl.Phys. 13 (1987) 135

TIME CHARACTERISTICS OF FISSION OF EXCITED HEAVY NUCLEI

Yu.V. Melikov, S.Yu. Platonov, A.F. Tulinov,
O.V. Fotina, and O.A. Yuminov
Institute of Nuclear Physics, Moscow State University, Moscow 119899, USSR

A method based on the blocking technique is used to study the fission lifetime of the excited $^{235,236,238,239}Np$ nucleus. It is demonstrated that only the double-humped fission barrier model with the inclusion of the lifetime of excited shape isomers of the fissile nucleus can discribe well experimental results. Analysis of the level density in the second potential well was made .

Introduction

The direct measurements of the times characteristics of nuclear reactions on single crystals in the 10^{-19} - 10^{-15} s range by the blocking technique /1,2/ offer new possibilities of studying the dynamics of nuclear transformations including a composite process , such as the fission of heavy nuclei.

Strutinsky's discovery of the double-humped fission barrier with the second deep well between the humps accounts for many unsolved problems ,for example, the nature of spontaneously fissionable isomers , the resonance-like peaks in the dependence of the fission cross section upon sub-barrier energies ,etc. /3,4/. A detailed study of these phenomena gives information about the low-lying excited states in the second potential well.

At higher excitation energies of a fissile nucleus in the above- barrier region, the barrier penetrability is close to unity, and the quasistationary states in the second potential well do not influence the energy behaviour of the time-intergrated characteristics (such as cross sections) of the decay of excited nucleus, However, these states can influence strongly the fission dynamics and increase the mean duration of the decay through the fission channel /5/. In the present paper we analyze the time characteristics of the decay of heavy nuclei possessing two classes of excited states and also present the results of measurement of the nuclear fission time of neptunium isotopes.

1. Two classes of excited states and the fission time

Figure 1 presents the scheme of population and decay of the excited nuclei possessing two classes of excited states.

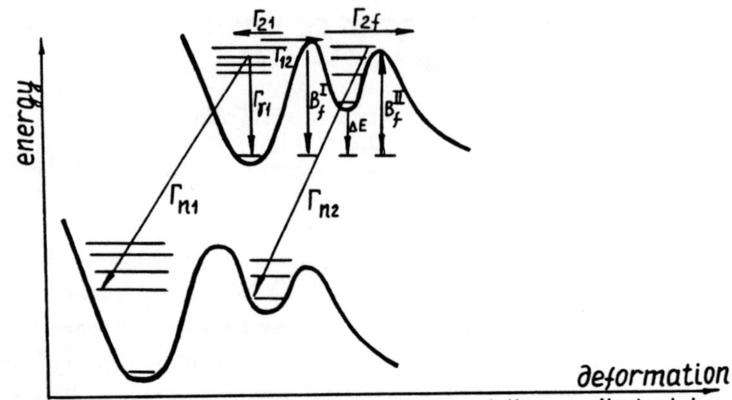

FIG.1. Scheme of population and decay of the excited states in the framework of the double-humped fission barrier model.

The excited state formed under equilibrium deformation, decays either through the emission of particles or γ-quanta or through the transition to the state in the second potential well. The excted state in the second potential well decays via the emission of particles or γ-quanta, or via the non-radiative transition across the inner barrier into the state with equilibrium deformation, or via the transition across the outer barrier, i.e. fission. In this case, the populations of the states in the first and the second potential wells $n_1(t)$ and $n_2(t)$ are described by the equations /3/ ($\hbar = 1$):

$$\frac{dn_1(t)}{dt} = -\Gamma_1 n_1(t) + \Gamma_{21} n_2(t) , \tag{1}$$

$$\frac{dn_2(t)}{dt} = -\Gamma_2 n_2(t) + \Gamma_{12} n_1(t) \tag{2}$$

where Γ_1 and Γ_2 are the total decay widths for the first- and second-well states:

$$\Gamma_1 = \sum_i \Gamma_{1i} + \Gamma_{12} \tag{3}$$

$$\Gamma_2 = \sum \Gamma_{2i} + \Gamma_{21} + \Gamma_{2f} \tag{4}$$

Γ_{1i} and Γ_{2i} are the partial widths of the emission of particles and γ-quanta; Γ_{12} and Γ_{21} are the width of the non-radiative transitions through the inner barrier, Γ_{2f} is the fissbn width. Under the initial conditions $n_1(0)=1$, $n_2(0)=0$ the solution of Eqs.(2)-(3) has the form

$$n_1(t) = \frac{S_2+\Gamma_1}{S_2-S_1} \exp(S_1 t) - \frac{(S_1+\Gamma_1)}{S_2-S_1} \exp(S_2 t) \tag{5}$$

$$n_2(t) = \Gamma_{12}\left\{\exp(S_2 t) - \exp(S_1 t)\right\} / (S_2-S_1) , \tag{6}$$

where $S_{1,2} = -\frac{1}{2}\left\{(\Gamma_1+\Gamma_2) \pm \left[(\Gamma_1+\Gamma_2)^2 - 4(\Gamma_1\Gamma_2 - \Gamma_{12}\Gamma_{21})\right]^{1/2}\right\}$ (7)

The decay rates in the fission channel and in the channel with the emission of a particle or a γ -quantum are given by the expressions:

$$\frac{dn_f(t)}{dt} = \Gamma_{2f}\, n_2(t) \tag{8}$$

$$\frac{dn_i(t)}{dt} = \Gamma_{1i}\, n_1(t) + \Gamma_{2i}\, n_2(t) \tag{9}$$

From (6)-(10) it follows that the two-humped fission-barrier structure and the transition of a fissile nucleus through the intermediate states of the second potential well lead to a non-exponential time-dependence of the yield of the decay of excited nuclei.

The mean decay time is also affected by the intermediate states of a strongly deformed nucleus. The mean decay time in the fission channel τ_f differs from that in any other channel - τ_i /5/:

$$\tau_f = \hbar \int_0^\infty t\, \frac{dn_f}{dt}\, dt \Big/ \int_0^\infty \frac{dn_f}{dt}\, dt = \hbar\, \frac{\Gamma_1 + \Gamma_2}{\Gamma_1 \Gamma_2 - \Gamma_{12}\Gamma_{21}} \tag{10}$$

and analogously

$$\tau_f = \hbar \left[\frac{\Gamma_1 + \Gamma_2}{\Gamma_1 \Gamma_2 - \Gamma_{12}\Gamma_{21}} \qquad \frac{\Gamma_{1i}}{\Gamma_{1i}\Gamma_2 + \Gamma_{2i}\Gamma_{12}} \right] \tag{11}$$

This difference is especially large if the coupling of the excited states in the first and the second potential well is weak, i.e. $\Gamma_{21} \ll \Gamma_2$, and under conditions - $\Gamma_{21} \ll \Gamma_{1i}$

$$\tau_f \simeq \frac{\hbar}{\Gamma_1} + \frac{\hbar}{\Gamma_2} \tag{10'}$$

$$\tau_i \simeq \frac{\hbar}{\Gamma_1}\,, \tag{11'}$$

i.e. the mean decay time in the fission channel is a sum of the mean lifetimes of the states in the first and second potential wells. From (10') and (11') it is also seen that the "delay" in the fission channel compared, for example, with the neutron channel, contains the direct information about the excited states in the second potential well. As it follows from the expressions (10')-(11'), the experimental data on the fission time convey information about the nuclear level density in the second potential well - ρ_2. The ρ_2-value can be estimated on the basis of the data on τ_t using the expression:

$$\rho_2 = \frac{(\tau_f - \tau_i) N_2}{2\pi\hbar} \tag{12}$$

where N_2 is the number of open channels of decay of the excited states in the second potential well. The values of N_2 and τ_i can be estimated on the basis of the data on the fissionability of nuclei and the level density in the first potential well.

2 . Experimental results . Discussion

The influence of the lifetime of excited states of the nucleus in the second potential well upon the fission time was studied by us for a number of Np isotopes with the parameters of the two-humped fission barrier most favourable for detection of the forecasted effects.

Figure 2 shows the results of measurement of the fission time of excited $^{235,236,238,239}Np$ nuclei produced as a result of confluence of $^{235,238}U$ nuclei with deuterons and protons and the subsequent emission of one or two neutrons.

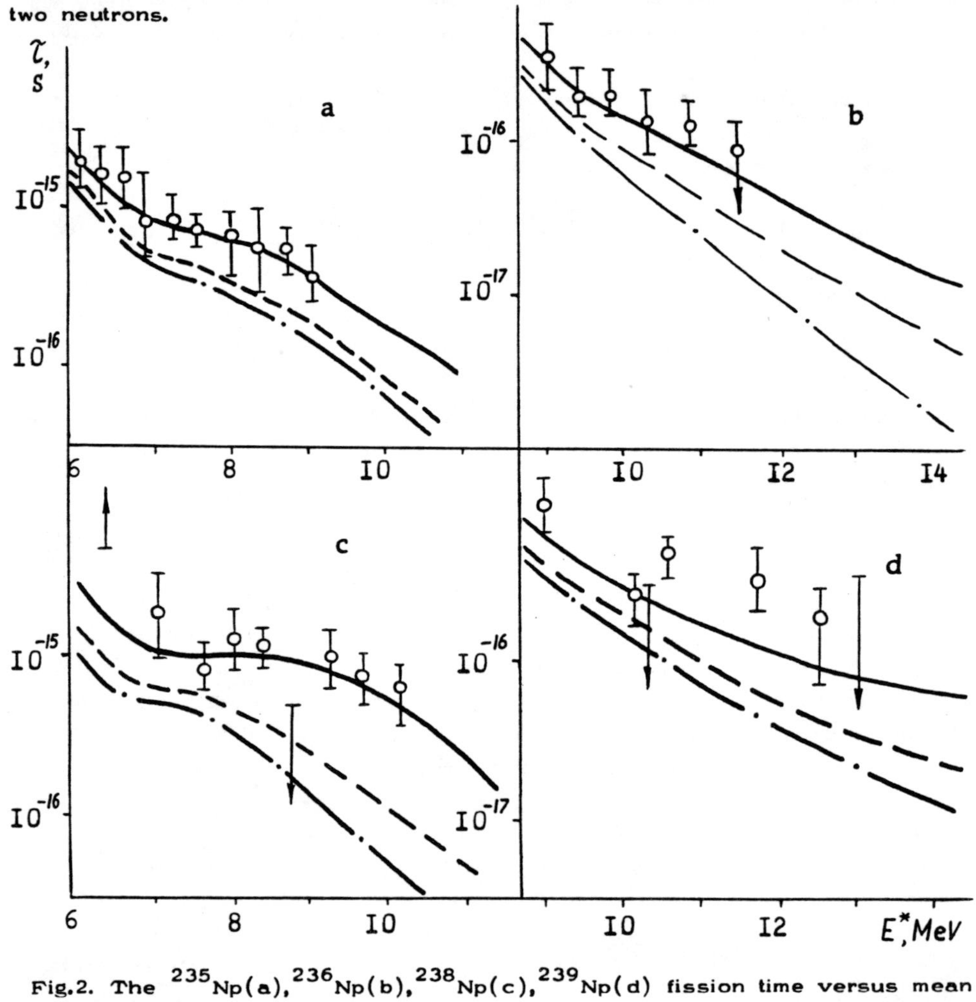

Fig.2. The ^{235}Np(a), ^{236}Np(b), ^{238}Np(c), ^{239}Np(d) fission time versus mean excitation energy. The dots are the (d, xnf) experimental values and open circles are the (p,nf) experimental values. The solid line curve represents the fission time values calculated with the inclusion of the lifetime of the excited state in the second potential well. The dash-dotted line curve shows the results of calculations of the fission time with the neglect of the lifetime of the excited states in the second potential well. The broken line curve

represents the calculations of the nuclear decay time in the neutron channel with the same level-density parameters as in the case of the solid line curve.

The experiment was made using the blocking technique. Setting up an experiment, data processing and analysis are described in detail in the paper /6/.

In the calculation of the Γ -widths entering into the expressions (10)-(11) we used the level density in the Fermi-gas model. The values of the parameters a and Δ under equilibrium deformation were taken from the systematics /7,8/. The parametrs in the first and second saddle points were assumed to be equal $a_{f1} = a_{f2}$, $\Delta_{f1} = \Delta_{f2}$ and were chosen as to provide the best descrption of the fissility of the isotopes in question, and the fission barrier parameters of these nuclei were taken from the review /9/. The level density parameter a_2 in the second potential well was regarded as free and its values were varied to describe satisfactorily the present experimental data. Table 1 lists the values of the level density-parameters used in the calculations.

Table 1

Level density parameters

Isotopes	a_1 (MeV^{-1})	Δ_1 (MeV)	$a_{f1}=a_{f2}$ (MeV^{-1})	Δf (MeV)	a_2 (MeV^{-1})	Δ_2 (MeV)
^{240}Np	28.24	0.0	30.90	0.0	-	-
^{239}Np	27.87	0.43	30.00	0.43	35.60	0.43
^{238}Np	25.47	0.0	30.00	0.0	35.11	0.0
^{237}Np	26.60	0.28	29.26	0.28	-	-
^{236}Np	26.50	-0.43	29.50	-0.20	36.00	0.37
^{235}Np	26.50	0.37	31.00	0.37	36.50	0.37
^{234}Np	26.50	-0.43	26.50	-0.43	-	-

Taking account of the intermediate states in the second potential well, produced in the course of the nuclear fission, enables one to describe satisfactorily the experimental dependences of τ_f for nuclei of neptunium isotopes. In this case the values of the level density parameter a_2 in the second cond well are ~ 1.3 times as large as the values of the parameter a_1 in the first well for all the examined neptunium isotopes. In Figure 2 one can see the extent to which the calculations agree with experiment.

To analyze the behaviour of the level density in the second potential well, we obtained the values of ρ_2 from the expression (12).

Figure 3 shows the ρ_2 -values for neptunium isotopes. There is also given (open circle) the value of the level density of ^{239}Np - ρ_1, in the

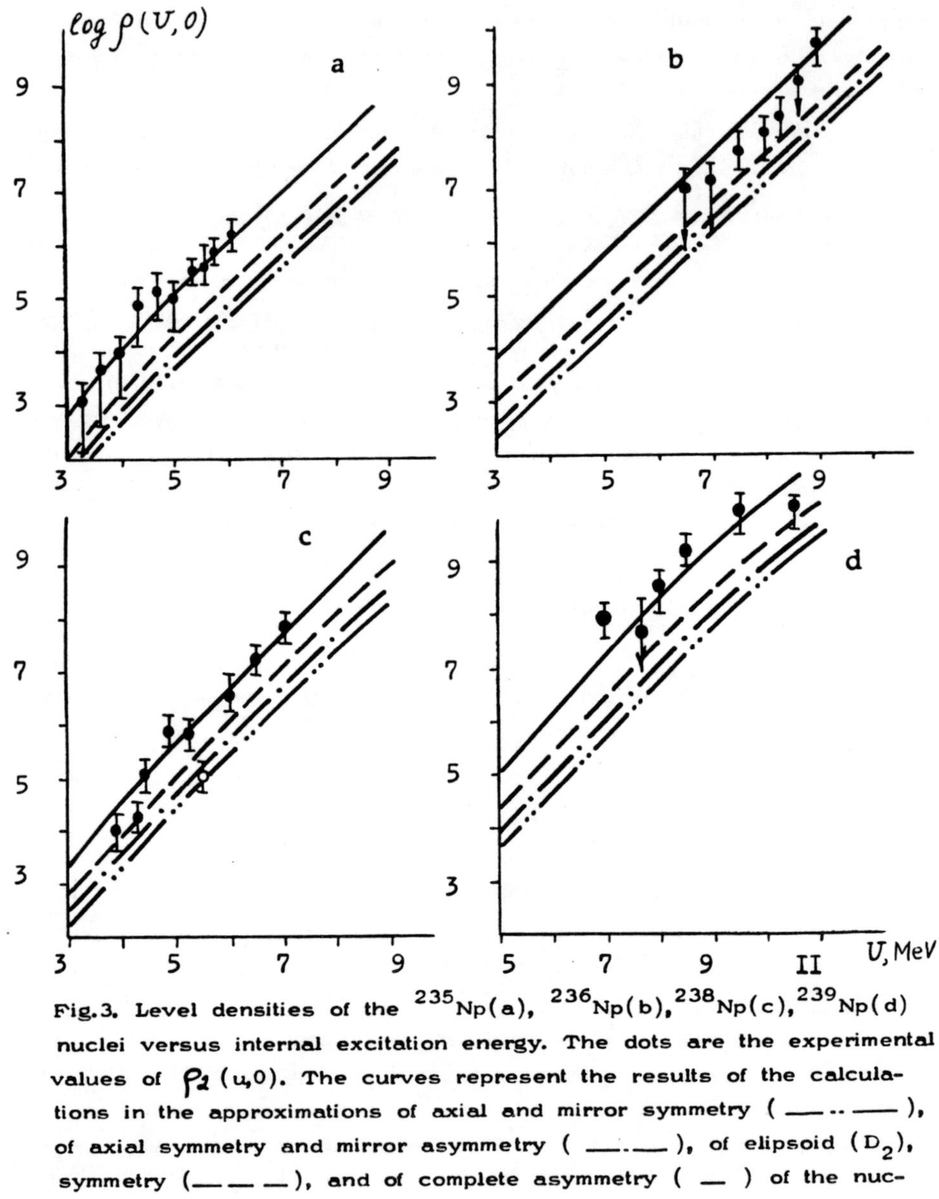

Fig.3. Level densities of the ^{235}Np(a), ^{236}Np(b), ^{238}Np(c), ^{239}Np(d) nuclei versus internal excitation energy. The dots are the experimental values of ρ_2 (u,0). The curves represent the results of the calculations in the approximations of axial and mirror symmetry (——··——), of axial symmetry and mirror asymmetry (——·——), of elipsoid (D_2), symmetry (—— —— ——), and of complete asymmetry (——) of the nuclear shape.

first potential well. Note that ρ_2 is much larger than ρ_1 at the same internal excitation energy. The analysis of ρ_2 made in the phenomenological model of the level density with allowance for the collective enhancement of level density /10/ showed that this is the result of violation of the axial and mirror symmetry of the nuclear shape in the second potential well (see Fig.3).

Thus, the studies of the nuclear fission time give information about the fission dynamics and the characteristics of nuclei in the strongly deformed excited states.

REFERENCES

1. Tulinov A.F. Dokl. Akad.Nauk SSSR, 162 (1965) p.546
2. Gemmel D.S., Holland R.E. Phys.Rev.Lett. 14(1965) p.945
3. Strutinsky V.M., Bjørnholm S. Nucl.Phys., v.A136(1969)p.1.
4. Polikanov S.M. Uspekh. Fiz.Nauk, v.107,issue 4(1972)p.686
5. Yuminov O.A. Proc. of the All-Union Meeting on the Physics of Interactions of Charged Particles with Solids (1985) p.68
6. Platonov S.Yu. et al. Proc. of Intern. Symp. on Modern Development in Nucl.Phys.,(1987)p.631
7. Gilbert A, Cameron A.G.W. Canad.J.Phys.,v.43(1965) p.1146
8. Dilg W. et al. Nucl.Phys.,v.A217(1973) p.269
9. Lynn J.E., Bjørnholm S. Rev.Mod.Phys.,v.52(1980) p.725
10. Ignatyuk A. V. Sov.J. of Nucl.Phys.,v.21(1975) p.485

URANIUM NUCLEI INVESTIGATED IN NEUTRON RESONANCE STATES BEFORE FISSION

A. Meister, D. Seeliger, and K. Seidel
Technische Universität, Dresden, Germany

ABSTRACT

Measuring chemically induced shifts of low-energy neutron resonances at ^{234}U, ^{235}U, and ^{238}U, mean-square charge radii of the nuclei excited in compound nucleus states are determined.

1. Introduction

The fission probability of an excited nucleus should depend on the structure of the states excited, and especially on the shape of the nucleus in that states. Some years ago, experimental determination of the mean-square charge radius of an uranium nucleus excited by neutron resonance capture to an isolated compound nucleus state succeeded by measuring resonance shifts chemically induced /1,2/. These experiments have been extended to uranium resonance states with large fission width. In the present work the results are presented and discussed.

2. EXPERIMENT

Any change of shape of a nucleus is connected with a weak shift of the nuclear state, caused by electrical hyperfine interaction between the charges of the nucleus and the atomic electrons. If the nucleus is investigated in chemical compounds having different electron densities in the nucleus region, $\Delta|\Psi_e(0)|^2 = |\Psi_{eI}(0)|^2 - |\Psi_{eII}(0)|^2$, the energy shift ΔE_0 of the nuclear state is

$$\Delta E_0 = \frac{Ze^2}{6\,\varepsilon_0} \cdot \Delta|\Psi_e(0)|^2 \cdot \Delta\langle r^2\rangle$$

where Z is the proton number and $\Delta\langle r^2\rangle$ is in our case the change of the mean-square charge radius due to neutron absorption. If $\Delta|\Psi_e(0)|^2$ and $\langle r^2\rangle_g$ for the ground state are known, the measurement of ΔE_0 yields $\langle r^2\rangle$ for the compound nucleus state. The shifts are in the order of 100 µeV or less. That is small in comparison to the total widths of low-energy neutron resonances, which are typically about 100 meV. Nevertheless, these shifts are measurable by means of high-resolution neutron time-of-flight spectroscopy at an efficient neutron source.

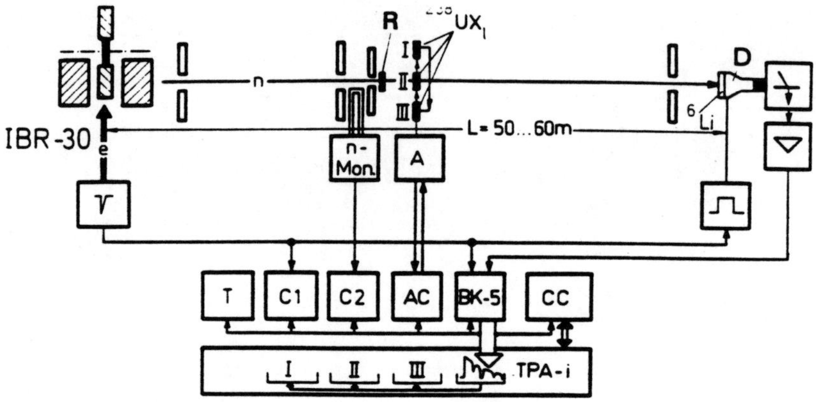

Fig.1: Time-of-flight spectrometer at the Dubna pulsed reactor IBR-30
n: neutron beam
I, II, III: uranium samples
R: reference sample
D: neutron detector.

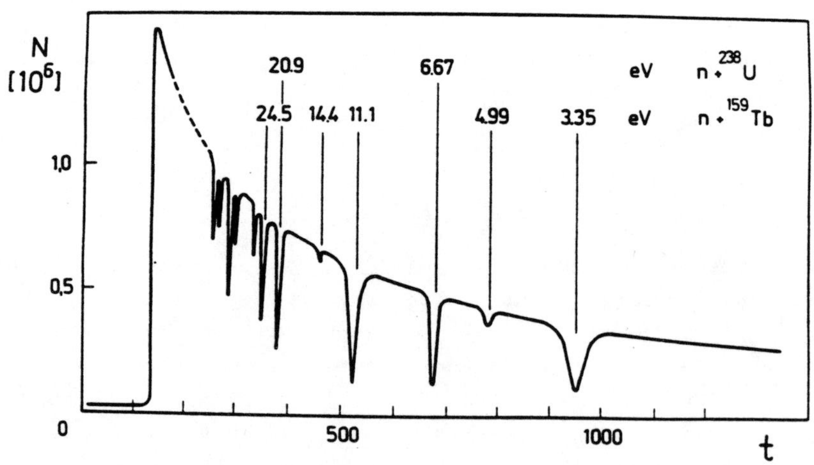

Fig.2: Time-of-flight spectrum for the low-energy resonances of ^{238}U.

The experimental arrangement used is shown in Fig. 1. The procedure for taking transmission spectra is concentrated on equal experimental conditions for the three samples containing different chemical compounds of the nuclide under investigation /1,2/. A typical time-of-flihgt spectrum for one of the samples is presented in Fig. 2. The resonances of ^{159}Tb serve as references. Experimental investigations are carried out for eight low-energy neutron resonances of ^{235}U, for the 5.2 eV resonance of ^{234}U and for the 6.7 eV resonance of ^{238}U. The analysis of the experimental transmission spectra must include the different Doppler broadenings of the neutron resonances in the samples. The treatement of this problem is discussed in Ref. /1-4/. With $\Delta|\Psi_e|^2$ obtained from relativistic Hartree-Fock-Slater calculations and from experimental data on chemical X-ray shifts and on Moessbauer isomer shifts, the $\Delta\langle r^2\rangle$ are deduced. As an example, the data obtained for the 6.7 eV resonance are shown in Fig. 3.

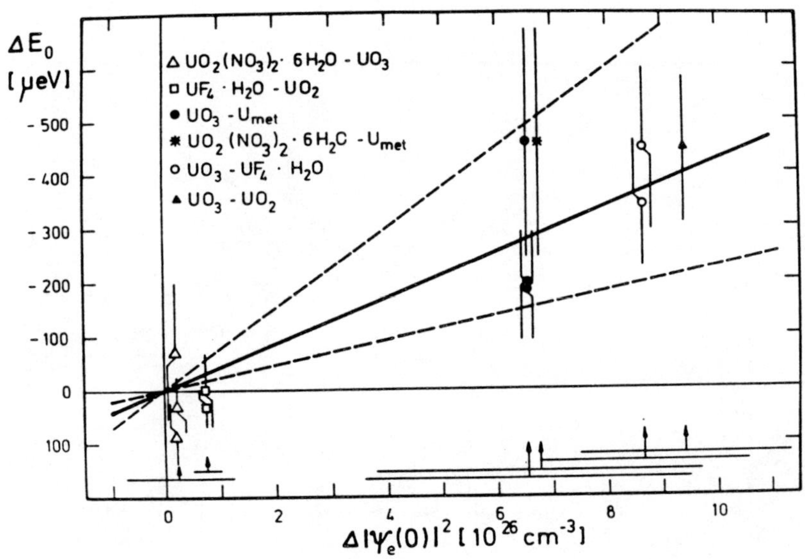

3. RESULTS AND DISCUSSION

The data on $\Delta\langle r^2\rangle$ determined in the experiment are corrected for isotope shift /5/ to have the difference of $\langle r^2\rangle$ between excited compound nucleus state and ground state of the same nucleus $\Delta\langle r^2\rangle_n$. They are shown in Fig. 4. The absolute values $\Delta\langle r^2\rangle_n$ are of the order of some per cent of the ground state mean-square radius $\langle r^2\rangle_g$ = 34 fm^2 /5/. It is obvious that on average the $\Delta\langle r^2\rangle_n$ are negative, not only for resonance states in ^{235}U and ^{239}U having a very small fission probability, but also for states in ^{236}U excited above the fission barrier. This unexpected mean behaviour can be interpreted to be caused by the release of pairing interaction at nucleus excitation /6/. The fission process as a large amplitude collective motion is suggested to be correlated with the mean size of the excited nucleus before fission. Indeed, in Fig. 4 a dependence of $\Delta\langle r^2\rangle_n$ on Γ_f seems to be indicated. If it is approximated by a straight line for the data from ^{235}U+n, also the two measured $\Delta\langle r^2\rangle_n$ from even-even target nuclei follow this tendency.

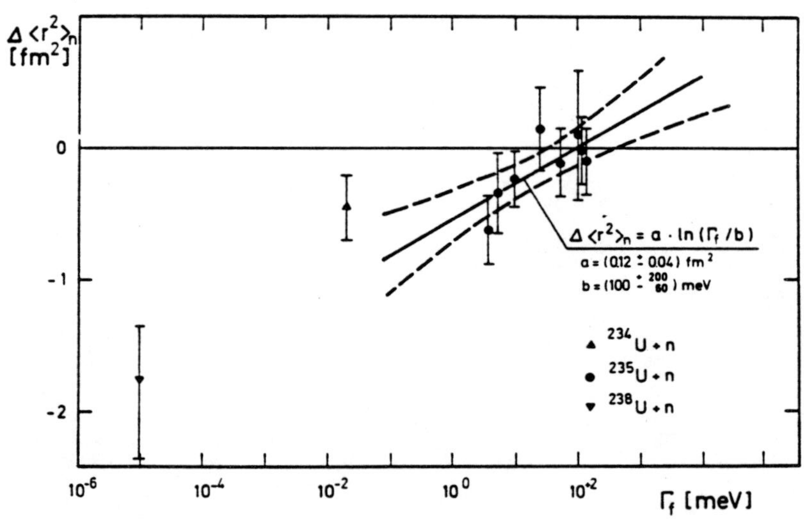

Fig.4: Difference of mean-square charge radii between compound nucleus state and ground state of nuclei ^{235}U, ^{236}U, and ^{239}U as function of the fission width /7/ of the excited state.

4. CONCLUSIONS

With the measurement of chemically induced shifts of neutron resonances a new method has been found for investigation of mean-square charge radii of nuclei in isolated compound nucleus states with excitation energies near the neutron binding energy. It was applied to uranium isotope resonance states.

The structure of sub-barrier fission cross sections is understood in the frame of the double-humped fission barrier /8/ as admixture of so called class II configurations having large deformation parameters and determining the fission probability. For states above fission barrier this approach should lose his validity. But, a correlation between fission width and mean-square radius, as indicated for ^{235}U+n by the present results, can be attributed to admixture of configurations with large deformation /9/. This is in accordance with the presence of intermediate structures in the ^{235}U+n fission cross section, reported about in Ref. /10/.

Further investigation of low-energy resonances of actinides, that have fission widths varying over orders of magnitude and have excitation energies below and above the fission barrier, should yield more insight into the starting point of the fission process.

REFERENCES

/1/ K. Seidel, A. Meister, D. Pabst, L.B. Pikelner, Dokl. Akad. Nauk USSR 256 (1981) 360; A. Meister, D. Pabst, L.B. Pikelner, K. Seidel, Nucl. Phys. A362 (1981) 18.

/2/ S. Mittag, D. Pabst, L.B. Pikelner, W. Pilz, R. Tschammer, A. Meister, D. Seeliger, K. Seidel, Nucl. Phys. A435 (1985) 97.

/3/ K. Seidel, D. Seeliger, A. Meister, S. Mittag, W. Pilz, Particles and Nuclei 19 (1988) 307.

/4/ K. Seidel, D. Seeliger, V.K. Ignatovich, A. Meister, S. Mittag, W. Pilz, Yad. Fiz. 42 (1985) 1041.

/5/ K. Heilig, A. Steudel, Atomic Data and Nuclear Data Tables 14 (1974) 613.

/6/ G.G. Bunatian, Yad. Fiz. 35 (1982) 16.

/7/ M.S. Moore et al., Phys. Rev. C18 (1978) 1328; G.D. James et al., Phys. Rev. C15 (1977) 2083; F.C. Difilippo et al., Phys. Rev. C21 (1980) 1400.

/8/ V.M. Strutinsky, Nucl. Phys. A95 (1967) 420.

/9/ A. Meister, S. Mittag, L.B. Pikelner, W. Pilz, D. Seeliger, K. Seidel, Proc. XVth Int. Symp. Nuclear Physics, Gaussig, 1985, ZfK-592 (1986) 70.

/10/ M.S. Moore et al., Phys. Rev. C30 (1984) 214.

III. THEORY OF NUCLEAR FISSION

SOME PECULIARITIES OF THE POTENTIAL ENERGY SURFACE IN NUCLEI AT LARGE DEFORMATION

V.V. Pashkevich
Joint Institute for Nuclear Research, 141980 Dubna, USSR

Abstract

The termination of the fission valleys of the deformation potential energy surface is discussed in connection with the multi-component mass and energy distributions of fission fragments. The results calculated using the Strutinsky method for ^{213}At are compared with the new experimental data obtained at the Alma-Ata cyclotron.

1. INTRODUCTION. The theoretical description of the fission fragment mass and energy distributions is far from being satisfactory. Neither statistical [1-3] nor dynamical models without taking viscosity into account [4,5] describe the main features of the distributions. The approach based on the Fokker-Planck equations for the distribution function of the collective variables with phenomenological viscosity was so far applied only to the description of induced fission at sufficiently high excitation energies[6-8]. In this connection any attempts to understand the main features of the distributions by using a comparatively simple statistical approach are of some interest. Recently a breakthrough has taken place which is connected with investigations of valleys on the deformation potential energy surface (DPES). In accordance with the classical concepts of the liquid drop model only one valley exists on the DPES and the nucleus following this valley gets over a smooth and wide barrier to attain the point of scission to two fragments.

However, the growing amount of experimental data on the mass and energy distributions of fission fragments provides evidence that in favorable circumstances not one but at least two fission valleys can manifest themselves in nuclear fission. The earliest phenomenological hypothesis concerning two ways of fission [9] was formulated in the course of the analysis of the fission fragment mass distribution and was not explicitly connected with specific valleys on the DPES.

The experimental evidence supporting this connection was first obtained in a series of experiments using the Alma-Ata cyclotron. In those studies in the lead region at low excitation energies two- and even three-component yields were observed[10-14] in the mass and energy distributions of fission fragments. The yields become single-component ones at both high and very low excitation energies of the compound nucleus. In the former case the liquid-drop description is applicable and in the latter one the yields from all valleys but the lowest one are fading down /13,14/. An especially wide discussion of the connection of various fission modes with the fission valleys on the DPES takes place after the publication of a paper /15/ on the bimodal fission of very heavy nuclei in the fermium region /16-21/. The theoretical investigation of the fission vallyes on the DPES encounters some difficulties of both fundamental and practical nature. The present work is devoted to a discussion of some of these problems and to the possible ways of coping with them. In particular, the numerical results of preliminary calculations of fission valleys in ^{213}At are given and compared with the experimental data of refs. /8-14/.

2. FISSION VALLEYS. The study of the valley on the DPES is complicated because the very notion of the valley in the multi-dimensional space of collective variables has no coordinate-independent definition. For example, by transforming the coordinate system one can convert a valley into a "ridge" /22/. Strictly speaking, the results in one parametrization on the position of a valley in the space of the collective parameters defining the nuclear shape cannot be compared to the results obtained in the framework of another parametrization.
But the difficulties arise only in the statical approach. In the dynamical approach one can quite definitely speak about the "path to fission". But in an attempt to conserve the simplicity of the statical approach one can make the minimal dynamical assumption that the change in the distance between the centres of gravity of future fragments is connected with the displacement of large masses and thereby takes place rather slowly in the process of fission, so that all the rest of the coordinates have sufficient time for reaching minimum values. This assumption can be justified in the dynamical calculations, in which the description of nuclear fission was made by solving an equation of the Fokker-Planck type. The calculation shows that after the transition from the coordinates $\{c, h, \alpha\}$ introduced in ref. /23/ to $\{R_{CM}, h, \alpha\}$ (where R_{CM} is a half of the distance between the centres of mass of the future fragments, all other coordinates remaining unchanged), the fission valley defined as

$$\{\partial V/\partial h = 0, \quad \partial V/\partial \alpha = 0\}$$

at constant c in the first case and at constant R_{CM} in the second one appears to be situated much closer to the classical trajectory of the descent from the saddle point to scission /6,8/ in the latter case, as compared with the former one.

It is also interesting to note that after such a transition the mass tensor matrix calculated by the Werner-Wheeler method becomes essentially closer to the diagonal one.
Since it is sometimes inconvenient to introduce a coordinate system with R_{CM} as one of the parameters, in order to locate the valley one has to carry out a constrained calculation, minimizing the potential energy V at constant R_{CM}.
As is shown in ref. /24/, in the liquid drop model the DPES has the following properties. For the R_{CM} values greater than some critical one, R_{CM}^{crit}, the minimization proposed above is impossible because the potential energy surface has no minimum at fixed $R_{CM} > R_{CM}^{crit}$. The nucleus becomes unstable against a decrease in the neck radius at constant R_{CM}. The magnitude of R_{CM}^{crit} was taken as a natural definition of the nuclear scission point. At the lower values of $R_{CM} \leq R_{CM}^{crit}$ the nucleus possesses an equilibrium shape with quite a thick neck with the radius R_{neck}^{crit} at $R_{CM} = R_{CM}^{crit}$.
In the calculations /19/ of the fission valleys in nuclei lying in the region of Z=100, which take shell corrections into account, the nuclear shape with a neck radius equal to R_{neck}^{crit} /24/ was assumed to be the scission point. It seems to be more consistent to define the critical shape of the fissioning nucleus in calculations with shell corrections in the same way, as it was done in ref. /24/, i.e. by determining the value of R_{CM} at which the nucleus loses its stability against decreasing neck radius. The numerical calculations of the critical shapes corresponding to the termination of the three valleys in ^{213}At will be discussed in the next section. To conclude this section we note that the nuclear instability against decreasing neck radius, i.e. the appearance of the second negative eigenvalue of the matrix of the second derivatives of potential energy with respect to the deformation parameters, in addition to the first negative eigenvalue corresponding to the loss of stability against nuclear elongation, has a wider geometrical meaning than the notion of the fission valley: namely, it is invariant with respect to the linear change of the variables. Indeed, after the transition from the old coordinates $x_i (i=1,n)$ to the new ones Z_k (k=1,n) the second derivatives matrix is transformed as follows

$$\frac{\partial^2 V}{\partial Z_k \partial Z_\ell} = \frac{\partial^2 V}{\partial x_j \partial x_i} \frac{\partial x_i}{\partial Z_k} \frac{\partial x_j}{\partial Z_\ell} + \frac{\partial V}{\partial x_i} \frac{\partial^2 x_i}{\partial Z_k \partial Z_\ell} . \qquad (1)$$

It is seen that the tensor law of transformation described by the first term of the right-hand side is valid not only at the extreme points where the gradient $\partial V/\partial x_i$ vanishes, but also in the case of any linear change in variables, when $\partial^2 x_i/\partial Z_k \partial Z_\ell = 0$.

For comparison it should be noted that the definition of the scission point as a point where the stability against decreasing neck radius is lost is formulated in terms of the potential energy V only, while the consideration of the instability of the liquid drop filament /26/ is invoked for the description of scission in ref. /25/.

3. THE CALCULATED RESULTS FOR ^{213}At. The calculations of the valleys termination, i.e. the scission points for ^{213}At, were carried out by the Strutinsky method, refs. /27,28/. The parameters of the liquid drop model were taken from ref. /29/. The shell corrections were calculated on the basis of single-particle spectra in the Woods-Saxon-type potential with parameters taken from ref. /30/. The nuclear shapes were described in the coordinate system based on the Cassinian ovals, as it was done in ref. /31/. The calculational details are described also in ref. /19/. Before the description of the main results which, because of the shortage of computer time, were obtained only for a limited number of points, lets us consider a model example in detail. In the vicinity of the scission point the DPES is characterized by the convergence of the valley defined as $\partial V/\partial R_{neck}=0$, $\partial^2 V/\partial R^2_{neck} > 0$ and the ridge on which $\partial V/\partial R_{neck}=0$, $\partial^2 V/\partial R^2_{neck} < 0$. So, at the very scission point $\partial V/\partial R_{neck}=0$ and $\partial^2 V/\partial R^2_{neck}=0$. Leaving only the most important terms in Taylor's expansion at the scission point we obtain that in the vicinity of the scission point the function V, in terms of $x=R_{neck}-R^{crit}_{neck}$ and $y = R_{CM}-R^{crit}_{CM}$, looks like

---

Fig. 1. a) A perspective view of the model surface that demonstrates the disappearance of the valley. A surface equation is given at the top. The x-axis (the neck thickness) passes from the centre of the figure to the left, the y-axis (the distance between the fission fragment mass centres) passes from the centre deep into the right of the figure. Fission proceeds along the valley from the depth of the figure towards the reader. The bottom of the valley and the top of the ridge are marked by a curve (a parabola (3)). The scission point is considered to be the origin of coordinates (x=0, y=0). It is situated at the top of the parabola, where the valley passes into the ridge.
b) Iso-lines of the same model surface, as in fig. 1a). The connections of the coordinate Rneck with x and RCM with y are given in the text. The projection of the valley's bottom and ridge's top on the plane (R_{neck}, R_{CM}) is also given. The point of connection between the valley and ridge is marked by a diamond.
c) A side view along the R_{neck}-axis of the curve that passes along the bottom of the valley (lower part) and along the top of the ridge (upper part). The point of their connection is marked by a diamond.
d) A side view of the same curve, as in fig. 1c), along the R_{CM}-axis.

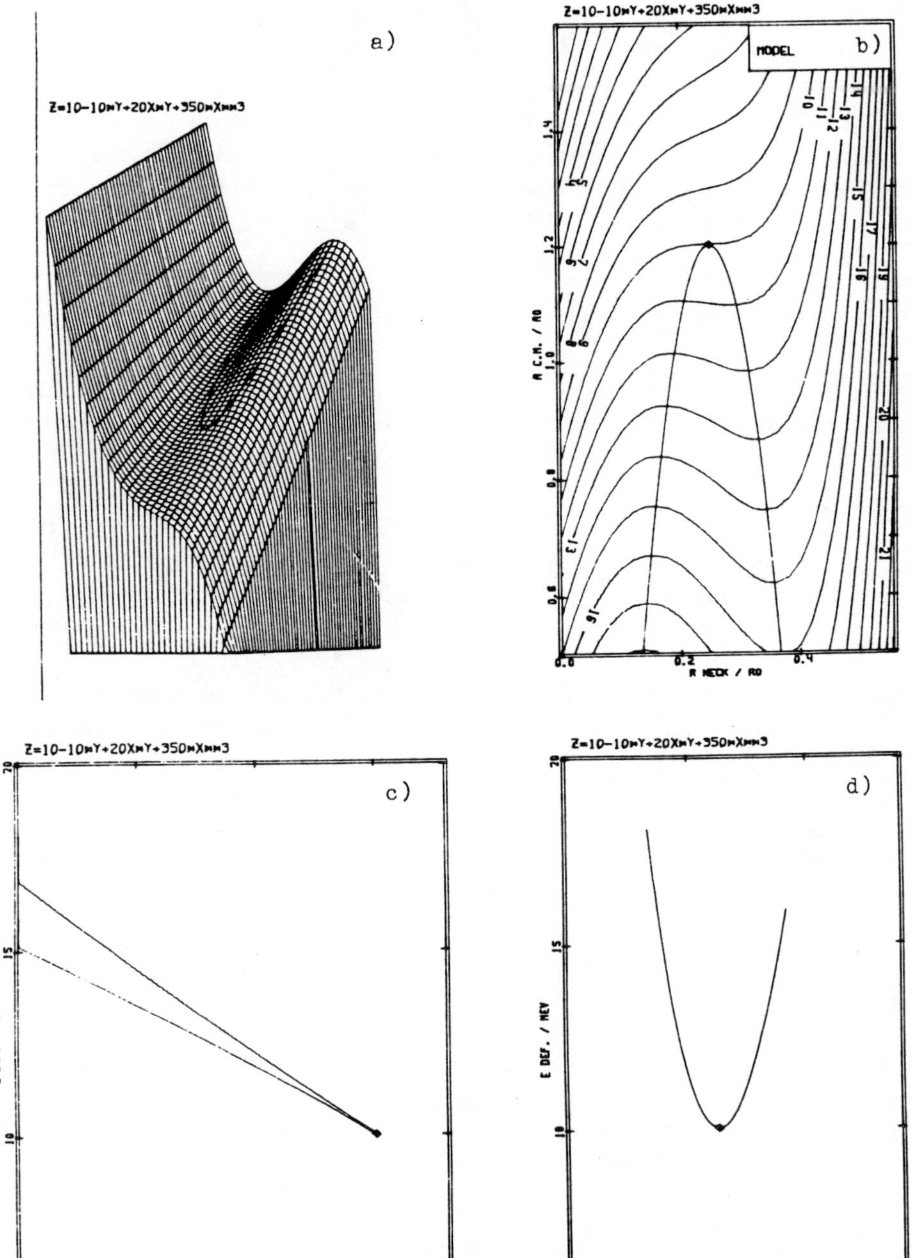
a)
Z=10-10*Y+20X*Y+350*X**3
b)
MODEL
R C.M. / RO
R NECK / RO
c)
E DEF. / MEV
R C.M. / RO
d)
E DEF. / MEV
R NECK / RO

Fig. 1

$$V = V_0 + \frac{\partial V}{\partial y} y + \frac{\partial^2 V}{\partial x \partial y} x \cdot y + \frac{1}{3!} \frac{\partial^3 V}{\partial x^3} x^3 . \qquad (2)$$

This kind of a surface with all specific parameters close to the real ones in ^{213}At, except for $\partial V/\partial y$, which is chosen to be considerably smaller in absolute value than in the real case, is shown as a perspective view in fig. 1a and as a map of iso-lines in fig. 1b. In the same figures the line $\partial V/\partial R_{neck}=0$ is also shown (a valley and a ridge), which is defined by the equation

$$\frac{\partial V}{\partial x} = \frac{\partial^2 V}{\partial x \partial y} y + \frac{1}{2!} \frac{\partial^3 V}{\partial x^3} x^2 = 0. \qquad (3)$$

At the point marked with a diamond the valley passes into the ridge. It is seen that in the most general case the valley/ridge line is a parabola rather than two intersecting straight lines, as is shown schematically in ref./32/. The value of the energy V along the valley/ridge line is demonstrated in figs. 1c and 1d as a function of R_{neck} and R_{CM}, respectively, in complete accordance with the schematic presentation /32/ and the exact calculations of ref. /24/. The full-scale calculations show that ^{213}At has, in the vicinity of the scission point, three valleys leading to binary fission, in agreement with experimental data /10-14/. The projection of the valleys on the (R_{CM}, R_{neck}) plane is shown in fig. 2 in which the fragmentary projections of the corresponding ridges are also shown (cf. the model example in fig. 1). At a large number of points both the valley and the ridge are calculated only in the liquid drop model and the value of $R_{CM}^{crit} = 1.167\ R_0$ is in good agreement with the exact calculation of ref./24/. In all the calculations presented in fig. 2 the nuclear shape was described in the coordinate system based on the Cassinian ovals, a deviation from the ovals being described as a series of the Legendre polynomials through the 10th order. It is seen that the two valleys that correspond to asymmetrical fission terminate for a more compact configuration and a smaller thickness of the neck, as compared with the valley of almost symmetric fission. The corresponding energies of the Coulomb interaction of the fragments, E_{int}^{Coul}, are compared in table 1 with experimental data on the fragment total kinetic energy, $\overline{E}_{kin}$, averaged over the appropriate mass intervals of the heavy fragments, ΔM. The surplus of $\overline{E}_{kin}$ over E_{int}^{Coul} is to be identified as the prescission kinetic energy, which, if one takes the calculated data at their face values,

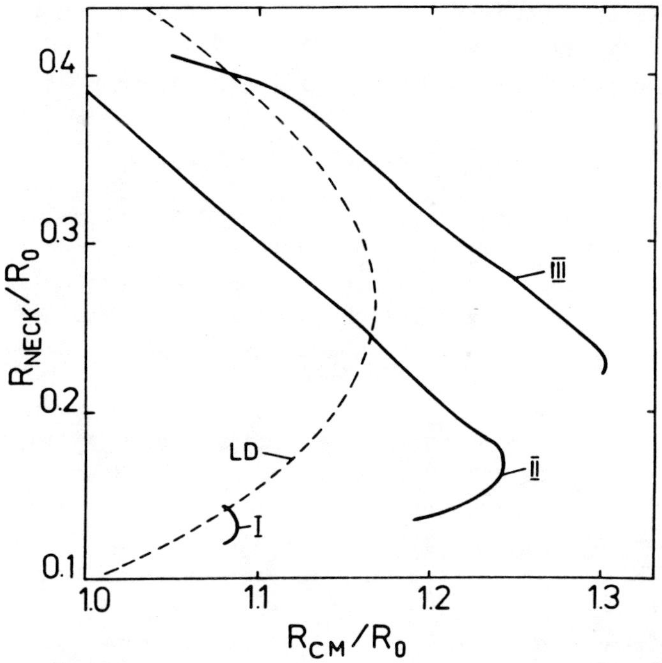

Fig. 2. The projection of the bottom of the valley and of the top of the ridge on the (R_{CM},R_{neck}) plane in the full-scale calculation for ^{213}At. The three fission valleys (full curves) are calculated taking the shell correction into account; also shown is the valley calculated in the liquid drop model (dashed curve).

Table 1

The most probable masses of heavy fragments, M, the fragment Coulomb interaction energy, E_{int}^{Coul}, and the fragment total kinetic energy $\bar{E}_{kin}$, averaged over the fragment mass range, ΔM, for different components of the mass-energy distribution.

Theory		Experiment /10-14/		
M_H (u)	E_{int}^{Coul} (MeV)	M_H (u)	ΔM (u)	$\bar{E}_{kin}$ (MeV)
113	145.3	107-113[a)]	113-113	146[a)]
133	158.5	132-134	132-133	160.8±1.4
135	139.0	137[a)]	134-137	149.2±0.8

[a)] The numbers are taken from the figure of the references given.

appears to be rather small for the very compact and very stretched, almost symmetrical configurations, of the order of 1-2 MeV, and moderately large for an intermediate configuration with a heavy fragment having a mass of about 135 u ($\approx$10 MeV). The small value of the prescission kinetic energy for the very compact configuration is possibly associated with the calculated result implying that the corresponding valley is very short: at $R_{CM} = 1.05\ R_o$ the minimization process goes away to the valley of the intermediate configuration. This means that the valley originates in this vicinity and terminates at $R_{CM} = (1.08 - 1.09)\ R_o$, where R_o is the radius of the sphere of the same volume.

Thus, the theory describes correctly the average values of the fragment masses and their kinetic energies for all the three experimentally observed components of the fragment mass and energy distributions. Moreover, theory explains the experimental evidence that the two asymmetric valleys have a common barrier and bifurcate only at the last stage of the descent /13-14/.

4. CONCLUSION. In the present paper a definition is given of the fission valley as a conditional minimum on the DPES with fixed R_{CM}, the location of the minimum being dependent on R_{CM} as a parameter. The point at which nuclear stability with fixed R_{CM} is lost is interpreted as a nuclear scission point in the calculations using shell corrections, as was done earlier in the liquid drop model /24/. The ratio between the volumes of the future fragments at the scission point is compared with the most probable ratio between the masses of the fission fragments and their Coulomb interaction energy is related to the total kinetic energy of fully separated fragments.

The different valleys and, correspondingly, the different scission points are considered to be related to different components of the mass and energy distributions of fission fragments. The calculated results for ^{213}At support this interpretation.

The author is grateful to G.N.Smirenkin for stimulating discussions.

REFERENCES

1. P.Fong, Statistical Theory of Nucl. Fission, N.Y., Gordon and Breach, 1969.
2. A.V.Ignatyuk, Yad. Fiz. 9 (1969) 357.
3. A.V.Ignatyuk and V.P.Krainov, Aktualnyie voprosy fiziki deleniya, M., MIFI (1983) 3.
4. J.R.Nix and W.J.Swiatecki, Nucl. Phys. 71 (1965) 1.
5. J.R.Nix, Nucl. Phys. A130 (1969) 241.
6. G.D.Adeev, I.I.Gonchar, V.V.Pashkevich, and N.I.Pischasov, Preprint JINR P4-86-552, Dubna (1986).
7. O.I.Serdyuk, G.D.Adeev, I.I.Gonchar, V.V.Pashkevich, and N.I.Pischasov, Yad. Fiz. 46 (1987) 710.
8. G.D.Adeev, I.I.Gonchar, V.V.Pashkevich, N.I.Pischasov, and O.I.Serdyuk, Particles and Nuclei 19 (1988) 1229.

9. A.Turkevich and J.B.Niday, Phys. Rev. 84 (1951) 52.
10. M.G.Itkis, V.N.Okolovich, A.Ya.Rusanov, and G.N.Smirenkin, Z.Phys. A320 (1985) 433.
11. M.G.Itkis, V.N.Okolovich, A.Ya.Rusanov, and G.N.Smirenkin, Yad. Fiz. 41 (1985) 849, 1109.
12. M.G.Itkis, J.V.Kotlov, S.I.Mulgin, V.N.Okolovich, A.Ya.Rusanov, G.N.Smirenkin, and V.N.Tolstikov, Europhys. Lett. 4 (1987) 275.
13. E.N.Gruzintsev, M.G.Itkis, Yu.V.Kotlov, V.N.Okolovich, A.Ya.Rusanov, and G.N.Smirenkin, Yad. Fiz. 47 (1988) 1201.
14. M.G.Itkis, V.N.Okolovich, A.Ya.Rusanov, and G.N.Smirenkin, Particles and Nuclei, 19 (1988) 701.
15. E.K.Hulet, J.F.Wild, R.J.Dougan, R.W.Lougheed, J.H.Landrum, A.D.Dougan, M.Schädel, R.L.Hahn, P.A.Baisden, C.M.Henderson, R.J.Dupzyk, K.Sümmerer, and G.R.Bethune, Phys. Rev. Lett. 56 (1986) 313.
16. U.Brosa, S.Grossmann, and A.Müller, Z. Phys. A325 (1986) 241; Z. Naturforsch. 41a (1986) 1341.
17. V.V.Pashkevich and A.Sǎndulescu, JINR Rapid Comm. no. 16-86 (Dubna,1986) p. 19.
18. P.Möller, J.R.Nix, and W.J.Swiatecki, Nucl. Phys. A469 (1987) 1.
19. V.V.Pashkevich, Nucl. Phys. A477 (1988) 1.
20. S.Cwiok, P.Rozmej, A.Sobiczewski, and Z.Patyk, Preprint GSI-88-41 (1988), submitted to Nucl. Phys. A.
21. P.Möller, J.R.Nix, and W.J.Swiatecki, Preprint Los Alamos National Laboratory, LA-UR-88-2266 (1988).
22. L.Wilets, Theories of Nuclear Fission, Clarendon Press, Oxford, 1964.
23. M.Brack, J.Damgaard, A.S.Jensen, H.C.Pauli, V.M.Strutinsky, and C.Y.Wong, Rev. Mod. Phys. 44 (1972) 320.
24. V.M.Strutinsky, N.Ya.Lyashchenko, and N.A.Popov, Nucl. Phys. 46 (1963) 639.
25. U.Brosa and S.Grossmann, Z. Physik A310 (1983) 177.
26. J.W.Rayleigh, Proc. London Math. Soc. X (1878) 4.
27. V.M.Strutinsky, Nucl. Phys. A95 (1967) 420.
28. V.M.Strutinsky, Nucl. Phys. A122 (1968) 1.
29. W.D.Myers and W.J.Swiatecki, Ark.Fys. 36 (1967) 343.
30. S.P.Ivanova, A.L.Komov, L.A.Malov, and V.G.Soloviev, Nucl. Phys. A135 (1969) 432.
31. V.V.Pashkevich, Nucl. Phys. A169 (1971) 275.
32. W.J.Swiatecki and S.Bjørnholm, Phys. Reports 4C (1972) 327.

ON THE ROLE OF THE SECOND WELL OF THE DEFORMATION POTENTIAL ENERGY IN NUCLEAR FISSION IN THE LEAD REGION

E.M. Rastopchin, G.N. Smirenkin
The Institute of Physics and Power Engineering, Obninsk, USSR
V.V. Pashkevich
Joint Institute for Nuclear Research, 141980 Dubna, USSR

Introduction .

The role of the second deformation energy minimum $V(\varepsilon)$ of heavy nuclei is known very well - it is considered to account for the origin of delayed fission, i. e. the spontaneously fissioning isomers /1,2/. Earlier /3/ it was shown that minimum in $V(\varepsilon)$ does not desappear in the lighter nuclei (up to Pb) though it is displaced in deformation $\varepsilon \cong 0.6$ to $\varepsilon \cong 0.4$ starting from Th and reaches the left-hand slope of the potential barrier of which the height rapidly increases with decreasing Z-value. This circumstance sets the lower limit of the region of spontaneously fissioning isomers (U-Cm). Observation of transitions from the second well to the first one also encounters great difficulties. Therefore, it would be of interest to try to search for other manifestations of a spontaneously fissioning isomers in preactinides, which would not be associated with their decay from the second well.

In this context our attention has been attracted by the island of spherical nuclei lying in the vicinity of the doubly magic nucleus of ^{208}Pb whose statistical properties look different aganist the background of deformed nuclei because of a conciderable difference between the coefficients of the rotational enhancement of level density /4,5/,i.e.

$$K_{rot} = \sigma_{\perp} \text{ for deformed nuclei, } 1 \text{ for spherical nuclei} \quad (1)$$

where $\sigma_{\perp}^2 = F_{\perp} \times T$ is the parameter representing the spin dependence of the level density which is larger than 10^2 for T=1 MeV; $F_{\perp}$ is the perpendicular moment of inertia and T is nuclear temperature.
The "jump" of K_{rot} in eq.(1) in passing from one nucleus to another considerably affects the energy dependence on nuclear fissility,$P_f(E)$ (i.e. the ratio between the fission cross section and that for compound nucleus formation) in the preactinide region /5,6/.

The present paper deals with a study of $P_f(E)$ as function of second well in V(E).The effect is expected to be the most considerable in spherical nuclei and manifests itself as follows.
With the excitation energy E of the compound nucleus increasing as a result of the difference between $\varepsilon_2 \cong 0.4$ and $\varepsilon_1 \cong 0$ leading to

$K^1_{rot} >> K^2_{rot}$, the level density in the second well can become equal to or even greater than that of the first well, despite the lower excitation energy $E - E_2$. This fact should lead to an increase in neutron emission and to a decrease in nuclear fissility. The smaller is the energy difference between the minima $E_2 = V(\varepsilon_2) - V(\varepsilon_1)$ the earlier the effect manifests itself. Evidently the expected effect of the second well will sharply decrease in the case of the nuclei deformed in the equilibrium state (the first potential well).

2. Attenuation of $K_{rot}(T,\varepsilon)$ and nuclear fissility .

Our analysis is based on experimental information and description of nuclear fissility /5,6/, i.e.

$$P_f(E) = \frac{\hbar^2}{4\mu r_0^2 A^{2/3}} \times \gamma(\bar{J}) \times \frac{\int_0^{E-E_f} \rho_f(U,0)\, dU}{\int_0^{E-B_n} \rho_n(U,0)\,(E-B_n-U)\, dU} \qquad (2)$$

where $\rho_f(U,J)$ and $\rho_n(U,J)$ - are the level densities of the fissioning nucleus at the saddle point and of the residual nucleus A-1 after emission of a neutron, as function of the excitation energy U and the angular momentum J respectivly; $\gamma(\bar{J})$ is the factor taking into account the J dependence of $\rho_i(U,J)$; E_f is the barrier height; and B_n is the neutron binding energy. In calculating $\rho_i(U,J)$ use was made of the superfluid model of the nucleus with the phenomenological inclusion of shell and collective effects. The model parametrs were in agreement whith the observed density of neutron resonances $\rho_{exp}(B_n,J)$ /5,7/. In what follows we shall turn to the only but very essential specification which we have introduced in the description of $\rho(U,J)$ and $P_f(E)$ /5-7/.

The adiabatic estimate of $K^{ad}_{rot} = \sigma^2_\perp$ for deformed nuclei made in eq.(1) is valid when it's possible to assume the independence of the single-partical modes of motion and rotation of the nucleus as a whole. Ref. /4/ gives temperature estimate

$$T = \hbar\omega_0 \times \varepsilon \cong 41 \times A^{1/3} \times \varepsilon \qquad (3)$$

which makes this assumption devoid of sense, namely (ε - is quadrupole deformation, as before; $\omega_0 \cong 41 \times A^{1/3} \times \hbar^{-1}$ MeV - is the average frequency of the anisotropic oscillator potential). The results of level density calculations /8/ taking into account the interactions between the rotatinal and internal degrees of freedom have confirmed the qualitative /4/ , in particular, they have shown that the attenuation factor which, as the nucleus gets heated, leads to a decrease as compared to K^{ad}_{rot}, can be approximately described using eq.(3).

The dependence $K_{rot}(T,\varepsilon)$ can be presented in the following form

$$K_{rot} = (K^{ad}_{rot} - 1)\times q(\varkappa) + 1 \ ; \ \varkappa = T/T_0 \cong 0.025\times A^{1/3}\times T/\varepsilon \qquad (4)$$

where $q(\varkappa) \to 1$ for small $\varkappa$ values ($K_{rot} \to K^{ad}_{rot}$), $q(\varkappa) \to 0$ for large $\varkappa$ ($K_{rot} \to 1$). A number of modifications of the analitical description of $q(\varkappa)$ is offered /8-10/. We have chosen the simpler one

$$q(\varkappa) = \exp(-\beta\times\varkappa^2) \qquad (5)$$

with $\beta = 1$ coinciding with $q(\varkappa)$ in ref. /9/. In what follows we employ the value $\beta = 1.37$ obtained by fitting the level density to $\rho_{exp}(B_n,J)$.

Thus, the attenuation of K_{rot} with energy depends substantially on nuclear deformation. This dependence is of major importance for nuclear since in $E < 100$ MeV in question the departures of K_{rot} from K^{ad}_{rot} in the neutron emission channel are considerable whereas they are negligibly small in the fission channel. These results contradict the previous assumptions /5,6/ concerning K^f_{rot} and K^n_{rot}. This leads to the consequences intimately related with the problem under study.

3. The necessity of including neutron emission in the second well.

In the lower part of fig.1 the results of calculating the fissility parameter in terms of more accurate model of $\rho(U,J)$ are compared with experimental data for three typical preactinide nuclei - spherical ^{212}Po, the deformed nucleus of ^{186}Os and ^{198}Hg considered to be an intermediate case in refs./5,6/.The top part of fig.1 shows the potential energies $V(\varepsilon)$ of these nuclei calculated using the shell correction method as in ref./3/. The $P_f(E)$ calculation was carried out using the barrier parameters from the phenomenological model /11/ and with equal asymptotic level density parameters in the fission and neutron emission channels, $\tilde{a}_f/\tilde{a}_n = 1$. As usual /5,6/, the E_f and $\tilde{a}_f/\tilde{a}_n$ values serve as free parametrs in fitting the calculated results to the experimental dependence. The given description does not involve these quantities (those not specified correspond to refs.(/5-7/). It should also be noted that in fig.1 all the curves were plotted taking into account fission processes preceded by neutron emission.

The dashed curves showed in fig.1 were calculated according to eq.(1) and using the traditional classification of nuclei according to the nature of the spectra of low - lying levels (^{212}Po and ^{198}Hg are spherical nuclei, and ^{186}Os is a deformed one). Following eqs.(4) - (5) the inclusion of the K_{rot} attenuation eliminates the disagreement with the experimental in the case of ^{186}Os, as regards the other two nuclei, taking the K_{rot} attenuation factor into account does not change the situation since K_{rot} = 1 for spherical nuclei. The latter are characterized by a growing deviation of the dashed curves from the experimental points with increasing E, this deviation occurring somewhat earlier ^{198}Hg than in ^{212}Po. In previous studies /5,6/ the discrepancies of results for ^{212}Po and adjacent nuclei were explained as being due to the K^f_{rot} deviations from the adiabatic estimate.These deviatons define the attenuation factor as $q(E-E_f) = P_f^{exp}(E)/P_f(E)$ which was valid also for describing the neutron emission channel, i.e. the K^n_{rot} value.

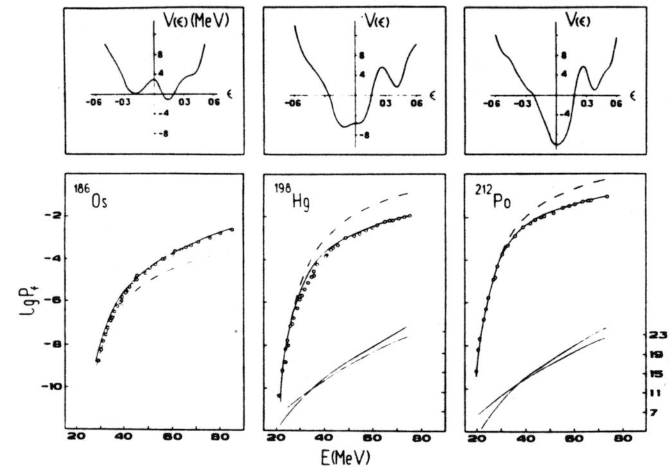

Fig.1. Top part: the potential deformatin energies $V(\varepsilon)$ of the ^{186}Os, ^{198}Hg and ^{212}Po nuclei. Bottom part: the energy dependences of fissility $P_f(E)$ for the same nuclei and the numbers of the residual nucleus final states accessible to neutron emission N_{n_i} for Hg and Po. The experimental data are taken from /5/, the calculations of $P_f(E)$ carried out without taking into account the attenuatin factor K_{rot} (dashed curve) and with taking into account the attenuation factor K_{rot} and neutron emmisson in the second well (full curve).

The results of the calculations of the fission probability for ^{212}Po and ^{198}Hg taking into account neutron emission in the second well ,which are shown by the full curves, to a considerable extent remove the disagreement. In these calculations the denominator (2) constituting the number of the residual nucleus final states accessible to neutron emission for J = 0, which has the form

$$N_{n_i} = \frac{4\mu r_0^2 A^{2/3}}{\hbar^2} \times \int_0^{U^i_{max}} \rho_{n_i}(U,0)\,(U^i_{max} - U)\,dU \qquad (6)$$

was replaced by the sum N_{n_i}, where i = 1 or 2 is the well index $U^1_{max} = E - B_n$, and $U^2_{max} = E - B_n - E_2$. This generalization (2) is a direct sequence to the statistical description of the decay probability for excited nuclei in terms of the two-humped barrier /2/.

The energy dependences N_{n_i} for spherical nuclei are given under the fissility parameters in fig.1. By comparing them one can see that the coming of the second well into play occurs in the vicinity of the interaction point $N_{n_1} = N_{n_2}$, as the result the nuclei in the question happen to fission like spherical ones at low energies and like deformed ones at high energies because of the predominance of neutron emission over fission in the second well. For the deformed ^{186}Os nucleus this is not essential because it has no second well and, moreover, even if it excisted, its role would be strongly suppressed since the difference between K^1_{rot} and K^2_{rot} is much smaller in this case than in spherical nuclei.

4. Conclusion

The inclusion of the dependence of K_{rot} on nuclear deformation and on neutron emission in the second well rules out the difficulties and disadvantages of the previous analysis of the fissility of preactinide nuclei /5,6/. We did not try to vary the parametrs in order to better describe the experimental data. It is worth doing after including the dynamical effects /12/.

The authors are grateful to A. V. Ignatyuk, G. A. Kudyaev and U. B. Ostapenko for useful discussions.

References.

1. V. M. Strutinsky, Nucl.Phys., 1967, v.A95, p.420.
2. S. Bornholm et al., Nucl.Pys., 1969, v.A136, p.1.
3. V.V.Pashkevich , Preprint JINR P4-4383.Dubna, 1969; In: International School-Seminar on Heavy Ion Physics.Dubna, JINR, 1983, p.405.
4. S. Bjornholm et al., In: Physics and Chemeistry of Fission. IAEA, Vienna, 1974, v.1, p.367.
5. A.V. Ignatyuk et al., Particles and Nuclei , 1975, v.16, p.709.
6. A.V. Ignatyuk et al., In Physics and Chemistry of Fission. IAEA, Vienna, 1980, v.1, p.421.
7. A.V. Ignatyuk et al., Yad. Fiz., 1979. v.29, p.875.
8. G. Hansen et al., Nucl.Phys., 1983, v.A406, p.236.
9. K.H .Schmit et al., Z.Phys.A - Atoms and Nuclei 315, 1984, p.159.
10. S.E. Vigdor et al., Phys.Rev., v.C26, 1982, p.1068.
11. W.D. Myers et al., Arkiv. Fys., 1967, v.36, p.343.
12. U.B. Ostapenko et al. Preprint FEI-1848.Obninsk, 1987; Voprosy Atomnoi Nauki i Tehniki.ser.Yadernye Konstanty, 1988., vyp.1, p.45

STATISTICAL AND DYNAMICAL ASPECTS IN NUCLEAR FISSION

A.A. Goverdovskij and A.V. Ignatyuk
Institute of Physics and Power Engineering, Obninsk, USSR

The main trends and results of the experimental and theoretical studies focused on the statistical description of observed fission cross sections and mass-kinetic energy destributions of fission fragments are reviewed.

Description of fission cross sections

The experimental data on the energy dependence of the fission cross section σ_f and the fission probability (or fissility) σ_f/σ_c, where σ_c is the cross section for compound-nucleus formation, are widely used for the determination of fission barrier heights. These main characteristics of the fissile nuclei are very important for the analysis of many properties of nuclei: the deformation energies, the range of nuclear stability, the heavy ion interaction potentials etc. With the fissility of nuclei being measured in the near-threshold and sub-threshold regions the extracted fission barriers do not depend greatly on the model employed in the analisis. However, when analyzing an over-threshold region the description of fissilities is essentially dependent both on the excitation energy above barrier and on the model description of the corresponding level density. On this account the analysis of fissilities in a wide excitation energy range turns into an effective method to investigate statistical properties of excited nuclei.

The Fermi-gas model is often used in the analysis and systematics of different experimental data on the level density and other statistical characteristics of nuclei. The relations of this model are rather simple, which may well be the main reason of its wide-spread application. However, this model does not take into account shell inhomogeneities in the spectrum of singe-particle levels as well as superfluid and collective effects. The whole totality of effects can be considered if the calculations of the level density are made on the basis of consistent microscopic approaches using realistic schemes of single-particle levels /1/. Recently the phenomenological approaches have been also suggested, which take into account the main concepts of the theoretical description of highly-excited nuclei with simplified analytical relations convenient for practical applications /2,3/.

When the collective effects are taken into consideration the relation for the level density can be presented in the form:

$$\rho(U) = \rho_{qp}(U)\, K_{vibr}(U)\, K_{rot}(U)\ , \qquad (1)$$

where ρ_{qp} is the density of quasi-particle (non-collectivized) excitations; K_{vibr} and K_{rot} are the coefficients of the level density increase due to vibrational and rotational excitations, respectively. The density of non-collective excitations should be described by the relations of the superfluid nucleus model. The principal feature of this model is the phase transition from the low-energy superconducting phase to the normal Fermi-gas one at the excitation energies $U_{cr} = 6 - 8$ MeV. The shell effects can be described within the phenomenological approach using an energy dependent level-density parameter

$$\mathbf{a}(U) = \begin{cases} \tilde{\mathbf{a}}\left[1 + \delta E\,\dfrac{f(U - E_{cond})}{U - E_{cond}}\right] & \text{for } U \geq U_{cr} \\ \mathbf{a}\,(U_{cr}) & \text{for } U > U_{cr} \end{cases} \qquad (2)$$

where δE is the shell correction to the nuclear binding energy, $\tilde{\mathbf{a}}$ is the asymptotic value of the level-density parameter for high excitation energies; $f(U) = 1 - \exp(-\lambda U)$ is a "universal" function specifying energy variations of single-particle level density in the vicinity of Fermi surface; $E_{cond} = .322\ U_{cr}$ is the condensation energy which defines the energy gain due to transition from the normal state to the superconductive one.

The coefficients Krot required for the description of the observed neutron resonance density of heavy nuclei ($A > 150$) were analyzed in Ref./2/, and the coefficients Kvib for light and medium nuclei - in /3/. These papers also contain a more complete consideration of all relations of the developed approach.

Which conclusions can be drawn about the main components of the statistical description of excited nuclei from the analysis of the observed fission cross sections ? The investigation of a wide range of pre-actinide nuclei in the reactions with light charged particles have shown that /4/:

1) Shell effects are displayed rather distinctly in the characteristics of neutron channel for near-magic nuclei in the region of lead, and the energy dependence of neutron widths extracted from the observed fissility is in a good agreement with the predictions of theory about the shell variations of the level density parameters (2);

ii) Systematic differences of the fissilities for spherical and deformed nuclei with the same excitation energies above the fission barrier are the direct confirmation of the necessity for taking into account of the rotational increase the level density (1);

iii) The observed energy variations of the parameter K_o^2 ,which define the angular distribution anisotropy of fission fragments, are in a good agreement with the behaviour of the moments of inertia predicted by the superfluid model.

The whole totality of the above-mentioned effects can be considered as the conclusive evidence of the necessity to pass to more complicated description of statistical properties of nuclei, than the Fermi-gas. A justification of complications is on the one hand the close relationship between the considerations of highly excited nuclei and the description of the ground and low-lying nuclear states, and on the other hand a significantly better consistency of parameters extracted from the analysis of various channels of compound-nucleus decay in a wide range of excitation energies and nuclear masses.

Recently an application of the phenomenological approach developed has been actively initiated for the analysis of neutron-induced fission cross sections of heavy actinide nuclei /5/.Here the main problems are concerned with the need for taking into account major portion of preequilibrium processes in the entrance channel and with the correct determination of shell corrections in the fission channel. Fig.1 can serve as a typical illustration of the experimental data description.

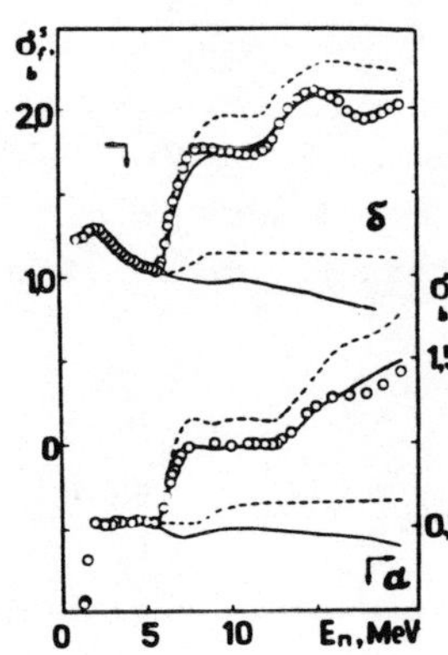

Fig. 1. Neutron-induced fisson cross section of ^{238}U (a) and ^{235}U (b). Solid curves are the calculations taking into account preequilibrium neutron emission and dashed ones - without preequilibrium decay.

Many uncertainties of the adopted model parameters are eliminated when the neutron spectra and the excitation functions of (n,xn)-reactions are analised together with the fission cross sections. The direct "experimental" determination of "first chance" fission cross sections, i.e. the fissility of initial compound nucleus, is of great interest for the definition of shell corrections in the fission channel. A more complete consideration of these problems is given in the poster presentation at this conference.

A preequilibrium component manifesting itself in neutron spectra is an example of the dynamical processes preceding the statistically equilibrium compound nucleus stage. The similar dynamic effects at high excitation energies ($U > 30$ MeV) seem to manifest themselves in the energy dependence of fission widths /6/. The theoretical and experimental analysis of the process mechanism in both predominating nuclear decay channels is of great interest for the further advance of the nuclear reaction theory as well as for the improvement of statistical description of the properties of heated nuclei. In particular, a reliable isolation of non-equilibrium processes is essential for analyzing the "damping" of level density increase coefficients at high excitation energies.

Mass distribution analysis

The question about the relationship of the statistical and dynamical effects is particularly urgent in the analysis of fission fragment yields, i.e. in the description of fragment yields and kinetic energy distributions observed . From the standpoint of the statistical approach the principal task is to determine the place of fragment distribution formation and their distortions with the final scission of the fissile nucleus and the divergence of fragments.

The study of the dispersion energy variation for the corresponding distributions is one of the direct empirical ways to solve the problem. For quasiequilibrium statistical formation of mass distributions the dispersion will be described by the relation

$$\sigma_M^{\ 2} = \frac{\hbar\omega}{2C_M} \,\mathrm{cth}\left(\frac{\hbar\omega}{t}\right) = \begin{cases} t/C_M & \text{for } t \gg \hbar\omega \\ \hbar\omega/C_M & \text{for } t \ll \hbar\omega \end{cases} \qquad (3)$$

where t is the nuclear temperature in the corresponding point; C_M is the stiffness coefficient for mass-asymmetric distortions; $\hbar\omega$ is the effective frequency of oscillations for these distortions. Determining the zero-point of the temperature scale we can localize the effective point of dispersion formation, and any deviations in the temperature dependence (3) could be indicative of the dynamic effects related to the formation point displacement and/or the contribution of supplementary non-equilibrium processes.

Detailed research of mass dispersions for pre-actinide nuclei shows that the place of their formation is a near-saddle point/7/.This result seems to be quite natural for saddle configurations with a well-pronounced dumb-bell like form, when the difference between saddle configurations and physical scission points (i.e. thick-neck configurations losing resistance to rupture on two fragments) being insignificant. The similar analysis is of great interest and importance for the actinide nuclei, whose differences in saddle configurations and physical scission points should be much greater.

Fig.2 represents the dependence of mass dispersions for actinides from the fissility parameter Z^2/A for spontaneous fission and for neutron-induced fission with the excitation energy U_f= 1.5 MeV above the barrier. The compilation of experimental data of Refs./8,9/ and the works quoted there has been used for the preparation of Fig 2. A significant dispersion of mass distributions for spontaneous fission as well as a dispersion growth with the increasing mass number indicate that the effective region of mass dispersion formation is located on the slope of barrier behind the exit point of subbarrier path for cold or spontaneous fission. If the dissipation energies obtained in Ref./10/ from the analysis the even-odd differences of fragment charge

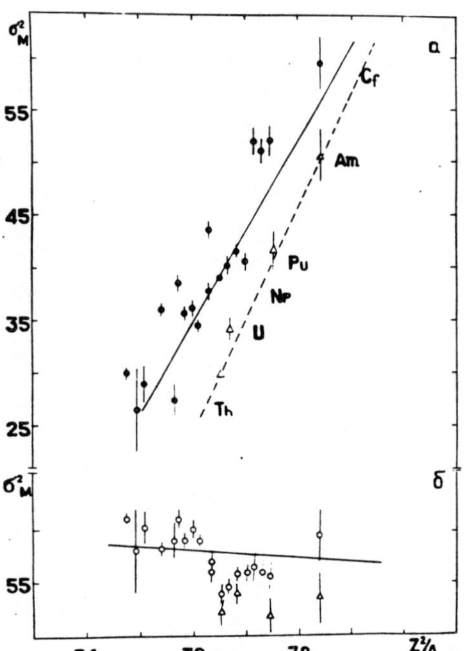

Fig.2a). Dispersions of actinide fragment yields with the excitations energy in the saddle point 1.5 MeV(●) and for the spontaneous fission (▲). б) The same is with the excitation energy 12.5 MeV in the effective scission point.

yield used to determine the excitation energies in the effective scission point on the slope of barrier, then the systematics of dispersions with the adopted energies $E_{dis}+U_f$ can be constructed. The results of this representation of dispersions can be treated as the determination of initial point on the scission excitation energy scale, i.e. the localization of formation place of mass distributions on the barrier slope.

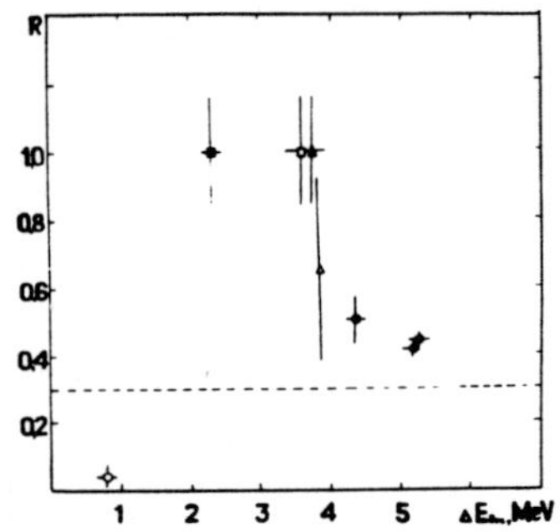

Fig. 3. The dependence of ratio **R** on ΔE_{def} /see the text/. The data on the following reactions were used: o-^{238}U(n,f); ■ -^{242m}Am(sf),^{241}Am(n,f); □ -^{250}Cf(sf),^{249}Cf(n,f); Δ -^{252}Cf(sf), ^{250}Cf(t,pf); Δ -^{238m}U(sf),^{238}U(sf); o -^{245}Cm(n,f),^{246}Cm(sf) o-^{239,241}Pu(n,f), ^{240,242}Pu(sf).

With the experimental data on mass distributions of near-barrier and subbarrier fission available, the dispersion systematics can be employed to solve the problem of the transformation of fissile nucleus deformation energy in thermal excitations. Fig.3 shows the ratio R obtained in this analysis for the thermal excitation energy difference determined from the observed dispersions to the total deformation energy ΔE_{def} accessible between the saddle point and the exit point of spontaneous fission on the barrier slope. The similar analysis results for spontaneously fissioning isomers are also included. The obtained data indicate that at the initial stage of descent virtually all deformation energy is converted into inner excitation. But at the next stage which is still far from the effective scission point only 40-50% of deformation energy is converted into thermal excitation, whereas the remaining energy is converted into kinetic energy of collective motion. The futher fate of this energy is still hardly understandable. However, the ratio R can hardly be expected to grow in the subsequent descent sections. Rather it will be approximately the same or even it will be reduced. As the kinetic energy of collective motion can be quite simply converted into the pre-scission kinetic energy of fragments, so the assumption on high values of the pre-scisson energy $E_s=(1-1.5)\Delta E_{def}$ becomes justified.

Note, that for the reaction ^{238}U(n,f) the mass yield dispersions are virtually unchanged with the decrease by 1 MeV below the fission threshold /11/. It indicates, that the conversion of deformation energy into inner excitation is hindered in the earliest part of descent for near-threshold excitations. The energy transformation is not initiated until the energy approximately comparable with the nucleon pairing energy $2\Delta_f$ is released.

This consideration of mass yield dispersions actually indicates the lack of thermodynamic equilibrium between thermal excitations and collective modes over a significant part of descent. It means that the consistent description of mass distributions may be obtained only within the quasiequilibrium statistical approach with considerable non-equilibrium kinetic energy for the main collective mode stretching deformation.

Variation of fragment kinetic energies

Kinetic energies of fragments are essentially defined by the energies of Coulomb repulsion of fragments at the scission moment, and the potential energy surface relief at the scission point defines the largest of the observed average kinetic energy dependence on mass split as well as the main changes of average kinetic energies with an increase of nucleus excitation energy. But recently the diverse fine effects have been observed in the energy variations kinetic energies whose nature is hard to be explained.

Fig.4 shows the results of average kinetic energy measurements for the neutron-induced fissin of even uranium isotopes /12/. The solid curves are the fission cross section for the corresponding isotopes. Systematic drops of the observed E_k by the value about 1 MeV is displayed in the region of vibrational resonances in fission cross sections. The contribution of non-resonance background being subtracted, the resonance variations of kinetic energy E_k related to intermediate vibrational states in a double-humped barrier can be evaluated. Fig.5 shows the dependence E_k obtained on the resonance energy measured from a well bottom. Together with the above data Fig.5 presents the analysis results for some other isotopes for which the corresponding experiments are available.

The research has shown that the kinetic energy variations are practically independent on mass split and are not accompanied by local variations of the number of emitted neutrons, i.e.the variations δE_k are not accompanied by total reaction energy changes. It points out that they are not caused by some variations of fragment's Coulomb energy at the scission moment, but they are directly associated with the variations of pre-scission kinetic energy of fragments. Hence, an

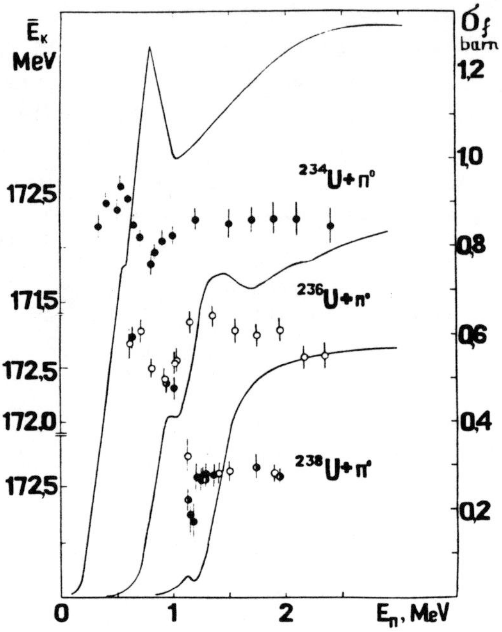

Fig. 4. The results of measurements for fragment average kinetic energies of near-barrier neutron-induced uranium fission /12/.

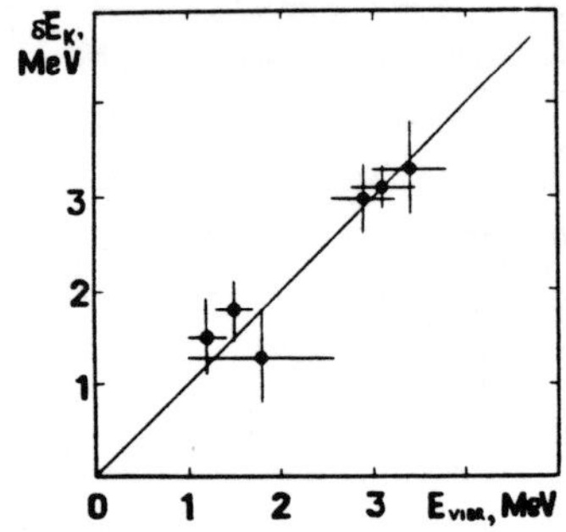

Fig. 5. Systematics of δE_k on vibrational resonance energies.

estimate of the lower limit for pre-scission kinetic energy of fragments 3.2±.3 MeV is obtained for uranium isotopes which is independent on the previous dispersion analysis. This value being compared with the dissipation energy E_{dis}=5-6 MeV obtained for near-threshold fission of uranium isotopes /10/, a fairly high value of pre-scission kinetic energy contribution to the total deformation energy released during the motion of nucleous from the saddle point to the scission point is obtained.

Possible advance in the statistical description of fission yields

The statistical theory of fragment yields gives a satisfactory explanation of the asymmetry of mass distributions, the dependence of average kinetic energies on mass split and the saw-toothed dependence of the number of prompt neutrons emitted by fragments /13/. At the same time this theory offers highly underestimated dispersions of the mass and kinetic energy distribution of fragments. The discussed experimental data enable us to conclude that the main fault of the statistical calculations is concerned with an assumption about thermodynamical equilibrium of stretching degree of freedom and inner excitations in the vicinity of geometrical scission point . Great pre-scission energy demands discarding this assumption and shifting the effective physical scission point upwards the barrier slope, so that this point could correspond to rather thick-neck shapes. When moving upwards the barrier slope the stiffness for mass-asymmetric distortions decreases, and consequently in terms of statistical description some higher dispersions of fragment yields can be obtained at the same temperature. The assumption on thermodynamical energy distribution over all the degrees of freedom seems to be discarded even for the thick-neck, and the statistical description should be formulated with the selected non-equilibrium stretching degree of freedom. For practical realisation of such calculations the parameters of a liquid drop model defining the steepness of barrier slope will be essential. The majority of widespread parameter sets offer a too steep slope, whereas numerous experimental data indicate lower energy changes between the saddle and the scission points. A special selection of model parameters correcting this defect is required /14/.

The quasi-equilibrium statistical description of fragment yields should certainly be considered as a simplified solution of a more general dynamical problem with fluctuational-dissipative motion from the saddle point to the scission one. Recently much progress has been achieved in the solution of this problem for highly heated nuclei whose potential energy surface and kinetic coefficients correspond to the liquid drop model /15/. The difficulties in employing the similar approach at the low and medium energies are concerned not only with the account of shell corrections in the calculations of the potential energies and thermodynamical characteristics, but also with the need for the equivalent shell determination of viscosity and diffusion of coefficients. Prior to these problems being completely overcome, the statistical description with the empirically localized effective scission point can be fairly fruitful for systematics of experimental data and tracing the relation of diverse characteristics of fragment yield.

References

1. Ignatyuk A.V. Statistical properties of excited atomic nuclei. M: Energoizdat, 1983.
2. Ignatyuk A.V. et al Sov.Nucl.Phys. 1979, v.29, p.54.
3. Blokhin A.I. et.al. Sov.Nucl.Phys. 1988, v.48, p.1012.
   Grudzevich O.T. et.al. -In book: Neutron physics.M: TsNIIAtominform, 1988, v.2, p.96.
4. Ignatyuk A.V. et al. ETchAYa, 1985, v.16, p.709.
5. Ignatyuk A.V. et al, Sov.Nucl.Phys.,1988, v.47, p.355; VANT, 1986, N 1, p.43.
6. Grande P. et.al. Phys.Rev.,1983, V.C. 27, p.2063.

7.Itkis M.G. et.al. ETchAYa, 1988, v.19, p.701.
8.Unik J.P. et.al. Phys.Chem.Fission. Vienna,IAEA, 1974, p.19.
9.Djachenko N.P. et.al. Proc.Int.Symp. on Interaction of Fast Neutrons with Nuclei. ZfK-410, Dresden, 1980, p.97.
10.Gonnenwein F. Proc.Symp. on Nucl.Fission, Obninsk, 1987.VANT, 1988, v.1, p.14.
11.Kuzminov B.D. et.al. Sov.Nucl.Phys., 1969, v.9, p.296.
12.Goverdovsky A.A. et.al. Sov.At.Energy, 1988, v.64, p.425.
13.Fong P. Statistical Theory of Nuclear Fission, N.Y.Gordon and Breach, 1969. Ignatyuk A.V.Sov.Nucl.Phys.,1969, v.9, p.357.
14.Hasse R.B. Ref.10, p.3.
15.Adeev G.L. et.ap. Ibid., p.49.

IV. FRAGMENT DE-EXCITATION

FRAGMENTS, NEUTRONS, AND GAMMAS IN THE FISSION OF ^{252}Cf : A UNIFIED AND PRECISE DESCRIPTION

U. Brosa
Phillipps Universität Marburg, Germany*
H.-H. Knitter
CBNM Geel, Belgium

Abstract: *Several recent discoveries concerning the spontaneous fission of ^{252}Cf made a new level of description feasible. To these discoveries belong:*

- *A very accurate formula for the yield $Y(A, TKE)$ as a function of the total kinetic energy TKE for fixed mass number A.*
- *The interpretation of $Y(A, TKE)$ in terms of multi-modal fission.*
- *Three sawteeth in the neutron multiplicity $\overline{\nu}(A)$.*
- *Shell effects on the deformability of a prescission nucleus.*
- *Evidence for scission neutrons.*
- *No sizable sawtooth in the gamma multiplicity.*
- *Some evidence for contraction gammas produced by fission via the super-long channel.*

* now at HLRZ/KFA Jülich, FRG

1. The Yield $Y(A,TKE)$

The end is a few-parameter description of the yield as a two-variable function. Therefore we established at every mass number A a very accurate description of the dependence on the total kinetic energy TKE. This function was then generalized to contain also the dependence on A, provided that fission goes through just one channel. Finally the yields from the various channels were added.

1.1 The TKE Dependence

When one fixes A and plots $Y(A,TKE)$ just as a function of TKE, it becomes evident that these curves are not really approximated by a gaussian distribution. The real functions are very skewed, and also their tails have little in common with a gaussian. We can offer now a representation which is better than the best gaussian by orders of magnitude.

At first one has to replace the actual value of TKE by the distance D between the centroids of the nascent fragments

$$TKE = \frac{Z'Ze^2}{D} + V_{nuc}. \tag{1}$$

The first term reflects the coulomb repulsion; Z and Z' are the charge numbers of the fragments. The second term is for the nuclear interaction. Although important for the Viola systematics, it does not affect the present problem. Thus we fixed this parameter at $V_{nuc} = 0$. It is convenient to use a shifted D, $T := D - d_{min}$, as variable

$$T = \left(\frac{Z_{cn}}{A_{cn}}\right)^2 \frac{A'Ae^2}{TKE} - d_{min}. \tag{2}$$

In this definition, Z_{cn} and A_{cn} fix charge and mass numbers of the compound nucleus, respectively, and A and A' denote the mass numbers of the fragments: $A + A' = A_{cn}$. Moreover, the uniform-charge-density relation $Z = (Z_{cn}/A_{cn})A$ was inserted. The desired formula is then

$$Y(TKE) = \left(\frac{200}{TKE}\right)^2 h \exp\left\{2\frac{d_{max} - d_{min}}{d_{dec}} - \frac{T}{d_{dec}} - \frac{(d_{max} - d_{min})^2}{T \;\; d_{dec}}\right\} \tag{3}$$

where $Y(TKE) := Y(A \text{ fixed}, TKE)$. The fit parameters

$$h, \quad d_{min}, \quad d_{max}, \quad d_{dec} \tag{4}$$

depend in this approach, of course, on A.

The first factor on the right-hand side of (3) stems from the differential associated with the change of variable from TKE to T, see eq. (2). "200" was included just for numerical convenience as most average TKE's in fission have about 200 MeV. The essential dependence is expressed by the exponential function. This function has its maximum at $T = d_{max} - d_{min}$ and takes there the value 1. h is thus a height.

All the parameters can be interpreted physically. The distance D is more or less the same as the half length of the prescission shape. The distribution (3), therefore, describes the fluctuations of that half length. According to equation (3) it is most probable that the prescission shape ruptures when its half length is d_{max}. It may also rupture at very large lengths, but the probability for this decreases exponentially with the length constant d_{dec}. There is, however, a shortest possible prescission shape beyond which the probability is exactly zero. The respective half length is expressed by d_{min}. It will be noted that the tails of the distribution (3) are in fact very different from those of a gaussian and that the skewness follows from the odd asymptotic behavior. The improvement over the gaussian distribution can be seen in Table 1.

Recently recorded data with good statistics (Budtz-Jørgensen and Knitter 1986 and 1988) made this discovery possible. The curious formula (3) arose when certain algorithms, which are not to be discussed, were systematically applied.

function	$h \exp \frac{-(TKE-\overline{TKE})^2}{2\sigma_{TKE}^2}$	$h \left(\frac{200}{TKE}\right)^2 \exp \frac{-(D-\overline{D})^2}{2\sigma_D^2}$	*eq.* (3)
params	$h = 15390$ $\overline{TKE}/MeV = 178$ $\sigma_{TKE}/MeV = 8.8$	$h = 12690$ $\overline{D}/fm = 18.6$ $\sigma_D/fm = 0.84$	$h = 12070$ $d_{max}/fm = 18.4$ $d_{dec}/fm = 0.35$ $d_{min}/fm = 13.5$
χ^2/dof	55	217	1.8

Table 1. Various representations of the yield $Y(TKE, A = 100)$ in the fission of ^{252}Cf. Given are the functions which were fitted to the data, the best parameters, and the minimum χ^2 values per degree of freedom. The data stem from Budtz's and Knitter's experiment (1988).

1.2 Multi-Mode Fission

Instead of putting the dependence on A into the fitting parameters (4) one should be inclined to consider them as constants for fission through one channel. To incorporate the dependence on A, one might take

$$Y_c(A,TKE) = Y(TKE)$$
$$\frac{1}{\sqrt{8\pi\sigma_A^2}}\left(\exp\frac{-(A-\overline{A})^2}{2\sigma_A^2} + \exp\frac{-(A-A_{cn}+\overline{A})^2}{2\sigma_A^2}\right) \tag{5}$$

with $Y(TKE)$ from eq. (3). We have now 6 fit parameters, 2 in addition to (4), namely the average mass number $\overline{A}$ and the standard deviation σ_A of the mass distribution. All these parameters should reflect certain properties of a definite fission channel: h describes the relative strength by which this channel is populated. d_{max} is the half length with the most favorable potential energy for scission (exactly such d_{max}'s were calculated by Brosa et al. 1986). d_{min} is the half length where the potential energy becomes too big to allow for any scission at all, and a similar interpretation is possible for d_{dec}. All the d's describe the extension of the fission valley in the direction of the length of the fissioning shape. $\overline{A}$ and σ_A, however, are associated with the characteristics of the fission valley in asymmetry.

The complete yield may then be written as

$$Y(A,TKE) = \sum_c Y_c(A,TKE) \tag{6}$$

If there are 5 fission channels, 30 parameters have to be accommodated. Compared with the thousands of data which are reproduced, this is a small number. Our results are given in Table 2.

The obtained numbers make sense: The properties of the super-long and super-asymmetric channels were cleanly isolated. Note the astonishing small value of d_{min} for the standard I channel. This channel is obviously a transitory mode from the standard I channel as observed in uranium (Hambsch 1988) to the super-short channel as we know it from the fermium and mendelevium isotopes (Hulet 1986). At present there seems to be no theoretical confirmation for the splitted standard channel, viz. standard II and III, and there is also no experimental argument why they should be considered as independent entities. It might be that a gaussian ansatz for the mass distribution is not better justified than a gaussian ansatz for the description of the TKE dependence.

channel name	h	d_{min}	d_{max}	d_{dec}	$\overline{A}$	σ_A
super long	2.02 10^4	16.8	18.7	0.521	126.9	11.62
standard I	1.47 10^5	10.9	17.1	0.125	134.9	3.16
standard II	8.70 10^5	13.7	17.9	0.270	142.5	4.99
standard III	3.15 10^5	13.5	18.7	0.317	148.5	7.14
super asymmetric	7.60 10^3	15.0	19.5	0.180	178.3	2.25

Table 2. The fission channels of ^{252}Cf. The numbers come from fitting formulas (3), (5), and (6) to Budtz's and Knitter's data (1988). Physically, only the relative sizes of the height h have significance. The absolute values appear here to indicate the experimental counts. The errors of the fit should have the size of the least digits, except for the super-asymmetrical channel where only the order of magnitude is claimed to be all right.

2. The Neutron Multiplicity $\overline{\nu}(A,TKE)$

It is still possible to discover remarkable features in the neutron multiplicity $\overline{\nu}(A)$ simply by extending the range of the fragments' mass number A. The modern trend, however, is to measure the multiplicity as a function of A *and* TKE. But it seems that the new information obtained in this way is finite: Apart from $\overline{\nu}(A)$ one finds one new one-variable function, the slope function $\partial\overline{\nu}(A)/\partial TKE$, which does not depend on TKE.

2.1 The Three Sawteeth

Fig.1. The three sawteeth of the neutron multiplicity in the spontaneous fission of ^{252}Cf. The curve is a prediction by Brosa (1988). The data were measured by Budtz-Jørgensen and Knitter (1988).

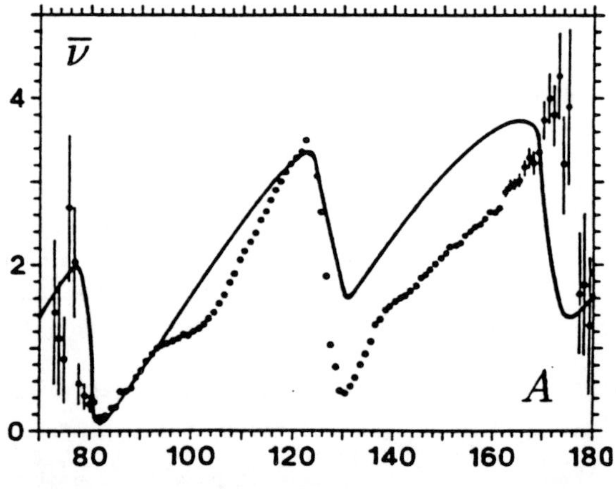

In a recent experiment (Budtz-Jørgensen and Knitter 1988) statistics and some other experimental conditions were so good that the multiplicity $\overline{\nu}(A)$ was measured for mass numbers below 78 and above 174. The result is presented in Fig. 1. Amazing are two additional sawteeth at very light and very heavy fragments. Although surprising at first sight, these features should have been anticipated after the reading of the previous chapter. Namely, when multimode fission shows up in the mass yield, it must be seen in the neutron multiplicities too. Unfortunately, the standard channels II and III are so powerful and their distributions so broad that they swamp the neutrons from the super-long and standard I channels. This is so for the presently available data, but it will change as soon as measurements of $\nu(A,TKE)$ without the averaging bar will be published. However, the super-asymmetric channel is so well separated from the others that its neutrons should be visible already now. This observation led to the computation of the drawn curve in Fig. 1. Shown is

$$\overline{\nu}(A) = \frac{\sum_c \overline{\nu}_c(A)\, Y_c(A)}{\sum_c Y_c(A)} \tag{7}$$

where $Y_c(A) = \int Y_c(A, TKE)\, dTKE$ with $Y_c(A, TKE)$ taken from eq. (5). The multiplicities $\overline{\nu}_c(A)$ were calculated using the random neck rupture model. One should note the preponderance of the contributions from the standard channels in the central part, whereas the outskirts are dominated by the super-asymmetric fission mode. Details can be found in (Brosa 1988).

2.2 The Straight-Line Representation

Experimentally is found that the detailed multiplicity may be represented as

$$\overline{\nu}(A, TKE) = \begin{cases} \frac{\partial \overline{\nu}}{\partial TKE}(A)\,\{TKE - TKE_{max}(A)\} & \text{if } TKE < TKE_{max} \\ 0 & \text{else.} \end{cases} \tag{8}$$

Two one-variable functions $TKE_{max}(A)$ and $\partial\overline{\nu}(A)/\partial TKE$ are thus sufficient to describe the two-variable function $\overline{\nu}(A, TKE)$ (Schmid-Fabian 1988, Budtz-Jørgensen 1988).

The linearity is just an approximation: Because of energy conservation, TKE has to pay for the excitation of fragments. This is a linear relation. But finally the excitation energy $E^*(A, TKE)$ has to go into $\overline{\nu}(A, TKE)$ evaporated neutrons, and this is not a linear act. The pertinent formula is

$$E^*(A, TKE) = \sum_{n=1}^{\overline{\nu}(A,TKE)} \{S_n(A, Z) + \overline{\eta}(A, TKE)\} + E_\gamma(A, TKE). \tag{9}$$

The average kinetic energy $\overline{\eta}$ of the neutrons and the energy E_γ removed by the gammas are small quantities. Eq. (9) shows that a linear connection between $E^*(A, TKE)$ and $\overline{\nu}(A, TKE)$ may be expected only if the separation energies S_n are constants. This, however, is not accurately the case. If n neutrons were just evaporated, it costs more work to separate the next. Therefore one has to expect a flattening of $\overline{\nu}(A, TKE)$ for decreasing TKE.

2.3 Inhomogeneous Elasticity of a Prescission Nucleus

Of the three functions

$$\overline{\nu}(A), \quad \frac{\partial \overline{\nu}}{\partial TKE}(A), \quad TKE_{max}(A) \tag{10}$$

one can be expressed by the others. To see this, one may insert eq. (8) in the trivial relation

$$\overline{\nu}(A) = \frac{\int_o^\infty \overline{\nu}(A, TKE)\, Y(A, TKE)\, d\,TKE}{\int_o^\infty Y(A, TKE)\, d\,TKE}. \tag{11}$$

An unimportant neglect at the integration limits gives

$$\overline{\nu}(A) = \frac{\partial \overline{\nu}}{\partial TKE}(A)\,\{\overline{TKE}(A) - TKE_{max}(A)\}. \tag{12}$$

The average kinetic energy $\overline{TKE}(A)$ is so well established that it is not free for disposition.

Equation (12) may be interpreted in the following manner: The "elastic energy" contained in the fragments just after scission depends on two factors. One of them is the "elasticity module" of that piece of the prescission nucleus from which the fragment is to be made. This module is expressed by the function $\partial\bar{\nu}(A)/\partial TKE$. In fact, one may identify the total kinetic energy as an inverse measure of the elongation of the prescission shape and the multiplicity as a measure of the elastic energy. The second factor is the deviation from the average elongation, expressed by the function $\overline{TKE}(A) - TKE_{max}(A)$. This factor is the more trivial one because it contains only the effects described by the random neck rupture model. It is therefore not surprising that it is again a sawtooth curve. But the sawtooth in the slope function is beyond expectation (see Fig.2). The discrepancy becomes understandable as soon as one recognizes that the model presumes homogeneous matter. If the prescission nucleus has a substructure, it must be wrong. Hence Fig. 2 may be interpreted by a prescission shape which has a soft neck and two hard heads. Stretching the shape is facilitated just by the the neck; the heads remain inert. The two minima in Fig. 2 correspond to the masses of these heads. Amousingly enough we find the mass numbers $A \approx 28 + 50$ and $A \approx 50 + 82$.

Fig.2. The slope function in the spontaneous fission of ^{252}Cf (Budtz-Jørgensen and Knitter 1988). The drawn line was obtained with the plain random neck rupture model (Brosa 1985).

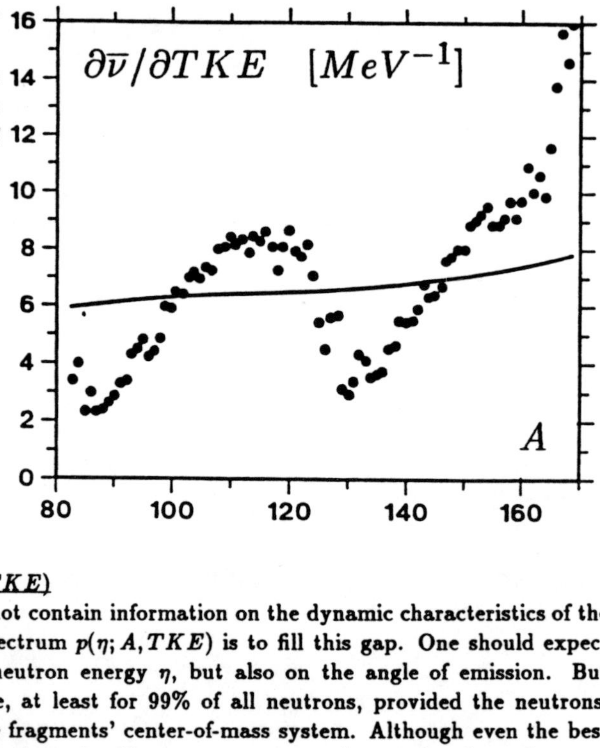

3. The Neutron Spectrum $p(\eta; A, TKE)$

The neutron multiplicities do not contain information on the dynamic characteristics of the emitted neutrons. The neutron spectrum $p(\eta; A, TKE)$ is to fill this gap. One should expect that it depends not only on the neutron energy η, but also on the angle of emission. But this turned out to be not the case, at least for 99% of all neutrons, provided the neutrons' characteristics are registered in the fragments' center-of-mass system. Although even the best data is still too ragged to give more than $p(\eta; A)$, we can construct the missing dependence on TKE from some elementary arguments.

3.1 Temperature $T(A, TKE)$ and Cascade Exponent $\lambda(A, TKE)$

Recent experimental work (Budtz-Jørgensen 1988) showed that the deviations from equilibrium are at least an order of magnitude smaller than what was derived from former experimental work (Bowman 1962). Nevertheless it is not justified to stipulate a simple maxwellian for $p(\eta, A)$. The first neutron already carries away so much excitation energy that the next neutron starts from a different environment. To take this into account, a simple generalization of Maxwell's law turns out to be sufficient:

$$p(\eta, A) = \frac{\eta^{\lambda(A)} \exp(-\eta/T(A))}{\Gamma(\lambda(A)+1)\, T(A)^{\lambda(A)+1}} \quad . \tag{13}$$

The denominator serves for normalization. One of the parameters is naturally the temperature $T(A)$. The other is the the cascade exponent $\lambda(A)$. Both are shown in Fig. 17 of Budtz-Jørgensen 1988. Remarkable is the tooth in the temperature $T(A)$ for A below 132. It can be associated with an anomaly of the level densities for the double-magic nuclei. According to theory (Lang 1964), the cascade exponent $\lambda(A)$ should take the value 5/11 when there is a proper neutron cascade, i.e. when several neutrons are evaporated. If at most one neutron comes out, $\lambda(A) \approx 1$ should be more appropriate, and this is indeed reasonably fulfilled (Budtz-Jørgensen 1988). We introduce hence the straightforward generalization:

$$\lambda(A, TKE) = \begin{cases} 1 & \text{if } \overline{\nu}(A, TKE) \leq 1 \\ 1/2 & \text{else.} \end{cases} \tag{14}$$

But it is also easy to establish a generalized $T(A, TKE)$:

$$T(A, TKE) = \sqrt{a(A)^{-1} E^*(A, TKE)}. \tag{15}$$

The level densities may be taken from Fig. 19 of Budtz-Jørgensen and Knitter's work (1988), and the excitation energy $E^*(A, TKE)$ can be computed from $\overline{\nu}(A, TKE)$ using an equation of the type (9) with the most trivial estimates for $\overline{\eta}$ and E_γ (see section 4 for better numbers). Inserting the functions (14) and (15) in (13) yields the desired spectrum $p(\eta; A, TKE)$.

3.2 How to Obtain the Fission Neutron Spectrum

The neutron spectrum $p(\eta; A, TKE)$ can only be measured in a highly coincident experiment. The most widely used quantity is the *fission neutron spectrum* $p(\epsilon)$ which is recorded in the lab frame without any notice on the properties of the fragments. So ϵ is the kinetic energy of the neutrons in the laboratory, whereas η refers to the fragment cms. Obviously it must be possible to find the coarser information from the finer one. We quote here just the result (Brosa 1988a):

$$p(\epsilon) = \sum_A \int_o^\infty d\,TKE\; \overline{\nu}(A, TKE)\; Y(A, TKE)$$
$$\frac{1}{4\pi} \int_o^{2\pi} d\varphi \int_o^{\pi} \sin\vartheta\, d\vartheta\; p(\eta; A, TKE) \sqrt{\frac{\epsilon}{\eta}} \;. \tag{16}$$

The neutron energies are transformed according to

$$\eta = \epsilon - 2\sqrt{\epsilon\,\kappa} \cos\vartheta + \kappa \tag{17}$$

with the kinetic energy κ of a fragment given by

$$\kappa = \frac{A_{cn} - A}{A} \frac{TKE}{A_{cn}} \quad . \tag{18}$$

It is of great use to know that two of these integrals can be done analytically:

$$\frac{1}{4\pi} \int_o^{2\pi} \int_o^{\pi} \sin\vartheta\, d\vartheta\; p(\eta; A, TKE) \sqrt{\frac{\epsilon}{\eta}} =$$
$$\frac{\Gamma(\lambda + 1/2)}{4\Gamma(\lambda+1)\sqrt{\kappa T}} \left\{ P\left(\lambda + \frac{1}{2}, \frac{(\sqrt{\epsilon} + \sqrt{\kappa})^2}{T}\right) - P\left(\lambda + \frac{1}{2}, \frac{(\sqrt{\epsilon} - \sqrt{\kappa})^2}{T}\right) \right\} \tag{19}$$

where

$$P(\lambda, x) := \frac{1}{\Gamma(\lambda)} \int_o^x t^{\lambda-1} e^{-t} dt \tag{20}$$

is the incomplete gamma function for which fast and accurate algorithms are available (Press 1986).

3.3 The Scission Neutrons

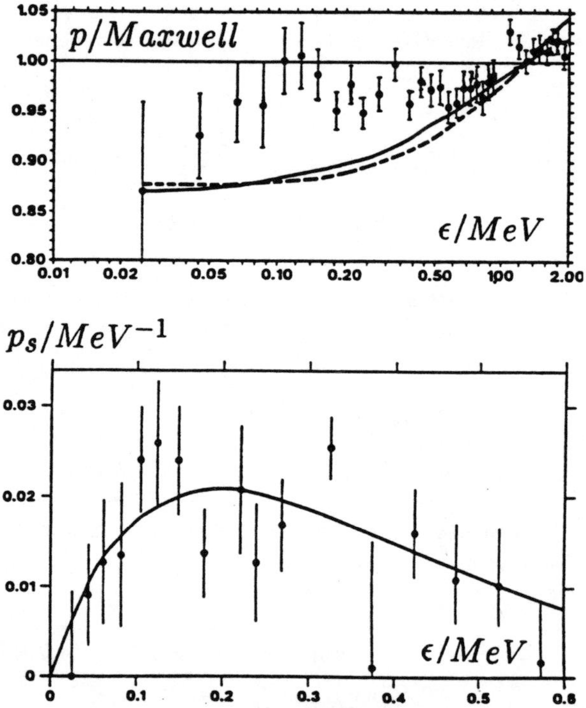

Fig.3. Upper part: The normalized fission neutron spectrum devided by $2\sqrt{\epsilon/(\pi(1.42)^3)}\exp(-\epsilon/1.42)$, the best maxwellian distribution. The direct data as compiled by Mannhart (1987) are entered as dots with error bars. The heavy line is the result from the calculation as described here. The dashed curves reflects the results by Batenko et al. (1988). Lower part: The dots with error bars come from subtracting the Brosa's (1988a) result from the experimental values in the upper part and are displayed without dividing something out.

The fission neutron spectrum is of outmost importance for applications. Much care was taken to record it directly. The indirect method just explained might seem inferior in every respect. However, it gathers cleanly the contributions from equilibrium evaporation (cf. eq. (13)). Thus it should be possible, by a comparison between the direct and the indirect evaluation, to detect even small deviations from equilibrium.

The comparison is shown in Fig. 3. The fat drawn line is our result, the dashed one comes from a somewhat different evaluation of different data by Batenko et al. (1988): The agreement is excellent. In contrast, the dots with error bars stem from direct measurements, but also here different types of counters in different experiments were used. Hence there is no doubt that the structure at 0.15 MeV is real and that it signalizes a deviation from boltzmannian statistics. The difference between these two sets is displayed in the lower part of the figure. The data were fitted by a function

$$p_s(\epsilon) = \frac{p_o}{T_s^2}\, \epsilon \, \exp\frac{-\epsilon}{T_s} \tag{21}$$

and the obtained parameters are

$$p_o = 0.01 \pm 0.003, \qquad T_s = (0.2 \pm 0.03)\ \mathrm{MeV}\ . \tag{22}$$

The present method is thus one order of magnitude more sensitive than previous ones (Budtz-Jørgensen 1988). Moreover, it is satisfactory to see that the pseudo temperature T_s is compatible with a cold prescission shape, but this is an argument which is by no means compulsory.

The low energy of the scission neutrons indicates that they come from a slow source. The width of the distribution (20) suggests that the source is small: radius about 5 fm. This all conforms with the idea of satellite droplets which are formed when the neck snaps. One can check this by looking for coincidences with small TKE since satellite droplets are preferentially produced when the prescission shape gets long. And in fact, inspection of such coincidences has shown that the respective angular distributions are not isotropic in the fragment-center-of-mass frame. However, the latter material presently is not conclusive enough to exclude other approaches (see Milek 1988 and the references therein).

4. Gamma Rays

A comparatively well-founded belief says that about 90% of the gammas are emitted after the neutrons. If this is true, the gammas only receive the excitation energy the neutrons can't spend. This is somewhat less than the separation energy of the neutron which just cannot escape. In first approximation, the neutron separation energies do not depend on the emitting isotope, and thus the gamma energy must be essentially constant.

In contrast to previous experiments, a recent one confirmed this point of view (Schmid-Fabian 1988). Looking at these data one may suggest

$$\overline{\nu}_\gamma(A,TKE) \approx 3.8, \quad E_\gamma(A,TKE) \approx 2.8 \text{ MeV} . \tag{23}$$

In this experiment, the TKE-dependence of the gamma multiplicity $\overline{\nu}_\gamma$ was not measured, and only sum energies were registered. So a part of eq. (23) is still an extrapolation, but hopefully a useful one in the context of eq. (9). Moreover, it is obvious from the data that sum energies $E_\gamma(A,TKE) + E_\gamma(A_{cn} - A, TKE)$ depend on TKE, but in comparison with the corresponding trends in $\overline{\nu}(A,TKE)$ this is a small effect, in keeping with the afore-mentioned statement of only 10% of the gammas being emitted in competition with the neutrons. Therefore also the values given in (23) should be considered as having errors of about 10%.

4.1 The Contraction Gammas

Fig.4. Contraction gammas in the spontaneous fission of ^{252}Cf. This is a multiplicity belonging to the pair of fragments A and $A' = A_{cn} - A$. Only those gamma are included which have an energy higher then 1.5 MeV. The data are from Schmid-Fabian's thesis (1988).

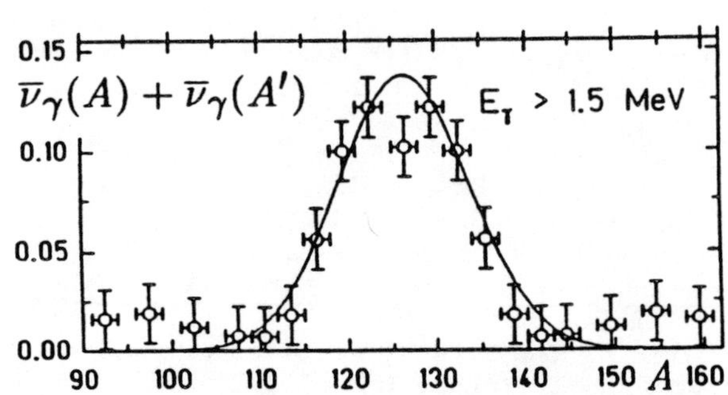

The spectrum of the gammas from fission should be a statistical one, describable with a Boltzmann distribution. This turns out to be case (Schmid-Fabian 1988), but with two significant deviations. First, below 1.5 MeV there are excess gammas, called E2 gammas, attributed to low-lying collective states of rotation. Seond, at about 4 MeV, in a broad bumb, extra gammas are found which are called contraction gammas. They are emitted only from fragments produced in symmetrical fission (see Fig. 4). These two properties are compatible with the following interpretation: The fragments are produced with large deformations. They have to contract. This necessitates, since the fragments are charged, emission of electromagnetic radiation. The energy of the vibrational modes is not in contradiction with the 4 MeV mentioned above. Especially deformed fragments are created when fission goes through the super-long channel. Thus one should expect the contraction gammas from fragments close at symmetry. Note the correct size of the width in Fig. 4 as compared with σ_A for super-long in Table 2.

5. Summary

How many numbers does one need to give a fairly accurate description of the fission of ^{252}Cf? For the yield $Y(A, TKE)$, eq. (6) and Table 2 are sufficient. For the neutron multiplicity $\overline{\nu}(A, TKE)$, tables of $\overline{\nu}(A)$ and $\partial\overline{\nu}(A)/\partial TKE$ are needed (see Figs. 11 and 14 in Budtz-Jørgensen 1988). The due extension can be done using the formulas (12) and (8) as the average kinetic energy $\overline{TKE}(A)$ follows from the yield. For the neutron spectrum, the level densities $a(A)$ (Fig. 19 in Budtz-Jørgensen 1988) and the neutron separation energies $S(A, Z)$ have to be provided. Temperature $T(A, TKE)$, cascade coefficient $\lambda(A, TKE)$ and neutron spectrum $p(\eta; A, TKE)$ come then from the equations (9), (15), (14), and (13). With eq. (23), the gammas experience a relatively sloppy treatment.

References

Batenko, O.I., Blinov, A.B., Blinov, M.V., and Smirnov, S.N. 1988, Proc. of IAEA Consultants Meeting on Physics of Neutron Emission in Fission, Mito Japan.

Bowman, H.R., Milton, J.C.D., Thompson, S.G., and Swiatecki, W,J. 1962, Phys. Rev. **126**, 2120.

Brosa, U. 1985, Phys. Rev. C **32**, 1438.

Brosa, U., Großmann, S., and Müller, A., 1986 Z. Naturforsch. **41a**, 1141.

Brosa, U. 1988, Phys. Rev. C **38**, 1944.

Brosa, U. 1988a, A new method to find the fission neutron spectrum, Internal Report CBNM Geel GE/R/VG/56/88.

Budtz-Jørgensen, C. and Knitter, H.-H 1986, in Proc. XVth Int. Symp. Nuclear Physics - Nuclear Fission - (eds. D. Seeliger, K. Seidel, and H. Märten), Akademie der Wissenschaften ZfK Rossendorf, p. 18.

Budtz-Jørgensen, C. and Knitter, H.-H. 1988, Simultaneous investigation of fission fragments and neutrons in ^{252}Cf (SF), Nucl. Phys. A.

Hambsch,F.J., Knitter, H.-H., Budtz-Jørgensen, and J.P. Theobald 1988, Fission mode fluctuations in the resonances of ^{235}U(n,f), Nucl. Phys. A.

Hulet, E.K., Wild, R.F., Dougan, R.J., Lougheed, R.W., Landrum, J.H., Dougan, A.D., Schädel, M., Hahn, R.L., Baisden, P.A., Henderson, C.M., Dupzyk, R.J., Sümmerer, K., and Bethune, G.R. 1986, Phys. Rev. Lett. **56**, 313.

Lang, D,W. 1964, Nucl. Phys. **53**, 113.

Mannhart, W. 1987, in Proc. of an IAEA Advisory Meeting, Leningrad 1986, IAEA-TECDOC-410, IAEA Vienna, p. 158.

Milek, B., Reif, R., and Révai, J. 1988, Phys. Rev. C **37**, 1077.

Press, H.W., Flannery, B.P., Teukolsky, S.A., and Vetterling, W.T. 1986, Numerical Recipes, Cambridge University Press.

Schmid-Fabian, R. 1988, Messung der spontanen Spaltung von ^{252}Cf am Darmstadt-Heidelberger-Kristallkugel-Spektrometer, Doktorarbeit an der Ruprecht-Karls-Universität Heidelberg.

FISSION FRAGMENT DE-EXCITATION

H. Märten
Technische Universität, Dresden, Germany

Abstract: Fission fragment de-excitation including the emission of neutrons (n), light charged particles (LCP), and γ-rays (γ) is studied in the framework of a complex statistical-model approach (SMA). Whereas nucleon emission (n and p) in fission, e.g. polar emission, can be well described by evaporation from the fragments, the release of heavier LCP (as α-particles) is mainly due to ternary fission. Recent experimental n data have been analysed in detail. The influence of both the intricate fragment occurrence probability $P(A,Z,E^x,TKE,J)$ (depending on mass and charge number, excitation energy, total kinetic energy, and angular momentum, respectively) and the emission model itself (including structure information and transition probabilities) on n spectrum shapes in the centre-of-mass system (CMS) is discussed. Conclusions are drawn concerning n emission from fragments in the crucial mass number region around 132-135. Neutron emission from fragments at low TKE, i.e. in the case of strongly deformed scission configurations, is probably suppressed due to high angular momenta of fragments. Considering n/γ-competition, which is much influenced by $P(\ldots,J)$, the correlation between the total γ emission energy E_γ and the average number of prompt neutrons $\bar{\nu}$ has been calculated.

1. Introduction

The investigation of mechanisms of particle emission in nuclear fission in conjunction with fission dynamics is a matter of great interest concerning fundamental and practical problems. Since the comprehensive review by Nifenecker[1] on n and γ emission in fission several new results have been published. Novel experimental methods have been applied to measure multiple-differential emission probabilities of n (cf. refs. 2-5), γ (cf. refs. 6-8), and LCP (cf. refs. 9,10). In order to study, whether a particle is created at a time close to scission (either from a triple scission configuration or just after scission from one of the fragments[11]) or later from one of the fully accelerated fragments, one should firstly apply a complex SMA to particle emission in comparison with experimental emission distributions depending at least on emission energy E and angle θ (defined as angle between the directions of the emitted particle and the light fragment).

Particle emission at scission (or ternary fission) has been treated within new theoretical approaches[12-14]. However, in spite of the progress achieved such calculations are much influenced by model parameters, e.g. assumptions concerning scission shapes and time dependence of potential changes. Thus, their combination with SMA applications should be reasonable to obtain a full-scale insight into particle release in fission.

The actual fragment de-excitation runs via n evaporation and γ emission mainly. These competing processes[15,16] depend on both $P(A,Z,E^x,TKE,J)$ (formulated as an asymptotic distribution[17], i.e. for conditions after fragment acceleration and deformation energy dissipation) and nuclear structure data. Based on recent SMA applications to n emission[18], the present work includes other de-excitation channels discussed above. A further study of n emission during fragment acceleration has been described in ref. 19.

2. Statistical-model approach (SMA)

As summarized in ref. 17, a complex SMA is based on the fragment distribution $P(A,Z,E^x,TKE,J,...)$. However, calculations are limited due to the lack of knowledge of this distribution. In the promising case of $^{252}Cf(sf)$, it can be deduced from experimental data with more or less accuracy in the following form[17,20]:

$$P(A,TKE,E^x),\ P(Z:A),\ P(J:A).$$

Applying standard evaporation theory to cascade particle emission by steps i (including γ emission) from a fragment for a given initial distribution $P_o(E^x,J)$, i.e. A, Z, and TKE fixed, we get the CMS spectrum of particle π as

$$\phi_\pi(\varepsilon_\pi) = \sum_i \int dE^x \sum_J P_i(E^x,J) \frac{\Gamma_\pi(\varepsilon_\pi,E^x,J)}{\sum_{\pi'} \Gamma_{\pi'}^{tot}(E^x,J) + \Gamma_\gamma^{tot}(E^x,J)} \tag{1}$$

$$\Gamma_\pi(\varepsilon_\pi,E^x,J) = (2\pi\ \rho(E^x,J))^{-1} \sum_{J'} \rho^\pi(U_\pi,J') \sum_{l_\pi,s_\pi} T_l^\pi(\varepsilon_\pi) \tag{2}$$

$$U_\pi = E^x - B_\pi - \varepsilon_\pi, \qquad \vec{J} = \vec{J'} + \vec{l}_\pi + \vec{s}_\pi \tag{3}$$

(Γ - emission width, Γ^{tot} - total decay width, ε_π - CMS emssion energy of particle π, $\rho(E^x,J)$ - nuclear level density, B_π - separation energy, U_π,J' - rest-nucleus excitation energy and angular momentum, respectively, l_π - orbital momentum, s_π - spin of particle π). Full-scale calculations have been performed for a simplified evaporation ansatz:

$$\Gamma_\pi(\varepsilon_\pi,E^x,\bar{J}) = (2\pi\ \rho(E^x,J{=}0))^{-1}\ \varepsilon_\pi\ \sigma_{inv}(\varepsilon_\pi)\ \rho(E^x,J'{=}0)\ C(E^x,\bar{J}) \tag{4}$$

(σ_{inv} - inverse cross section of compound-nucleus formation to be calculated in the framework of the optical model) with a correction factor

$$C(E^x,J) \sim (1 - \exp(-C_1(E^x - B_\pi)^{C_2})) \tag{5}$$

to account for neutron-γ-competition in respect of eq. (1). (C_1, C_2 – parameters depending on $\bar{J}$).

Note that the fragment occurrence probability $P_i(E^x,J)$ for i≥1 has to be deduced from $P_{i-1}(E^x,J)$ considering the spectra of emission step i-1. Further, the quantities as ρ, T_l (or σ_{inv}), and B have to be taken for relevant A values, which are diminished by steps 1 during cascade emission of neutrons. In particular, B_n rises within a n cascade.[21] Transformation of Eq. 1 into the laboratory system (LS)[17,20] taking into account CMS anisotropy due to fragment spin[22] yields the LS distribution $N_\pi(E,\theta{:}A,Z,TKE)$. Thus, the total distribution is given by

$$N_\pi(E,\theta)=\sum_{A,Z}\int dTKE\,N_\pi(E,\theta{:}A,Z,TKE)P(A,Z,TKE) \qquad (6)$$

3. LCP evaporation

The cascade evaporation model (CEM)[18,20] based on Eqs. (1,4-6), but using averages $\bar{Z}(A)$, $\bar{\rho}(A)$, $\bar{\sigma}_{inv}(A)$, and $\bar{B}_\pi(A)$, has been applied to p and α emission (in addition to n emission)[23]. Fig. 1 represents essential parameters of the calculation as function of fragment mass number. Yields are shown in fig. 2. Whereas the α yield is clearly underestimated, calculated p and n yields and spectra agree with experimental data (in particular, 4.90 10^{-5} p yield, $\bar{E}_p$ =7.99 MeV, cf. ref. 24 and references therein, 1.42 10^{-7} α yield, $\bar{E}_\alpha$ = 14.14 MeV).

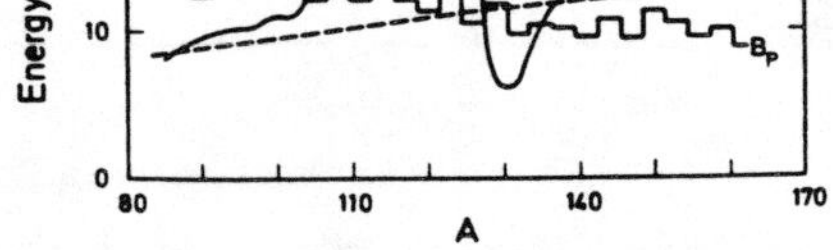

Fig. 1 Average values of E^x, p separation energy and effectice Coulomb barrier C for p and α.

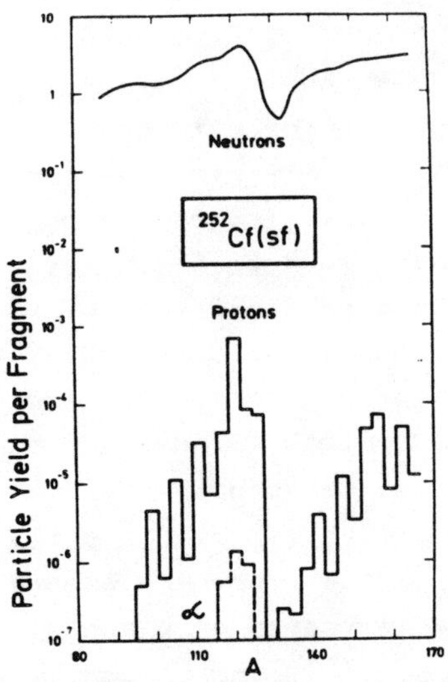

Fig. 2 Calculated particle yields of n, p, and α for 252-Cf(sf)

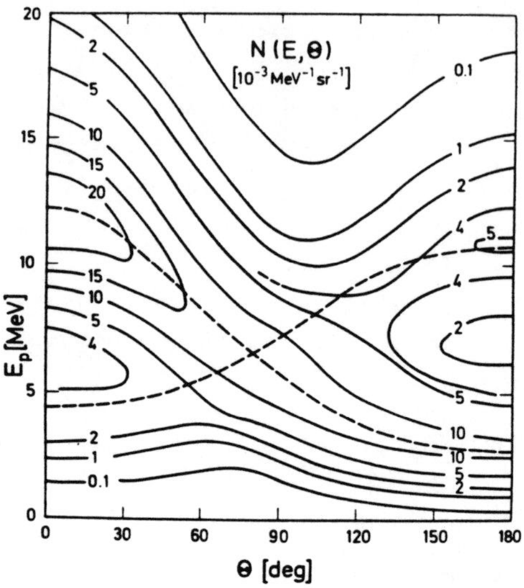

Fig. 3 Contour lines of N(E,θ) for p. The dashed lines represent the position of the average CMS p emission energiesin LS co-ordinates (shown for the light and heavy fragment groups)

As shown in fig. 1, LCP emission is strongly suppressed due to the Coulomb potential entering the optical potential for calculating σ_{inv}. This fact is reflected in the particle yields shown in fig. 2. LCP are predominantly emitted from fragments with high $\bar{E}^{x}(A)$, i.e. for 110<A<125 and A>150. Double-differential emission probabilities of p are represented in figs. 3-6 ($N_p(E_p,\theta)$ normalized to 1.). Obviously, these LS distributions exhibit a structure defined by emission kinematics mainly. The average CMS emission energies of p from the light and heavy fragment groups are 6.6 and 7.6 MeV, respectively. Polar and equarorial emission spectra of p from ^{252}Cf(sf) are represented in comparison with experimental data in figs. 4-6. Polar distributions indicate the influence of p emission from the complementary fragment, especially light-fragment p emission at backward angles (180 deg). Unfortunately, the relevant energy range is out of that measured by Nowicki et al.[25]. In contrast to the emission of p, α-particles are clearly pronounced in equatorial direction indicating that they are predominantly released in ternary fission. The polar emission of α can also be described within scission model approaches[12,14]. The present study shows that a considerable part of fission p is emitted from fully accelerated fragments.

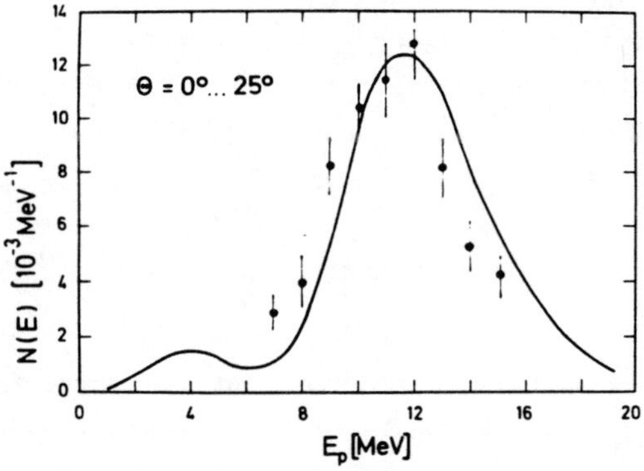

Fig. 4 Polar p spectrum for light-fragment direction (points-ref. 25, line - CEM)

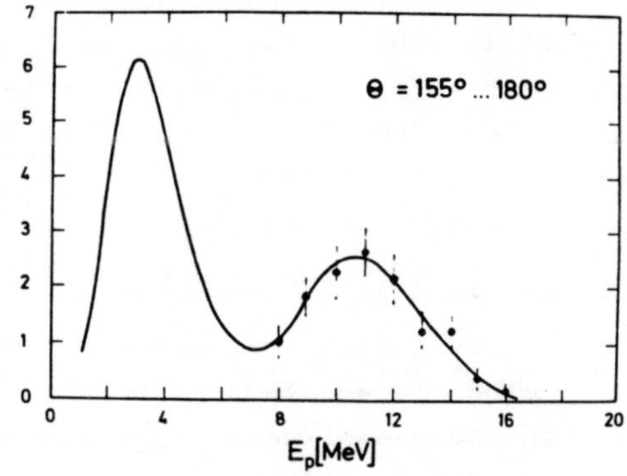

Fig. 5 Same as for fig. 4, but for heavy-fragment direction

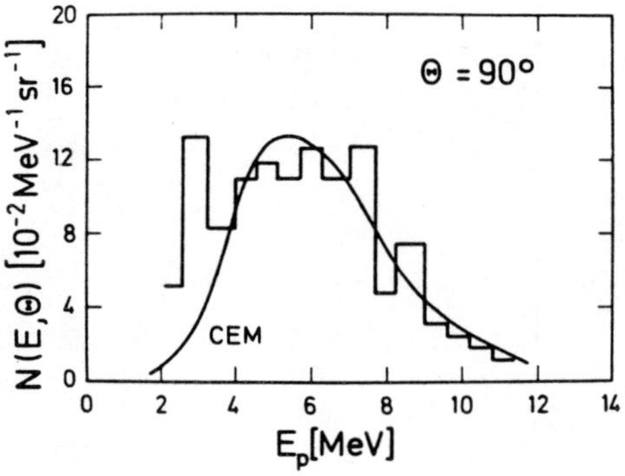

Fig. 6 Equatorial p spectrum (histogram - experimental data, ref. 26)

TABLE 1: p spectrum parameters for polar emission (PE) in the direction of the light (L) and heavy (H) fragment as well as the yield ratios forward/backward and polar/equatorial (average energies and FWHM in MeV)

Work	$\bar{E}_L$	$FWHM_L$	$\bar{E}_H$	$FWHM_H$	LE/HE	PE/EE
Exp.[25]	11.4±0.1	5.4	11.3±0.2	5.7	3.9±0.3	0.11±0.02
CEM	11.5	5.8	11.0	5.0	4.0	0.13

In contrast to the emission of p, α-particles are clearly pronounced in equatorial direction indicating that they are predominantly released in ternary fission. The polar emission of α can also be described within scission model approaches[12,14]. The present study shows that a considerable part of fission p is emitted from fully accelerated fragments.

4. Neutron emission in fission

As clearly shown in recent papers[17,18,26], prompt fission n are predominantly emitted (evaporated) from fully accelerated fragments. Within the experimental and theoretical uncertainties, energy and angular distributions can be well reproduced in the framework of CEM. The upper yield of scission n was estimated to be less than 5 %. An example is shown in fig. 7.

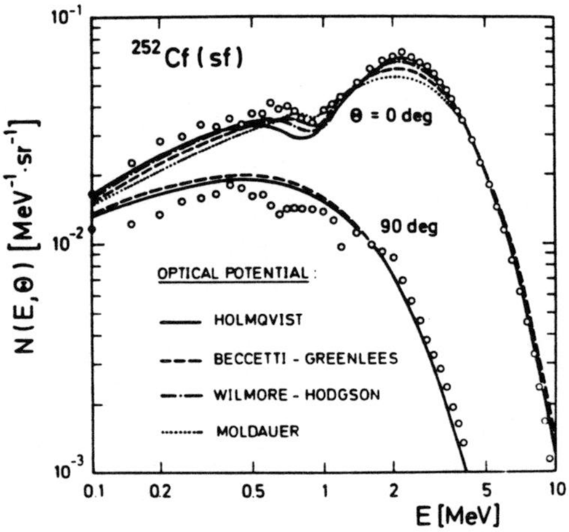

Fig. 7 Differential spectra of n from 252-Cf(sf). Experimental data (ref. 4) in comparison with CEM calculations based on different optical potentials as indicated.

A new CEM calculation has been performed on the basis of $\bar{E}^x(A,TKE)$ deduced by the help of recent $\bar{\nu}(A,TKE)$ data[27]. Earlier results could be confirmed. The CMS spectra have been analysed as in ref. 27, i.e. they have been fitted to the ansatz

$$\phi(\varepsilon) = [T^{\lambda+1}\,\Gamma(\lambda+1)]^{-1}\,\varepsilon^{\lambda}\,e^{-\lambda/T}. \tag{7}$$

In this case, both parameters T and λ define the average $\bar{\varepsilon} = (\lambda+1)T$, the variance $\sigma^2 = (\lambda+1)T^2$, the skewness (asymmetry) $A = 2/(\lambda+1)^{1/2}$, and the excess $E = 6/(\lambda+1)$ of the distribution. Figs. 8,9 show $\bar{\varepsilon}$ and λ as function of fragment mass number in comparison with experimental results[27]. λ(A) is reproduced by the CEM qualitatively. It has been found that the deviations are probably due to non-correct variances of the excitation energy distribution as function of A and TKE. The used data from ref. 1 deduced from multiplicity variances by statistical methods will be substituted by CEM fitted values in next future. Concerning $\bar{\varepsilon}(A)$, a clear deviation between experimental and calculated results appear for 128<A<138. This discrepancy can be interpreted as

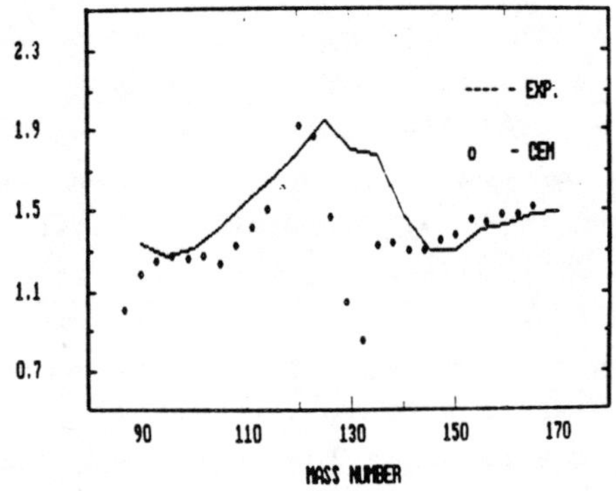

Fig. 8 Average CMS energy of n from 252-Cf(sf) as function of A (exp. data - ref. 27)

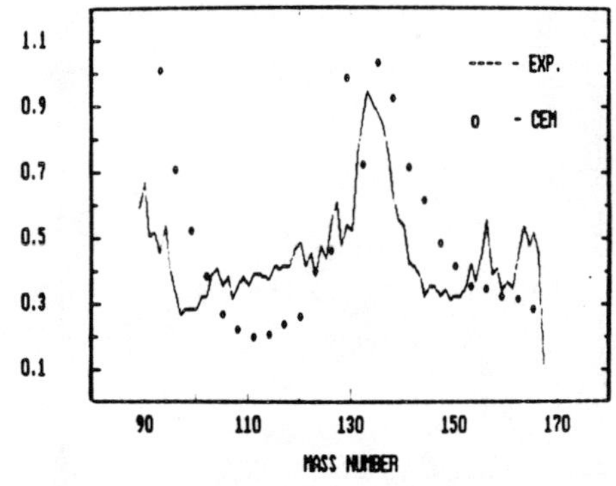

Fig. 9 The "shape" parameter λ versus A (experimental data - ref. 27)

due to the possible influence of the deformed closed-shell region H[28] corresponding to deformation energies of about 22 MeV. If one takes this value as input excitation energy for A around 132, the parameters T and λ as found in ref. 27 can be exactly reproduced. This confirms the previous assumption that a TKE dependent part of scission point configurations consists of the heavy fragment in a deformed state around shell region H. The other part consists of a heavy fragment within the double-magic shell region G^{28}, i.e. at deformation close to zero. It should be pointed out that a fit of $\phi(\varepsilon)$ to Eq. (7) in the whole energy scale yields λ and T parameters which don't correspond to the values deduced for the low-energy region (1-100 keV). As a consequence of the CMS-LS transformation relations

$$N(E,\theta) = (E/\varepsilon)^{1/2}\ \phi(\varepsilon,\delta), \qquad \varepsilon = E + E_f - 2(E\ E_f)^{1/2} \cos\theta \qquad (8)$$

(δ - CMS angle of n emission with reference to fragment direction, E_f - kinetic energy of the fragment per nucleon) the parameter λ defines the behaviour of $N(E,\theta)$ at the singular point $E \rightarrow E_f$ and $\theta \rightarrow 0$, i.e. $\varepsilon \rightarrow 0$. $\lambda>0.5$ ($\lambda<0.5$) corresponds to a dip (a pole) of $N(E,\theta)$ at the crucial point. As shown in fig. 7, the experimental data indicate a clear dip for n emission from the light fragments (as average).
λ has been calculated from CEM results for the low-energy region as well as the total ε range, see fig. 10. The "low-energy λ" is mainly influenced by the optical potential, by cascade emission, and n/γ competition. It doesn't agree with the "total λ". Therefore, eq. 7 is not suitable to describe the whole CMS spectra with sufficient accuracy. "Total λ" values give rise to false descriptions of angular distributions at E close to E_f. Note that the CEM reproduces measured $N(E,\theta)$ even for the crucial region mentioned above.

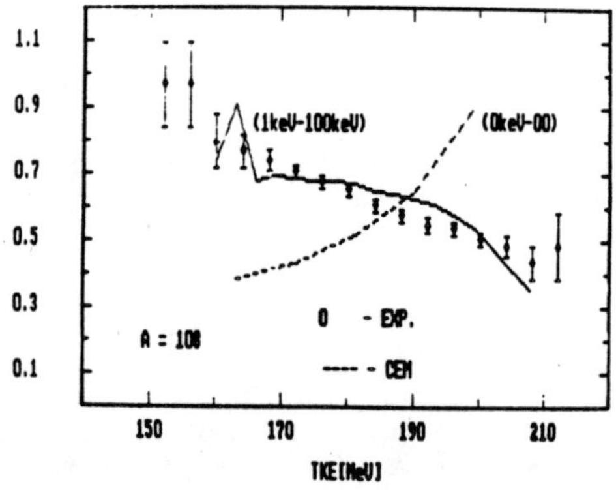

Fig 10 The "shape" parameter λ deduced from CMS spectra for A = 108. Experimental data have been taken from ref. 3

5. Neutron-γ competition of fragment de-excitation

The CEM has been applied to calculate average γ emission energies by the use of the relation

$$\bar{E}_\gamma = \sum_i \int_0^{B_n^i+\Delta} dE^x\, E^x\, P_i(E^x) + \sum_i \int_{B_n^i+\Delta^i}^{\infty} dE^x\, E^x\, P_i(E^x)\, \frac{\Gamma_\gamma^{tot}(E^x,J)}{\Gamma_n^{tot}(E^x,J)+\Gamma_\gamma^{tot}(E^{x-},J)}$$

(Δ - pairing correction energy). The first part, i.e. $\bar{E}_{\gamma,1}$, is a value close to $(B_n+\Delta)/2$. The second one is strongly influenced by the ratio Γ_n/Γ_γ. The total n width are rather low for high level density parameter, i.e. in the mass number regions $108<A<126$ and $A>145$ mainly. Further, they are diminished due to the influence of fragment angular momentum (cf. Eqs. 1-3). Large differences between J and J′ can only be realized for sufficiently high orbital momenta l of the neutrons. Due to the centrifugal barrier, large l give rise to small emission probabilities at small E^x.

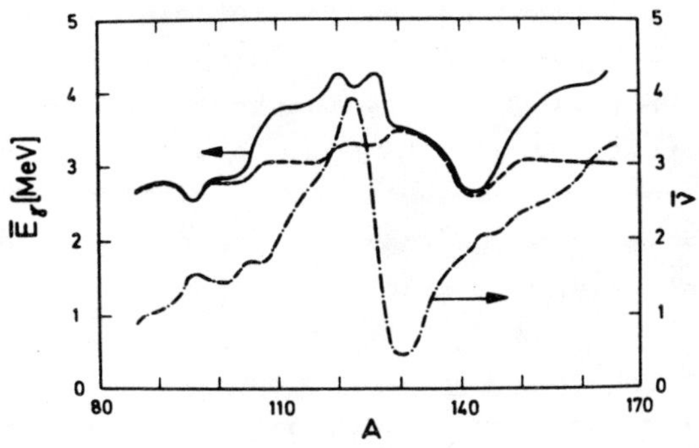

Fig. 11 $\bar{E}_\gamma$ and $\bar{\nu}$ versus A (CEM calculation). The dashed line represents $\bar{E}_{\gamma,1}(A)$

The correlation between $\bar{E}_\gamma$ and $\bar{\nu}$ is represented in fig 12. The CEM reproduces the systematics found in ref. 15. It is pointed out that the dependence of the average angular momentum on TKE is still an open question.[8] The increase of $\bar{E}_\gamma$ with decreasing TKE can be interpreted as due to the increase of B_n with increasing emission step (about 0.2-0.3 MeV per emitted neutron). The experimental $\bar{\nu}(A,TKE)$[1,27] exhibit remarkable deviations from a linear dependence on TKE for very low TKE, in particular for fragments with high level density parameter. This behaviour of experimental data can only be reproduced if one assumes high fragment angular momenta at low TKE ($J \gtrsim 10$), i.e. in this case n emission is suppressed. Fission configurations at very low TKE, interpreted as "cold-deformed"[29], are probably characterized by this effect giving rise to fine-structure effects in the n yield as discussed by Hasse[29].

6. Summary

A complex SMA, originally applied to n emission only, has been extended to account for LCP and γ emission from fission fragments. The effect of particle emission during fragment acceleretion depends on dissipation mechanism strongly (as studied in the case of n recently[19]). Whereas heavier LCP are mainly released in ternary fission, a considerable yield of fission p is due to simple fragment de-excitation. The CEM reproduces experimental data. However, further measurements seem to be useful (cf. figs. 4,5). The analysis of fission n distributions have shown that the deformed-shell region H has an remarkable influence on the initial distribution in E^x for A around 132-135. A similar effect has been found by measuring TKE distributions in Pu fission[30]. The spectrum parameter λ is important in defining the structure of n LS distributions at the crucial points corresponding to zero ε. However, CMS spectrum fits give totally different results for the low-energy range compared with a fit to the total distribution. Hence, one should be careful in interpreting such parameters. It is emphasized that the CEM reproduces experimental distributions within experimental and theoretical uncertainties. The same holds true for the average γ emission energy. Here, an important question concern the fragment angular momentum as function of both A and TKE.

References

1 H. Nifenecker, C. Signarbiex, R. Babinat, and J. Poitou, Proc. IAEA Symp. on Physics and Chemistry of Fission, Rochester, New York, 13 - 17 Aug. 1973 (IAEA, Vienna, 1974), Vol. II, 117

2 C. Budtz-Jorgenson and H.-H. Knitter, Proc. XV-th Int. Conf. Nucl. Phys. - Nuclear Fission -, Gaussig (GDR), 11 -15 Nov. 1985, ZfK-592 (1986) 18

3 E.A. Seregina and P.P. Dyachenko, Yad. Fiz. **42** (1985) 1337

4 H. Märten, D. Richter, D. Seeliger, W.D. Fromm, and W. Neubert, loc. cit. (2), p. 39; and Nucl. Instr. Meth. **A264** (1988) 375

5 O.I. Batenkov, A.B. Blinov, M.V. Blinov, and S.N. Smirnov, Proc. All-Union Conf. onNeutron Physics, Kiev, 1983 (Moscow, 1984), Vol. I, 334

6 K. Skarsvag, Phys. Rev. **C22** (1980) 638

7 W. Pilz, W. Neubert, K. Arnold, D. Lucas, and H. Märten, Proc. XVII-th Int. Conf. Nucl. Phys., Gaussig (GDR), 9 - 13 Nov. 1987, ZfK-646 (1988) 163

8 R. Schmid-Fabian, Thesis (Dissertation), Heidelberg, 1988

9 J.P. Theobald, Proc. Seminar on Fission, Castle of Pont d'Oye,

Habay-la-Neuve, Belgium, 1986, BLG-588 (1986) 63
10 L. Novicki et al., Nucl. Phys. **A375** (1982) 187
11 I. Rotter, J. Phys. G: Nucl. Phys. **15** (1989) 251
12 P. Mädler, Z. Phys. **A321** (1985) 343
13 B. Milek, R. Reif, and J. Revai, Phys. Rev. **C37** (1988) 1077
14 O. Tanimura and T. Fliessbach, Z. Phys. **A328** (1987) 475
15 H. Nifenecker et al., Nucl. Phys. **A189** (1972) 285
16 A.E. Savalev et al., Yad. Fiz. **24** (1976) 261
17 H. Märten, Proc. IAEA Advisory Group Meeting on Nuclear Theory for Fast Neutron Nuclear Data Evaluation, Beijing, 12 - 16 Oct. 1987, IAEA-TECDOC-**483** (1988) 148
18 H. Märten, A. Ruben, and D. Seeliger, Proc. Int. Conf. on Neutron Physics, Kiev, 14 - 18. Sept. 1987 (Moskow, 1988), Vol. III, 49
19 H. Märten and D. Seeliger, J. Phys. G: Nucl. Phys. **14** (1988) 211
20 H. Märten and D. Seeliger, J. Phys. G: Nucl. Phys. **14** (1986) 349
21 E. Nardi et al., Phys. Rev. **C8** (1973) 2293
22 T. Ericson and V. Strutinski, Nucl. Phys. **8** (1958) L84 (Errata: Nucl. Phys. **9** (1959) 689)
23 I. Dühring, unpublished report, TU Dresden (1988)
24 J.F. Wild et al., Phys. Rev. **C32** (1985) 488
25 L. Nowicki et al., Nucl. Phys. **A375** (1982) 187
26 G.M. Raisbeck and T.D. Thomas, Phys. Rev. **172** (1968) 1272
27 C. Budtz-Jorgenson and H.-H. Knitter, Proc. IAEA Consultants' Meeting on the Physics of Neutron Emission in Fission, Mito (Japan), 24 - 27 May 1988, in print
28 B. Wilkins et al., Phys. Rev. **C14** (1976) 1832
29 R. Hasse, Yad. Const. 1 (1988) 3
30 P. Schillebeeckx et al., loc. cit. (9), p. 124

AN EXPERIMENTAL STUDY OF THE PROMPT NEUTRON SPECTRUM OF U-235 FISSION INDUCED BY THERMAL NEUTRONS

Wang Yufeng, Bai Xixiang, Li Anli, Wang Xiaozhong, Li Jingwen, Meng Jiangchen, and Bao Zongyu
Institute of Atomic Energy, P.O. Box 275-46, Beijing, China

Abstract
The prompt neutron spectrum from the fission of U^{235} induced by thermal neutrons has been measured in the energy range 0.56 to 15.4 MeV with a TOF spectrometer consisting of a multisection fast fission chamber and an ST-451 liquid scintillation neutron detector. The gamma-ray background is suppressed by applying efficient pulse shape neutron gamma discrimination. The several systematical influences upon the measured TOF spectra are corrected in detail. Particularly, the correction for the distortion resulting from irrelative fission signals has been carried out meticulously. The experimental result demonstrates that the measured spectrum can not be described by Maxwellian distribution in the energy range given above. However, Madland-Nix model (MNM) calculation corresponds to the data fairly in a quite wide energy range.

1. Introduction

The prompt fission neutron spectra occupy a key position among the nuclear data required in reactor design; besides, a detailed knowledge of fission spectra is alse important for other applications. There are several cases of discrepancies between differential data and integral data measured as averages over fission spectra. More reliable and accurate fission spectrum data are necessary for clarifying the discrepancies. Moreover, U^{235} fission spectrum could be used as a standard for several categories of nuclear data. On the other hand, accurate data of fission spectra are significant for verifying the theory model and stadying fission mechanism further.

Most of the existing data on prompt neutron spectrum of U^{235} fission induced by thermal neutrons cover only a narrow energy range (1-10 MeV) and the data in the high energy range are very rare. In addition, the methods used in these measurements are mostly outmoded. The discrepancies among them are larger than the errors given by the authers. Obviously, there exist some systematic errors. In view of the facts mentioned above, more accurate measurement of this spectrum seems to be necessary.

2. Experimental arrangement

The prompt neutron spectrum from fission of U^{235} induced by thermal neutrons has been measured in the energy range 0.56 to 15.4 MeV with a TOF spectrometer consisting of a multisection fast fission chamber and an ST-451 liquid scintillation neutron detector. The high purity thermal neutron beam is supplied by the thermal column at the heavy water reactor of IAE[1]. A collimator and a single crystal silicon filter are utilized for upgrading the beam quality. The thermal neutron flux is $7.5 * 10^6 / cm^2.s$ and Cd ratio is 18000:1 for the thermal neutrons. The fission chamber contains 21 mg of enriched uranium (90% U^{235}). The fission plates are perpendicular to the thermal neutron beam and the fission rate is about $6 * 10^4/s$. The 12 fission plates in the chamber are divided into three sections to decrease capacity and keep fast rise-time of the fission signals. Fission fragment signals taken from the three sections of the fission plates are fed into three ORTEC 142B preamplifiers respectively, and the timing outputs of the 142B are mixed together by a Mixer through corresponding constant fraction discriminators.

The neutron detector consists of an ST-451 liquid scintillator, 10 cm in diameter and 5 cm thick, and a 58AVP photomultiplier. Pulse shape discrimination is applied to suppress gamma ray background efficiently. The neutron detector is heavy shielded with Pb, Fe, and paraffin and located at 44 degree to the direction of the thermal neutrons.The flight pass is 3.17 m. A collimator is built between the fission chamber and the neutron detector to prevent the neutrons emitted from fission events and scattered by surrounding material from hitting the neutron detector.

For obtaining suitable effect-background ratio in higher energy range, three different biases of the neutron detector are set. The contained energy ranges of the fission neutrons are 5 to 15.4 MeV for the high bias, 1.0 to 9 MeV for middle bias, and 0.56 to 3.2 MeV for low bias. The corresponding neutron TOF spectra have been measured by using three MCAs simultaneously. The fission neutron spectrum with a wide energy range has been obtained by merging these three spectra. The detection efficiencies of the neutron detector are determined by combining the (n,p) scattering experiments using 14.8 MeV neutrons on a neutron generator with Monte carlo calculation using the standard code NEFF4[2]. The results of the experiments and the calculations accord with each other in suitable accuracy.

3. Corrections

The measured neutron TOF spectrum is always disturbed and should be corrected. A more careful consideration of the corrections is given in the present work.

At first the correction for accidental coincidence of irrelative signals from neutron detector and fission chamber has been done. In the measurements the counting rate of the fission chamber is as high as $10^6/s$, only a few events per second are expected in the neutron detector because of the finite solid angle and the limited neutron detection efficiency.

In order to avoid deadtime loss in the time to amplitude converter (TAC), the time scale is inversed by starting with the neutron events and stopping with fission events. Both the statistical analysis and the measurements have indicated that this background is channel--dependent, and can be determined in full range by fitting the distribution beyond the prompt gamma peak of the TOF spectrum with a proper exponential form.

The distortions due to the stop signals of the TAC from irrelative fission events of random distribution instead of the associated fission events have been corrected by means of probability analysis and iteration teatment.

The scattered neutrons which were emitted from the fission events and scattered by the surrounding material has been measured with a shadow bar in length of 60 cm and corrected in the same way as mentioned above. Finally, it has been subtracted from the TOF spectra.

It is not always negligible to correct the experimental spectra for fission neutron intensity attenuation in air due to the absorption and scattering of the oxygen and nitrogen in the flight pass, even the correction is usually small; however, it becomes significant in the low energy resonance region of nitrogen and oxygen.

Other influences to the spectrum, such as incompletely thermalized neutrons, delayed gamma rays and delayed neutrons following fissions are negligible.

4. Results and discussions

As an approximate calculation, the prompt fission neutron spectra are described by Maxwellian distribution. The fitting has been carried out by means of least-squares fit. The Maxwellian temperature T and average energy E = 3T/2 can be obtained. The results of the fit indicate that Maxwellian distribution can not fairly describe the measured spectrum in the full energy range. For the optimum fitting, Maxwellian temperature T=1.321 MeV. Based on the standard nuclear evaporation theory, Madland and Nix calculated the prompt fission neutron spectrum N(E) as a function of both fissioning nucleus and its excitating energy[(3)]. We have compared the experimental results with MNM calculation. The exact theoretical calculation of the prompt fission neutron spectrum for the thermal neutron induced fission of U^{235} has been carried out by Madland[(4)]. The ratios of the experimental spectrum and the exact calculation to approximate calculation using a constant cross section in the model are shown in Fig.1. Meanwhile, the ratio of Maxwellian distribution to the approximate calculation is also shown in Fig. 1. The exact calculation improves the approximate calculation slightly. In the low energy range, the experimental data 12% higher than the exact calculation. This is a problem which needs to improve both experimental atudy and theoretical calculations.

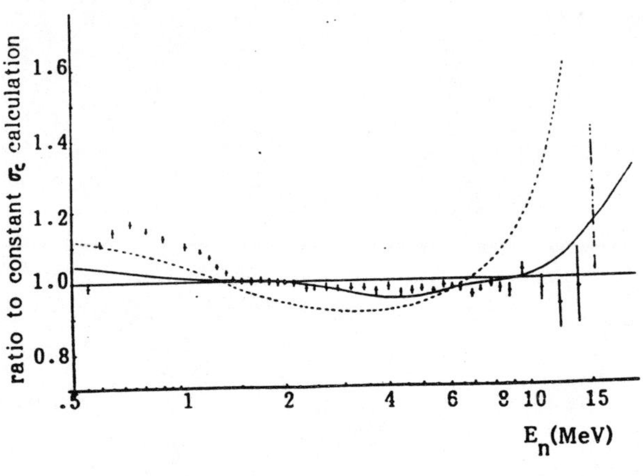

Fig.1 The experimental neutron spectrum from fission of U^{235} induced by thermal neutrons compared with calculations.
+ experimental data
— calculated by Madland[(4)]
--- Maxwellian distribution (T = 1.38 MeV)

Acknowledgements

We are grateful to Shi Zongren, Cao Zhong, and Yu Chunying for their technical assistances. Further ackowledgement for valuable suggestions goes to Prof. Huang Shengnian and Prof. wang Dahai. We would also like to thank Dr. G.Dietze for his help in supplying the Monte Carlo code NEFF4 and Dr. D.Madland for his useful help in the MNM calculation.

References

(1) Shi Zongren, et.al., A facility and design of high purity thermal neutron beam, to be published

(2) G.Dietze and H.Klein, PTB-ND-22

(3) D.Madland and J.Nix, New calculation of prompt fission neutron spectra and average prompt neutron multiplicities, Preprint LA-UR-81-2968

(4) D.Madland, Private letter

V. TERNARY FISSION

LOW ENERGY NUCLEAR TRIPARTITION

P. Heeg, M. Mutterer, J. Pannicke and J.P. Theobald
Institut für Kernphysik der Technischen Hochschule Darmstadt,
Schlossgartenstr. 9, D-6100 Darmstadt, Germany

Abstract

Based on our experimental work with the detector system DIOGENES different ways of interpreting ternary fission observables are discussed, in particular the width of the equatorial light particle angular distribution and its dependence on the assumed fragment configuration at scission. We consider the total kinetic energy TKE, the sum of the main fragment kinetic energy E_{ff} and the light particle energy, as the key quantity in this discussion.
For polar alpha particle accompanied fission, structures were observed in the main fragment mass distributions. We propose a possible explanation for the most prominent structure at a mass split of about $A_H/A_L = 132/116$.

1 Introduction

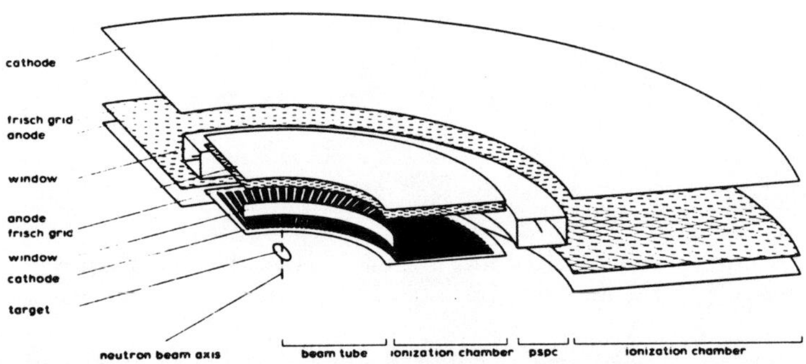

Figure 1: *Quarter section of the detector system DIOGENES*

Nuclear spectroscopy in the second potential minimum of the fissioning nucleus has revealed a quite consistent picture of the static fission barriers of the actinides. Experiments with the fission fragment spectrometers LOHENGRIN and COSI FAN TUTTE of the Institute von Laue - Langevin (ILL) as well as with twin ionization chambers have provided a wealth of accurate data on fragment mass, nuclear charge and energy distributions. However, there is a missing link between the experiments for the study of compound nuclear deformation up to the fission barriers and those on the fragment properties. The interplay between prescission kinetic energy and excitation energy for the advanced fission process, when the potential energy drops from the saddle to the scission point, is not well known, although there are recent important investigations on this subject)[1,2].

One of the few messages, which the fissioning nucleus releases in this late stage of the fission process is, in some permillage of events, carried by a light nucleus emitted in most cases nearly perpendicular and in some percent of cases nearly into the direction of the two main fission fragments. About 90 % are alpha particles called according to their observable final emission direction "equatorial" or "polar" alpha particles. This mode of ternary fission has been named long-range alpha-particle accompanied fission (LRAF).

Since its discovery in 1946 in Paris the interpretation of LRAF data, essentially the kinetic energies and the relative angular distributions of the three reaction products from which also the main fragment mass distribution can be deduced, has been given under three tacitly accepted assumptions:

1. The alpha particle emission coincides in space and time with the scission point.
2. The alpha particle is a probe for the late *binary* fission configuration. When the configuration is known, the repulsive Coulomb energy at the scission point can be calculated and from the final fragment kinetic energy the prescission energy determined.
3. It is possible in a univocal way to transform the above mentioned observables of LRAF into microscopic scission point parameters of the main (binary) fragments with the aid of trajectory calculations. Of particular interest have been the shape parameters of the nascent main fragments.

After a short description of our experimental setup and a presentation of some of our results we shall discuss the validity of these assumptions and cast doubts on conclusions which have been drawn from there.
In a final chapter we present some of our recent results on polar LRAF.

2 Experiments

Our experiments have been performed with the double-torus ionization chamber DIOGENES, a quarter section of which is shown in figure 1.

This is an angular-position sensitive detector system composed of concentric ionization chambers and proportional counters of high angular and energy resolution. DIOGENES is described in detail elsewhere)[3].
Kinetic energies and relative angular distributions have been recorded in coincidence for the three reaction products of LRAF for neutron induced fission of ^{235}U and ^{239}Pu and for ^{252}Cf(s.f.). The neutron induced reactions have been measured in the neutron flux of $5 \times 10^{8}\ s^{-1}cm^{-2}$ of a neutron guide tube at the High Flux Reactor HFR of the ILL, and the spontaneous fission experiment has been performed at the Institute for Nuclear Physics of the Technische Hochschule Darmstadt.

3 Some Results on Equatorial Emission

From the experimental data we have derived a great variety of correlations between alpha particle and fission fragment parameters. Only a few of them will be presented here, which are

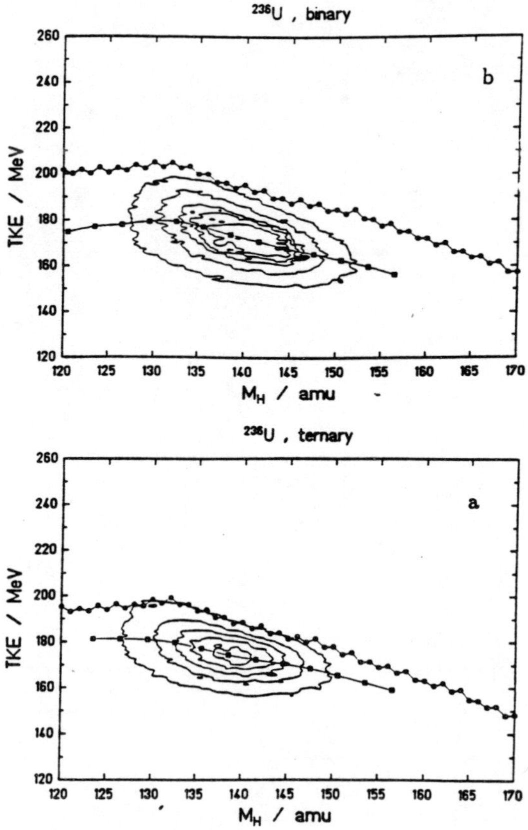

Figure 2:
a: The sum TKE of the main fragments and alpha particle energies as function of the heavy fragment mass, together with the average $< TKE >$ *and maximum Q values. b: The fragment kinetic energies* E_{ff} *together with* $< E_{ff} >$ *and* Q_{max}*, binary fission*

important for the following discussion.

In this chapter, data on the reaction ^{235}U(n,αf) will be interpreted. Figure 2a shows the total kinetic energy TKE = E_{ff} +E_α, the sum of the main fragments and alpha particle energies, as function of the heavy fragment mass together with the average $< TKE >$ and maximum Q values.

For comparison, E_{ff} is also displayed in figure 2b together with maximum Q values and $< E_{ff} >$ for binary fission. It is obvious that ternary main fragments are less excited than binary ones and that "true" cold fragmentation, for which the total Q value is tied up in kinetic energy, occurs in a broader fragment mass window than in binary fission. The investigation of ref.)[2] indicates that at the scission point only about 50 % of the excitation energy is intrinsic. It is therefore reasonable to assume that the fragments in LRA accompanied fission are also less deformed than binary fragments. The alpha particle is not directly a probe for binary scission point configurations and not simply an observer but a participant in the fission process.
The fact that the fragment properties in binary fission and LRAF are similar has caused in

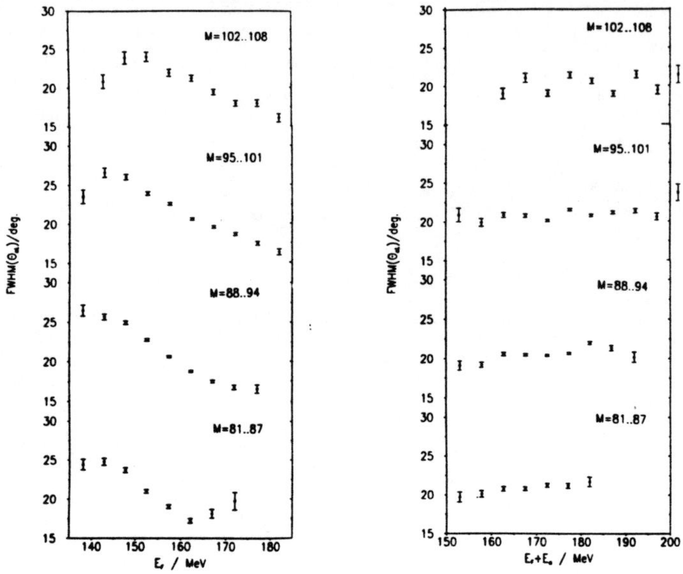

Figure 3:
a: The width $\Delta\theta_{\alpha l}$ of the alpha particle angular distribution relative to the direction of the light main fragment for different mass windows as function of the fragment kinetic energy E_{ff}
b: $\Delta\theta_{\alpha l}$ as function of the total kinetic energy $TKE = E_{ff} + E_\alpha$

the past some wrong conclusions from LRAF data, in particular from the width $\Delta\theta_{\alpha l}$ of the alpha particle angular distribution. This width decreases with increasing main fragment kinetic energy as it is plotted in figure 3a.

The conclusion was that compact scission configurations, which release high energetic fragments, have a narrow equatorial alpha particle focussing)[4]. However, $\Delta\theta_{\alpha l}$ is independent of TKE (figure 3b).
One could argue that compact scission configurations are also associated with high alpha particle kinetic energies, which would prohibit stronger focussing, but, as shown in figure 4, fragment and alpha particle energies are anticorrelated and add up to a nearly constant $< TKE >$.

It is evident that the angular width $\Delta\theta_{\alpha l}$ is not correlated in a simple way with the distance d of nascent fragments at the scission point. This conclusion is confirmed by the observation that $\Delta\theta_{\alpha l}$ is also constant as function of the main fragments' mass ratio R as shown in figure 5, and does even not dependent on the fissioning systems ^{236}U, ^{240}Pu and ^{252}Cf. There is, however, an indication that $\Delta\theta_{\alpha l}$ increases towards symmetric fission (figure 5). Fragment shape fluctuations at symmetric mass splits would broaden the alpha particle angular distribution as is observed in the experiment.

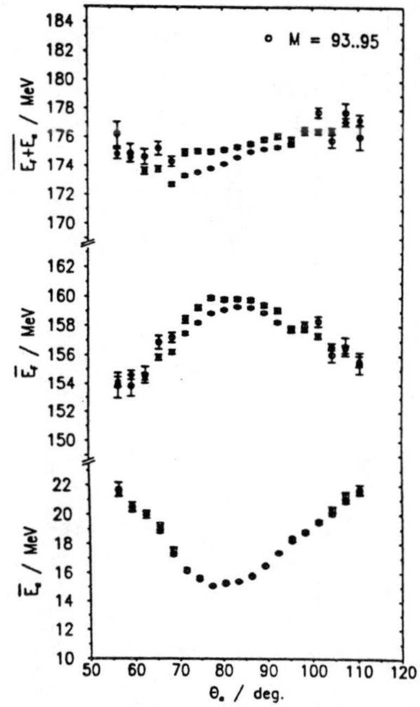

Figure 4: *Average alpha particle energy $< E_\alpha >$, average main fragment kinetic energy $< E_f >$ and average total kinetic energy $< TKE > = < E_{ff} + E_\alpha >$ as function of the alpha particle emission angle $\theta_{\alpha l}$ integrated over all mass splits and over mass splits in the window M = 93–95.*

4 Discussion of these Results on Equatorial Alpha Emission

The results presented in the last chapter have motivated us to discuss LRAF in the light of the quantity TKE of the three reaction products, which is independent on the angular distribution and is directly related to the total excitation energy of the ternary fissioning system TXE = Q_{max} −TKE. Decreasing fragment deformation can thus be defined by decreasing TXE and cold ternary fission by TXE → 0. This definition is more appropriate than the previous one by relatively large main fragment kinetic energies E_{ff}, which in the past, like in the case of binary fission, was taken as a measure for the excitation energy.

The independence of the angular width $\Delta\theta_{\alpha l}$ of the total fragment excitation energy TXE as well as from the fragment mass ratio can be understood as an indication for particle emission prior to scission. Nascent fragments have a very small influence on the alpha particle observable $\Delta\theta_{\alpha l}$. O. Tanimura and T. Fliessbach)[5] have shown, using an idea of N. Carjan)[6] that alpha particles can be emitted as soon as they have gained sufficient energy by collisions with the shrinking potential wells in the neck of the fissioning nucleus. In this context it is interesting to compare main fragment mass distributions of charged particle accompanied fission of a nucleus with mass number A and nuclear charge number Z with the binary fragment mass distribution of a nucleus with the corresponding numbers (A − A_p) and (Z − Z_p), where (A_p Z_p) characterize the emitted light charged particle.

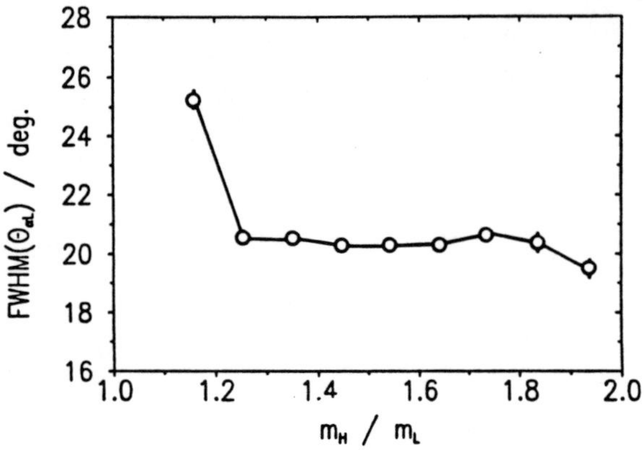

Figure 5: $\Delta\theta_{\alpha l}$ *as function of the main fragments' mass ratio* $R = m_H/m_L$

An attempt in this direction has given the result presented in figure 6 for light charged particle accompanied spontaneous fission of ^{252}Cf with light particle nuclear charge numbers $Z_p > 6$ and $E_p > 25$ MeV)[7]. Here, this main fragment mass distribution is compared with the

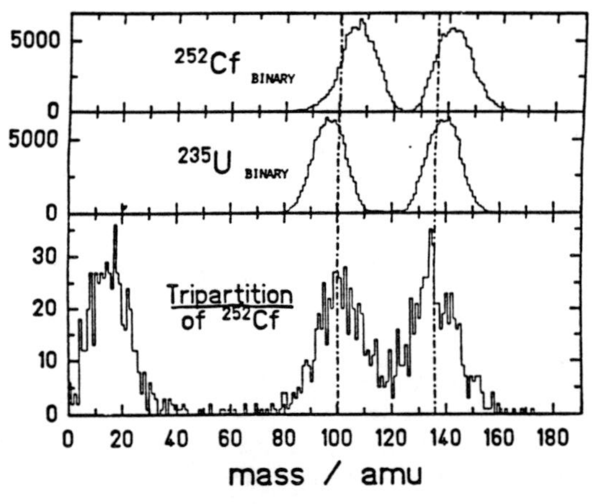

Figure 6: *Mass spectra of* ^{252}*Cf spontaneous binary fission, of* ^{235}*U neutron induced binary fission, and of charged particle associated spontaneous fission of* ^{252}*Cf*

binary distribution of ^{235}U(n,f). It is obvious that the two distributions are very different, in particular the distance between the mass of the most probable heavy and light fission fragment. Obviously the charged particle is not emitted at such an early stage of the fission process that the fissioning nucleus is converted before scission into the corresponding lighter fissioning system. On the other hand, the light charged particle must not necessarily be emitted at the instant of scission.

Also the assumption that trajectory calculations allow the univocal transformation of LRAF observables into microscopic scission point configuration parameters is debatable as we have shown in a previous paper)[8] and as was demonstrated already by Pik–Pichak)[9]. For a correct interpretation of LRAF it is necessary to develop a model, which takes the generation process of the alpha particle into account, in order to start the trajectory calculations with initial parameters which have a sound physical basis. Attempts in this sense are described in ref.)[5,10]

It should be mentioned that, although the LRA particle is not a probe for binary scission point configurations, it is a message of a nucleus with large amplitude nuclear deformations.

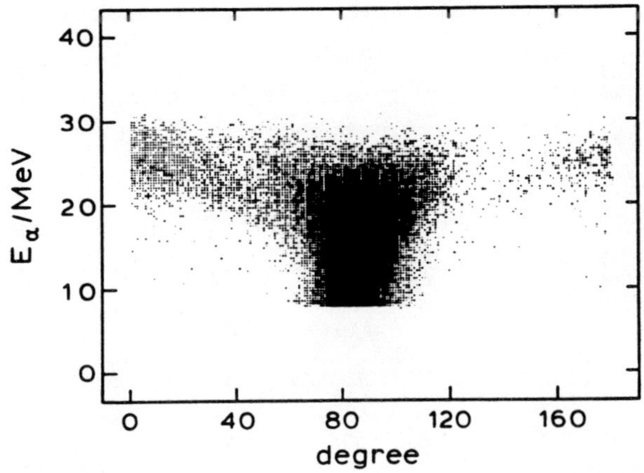

Figure 7: *Two dimensional alpha particle yield distribution as function of ($\theta_{\alpha l}, E_\alpha$)*

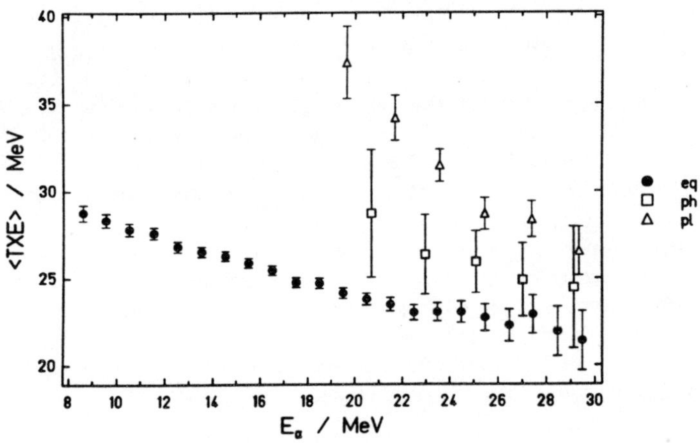

Figure 8: *Total excitation energy $< TXE > = < Q_{max} - TKE >$ as function of alpha particle energy E_α for the three emission regimes: equatorial, polar–light and polar–heavy*

5 Polar Alpha Particle Emission

This chapter is based on our LRAF data on spontaneous fission of ^{252}Cf. Figure 7 is a plot of the two dimensional alpha particle yield distribution as function of emission angle $\theta_{\alpha l}$ and alpha particle kinetic energy E_α. Polar emitted alpha particles form separated spots in this two–dimensional display. If we see this polar emission process in the light of the $< TXE > = < Q_{max} - TKE >$, we realize that $< TXE >$ has a different dependence on the alpha particle energy E_α for polar emission in direction of the light and the heavy fragments flight paths, respectively, as shown in figure 8. Polar emission into the direction of the heavy fragment costs more excitatiton energy than the emission into the direction of the light fragment (figure 9).

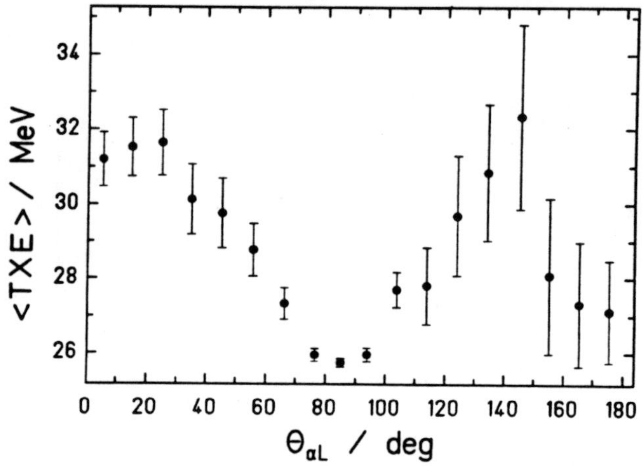

Figure 9: $< TXE > = < Q_{max} - TKE >$ *as function of the alpha particle emission angle* $\theta_{\alpha l}$

In the model of O. Tanimura and T. Fliessbach)[5] polar particles are emitted later than equatorial ones, when the movement of the potential wells in the shrinking and closing neck is very fast and going into the direction of the nascent fragments. In order to be emitted the alpha particles have to overcome the potential barriers at the poles of the fissioning dinuclear system. These barriers are of different heights. Therefore the nuclear system is left with different TXE after alpha emission in the two cases of polar emission regimes as shown in figure 9.

Of particular interest are the main fragment mass distributions for polar LRAF, which show distinct fine structures as can be seen in figure 10. With increasing TKE these structures are relatively more pronounced, particularly those at the fragment mass numbers A $\simeq$ 90 and A $\simeq$ 132, which have minimum neutron emission, i.e. which are relatively cold or close to their ground states. This property seems to be enhanced by polar alpha particle emission. For equatorial alpha particle emission, no fine structures are observed in the energy integrated mass spectra, although the average excitation energy is smaller in this case.

At present, we are investigating these fine structures. Here we only try to give an interpretation of the most prominent structure in the main fission fragment mass spectrum at A $\simeq$ 132 for polar emission of alpha particles into the direction of the light main fission fragments. In the binary fragment mass distribution for *cold* fission of ^{252}Cf the mass split A_H / A_L = 132/120 is enhanced relative to neighbouring fragmentations)[11]. If we assume that in polar LRAF also the cold double magic heavy fragment is formed the corresponding light fragment, with A_L= 116 after alpha particle emission, has to take over the total excitation energy of approximately 30 MeV (figure 9) and consequently emits about 3 neutrons. The corresponding mass split is then $^{132}Sn^{82}_{50}$ / $^{113}Pd^{67}_{46}$ compared to $^{132}Sn^{82}_{50}$ / $^{116}Cd^{68}_{48}$ in binary fission. The other fine structures in the main fragment mass spectra of polar LRAF may be fingerprints of the proton odd–even effect, which seems also to be enhanced in polar LRAF. However, interpretation is limited by the mass resolution and the insufficient statistical significance in the case of the rare polar events.

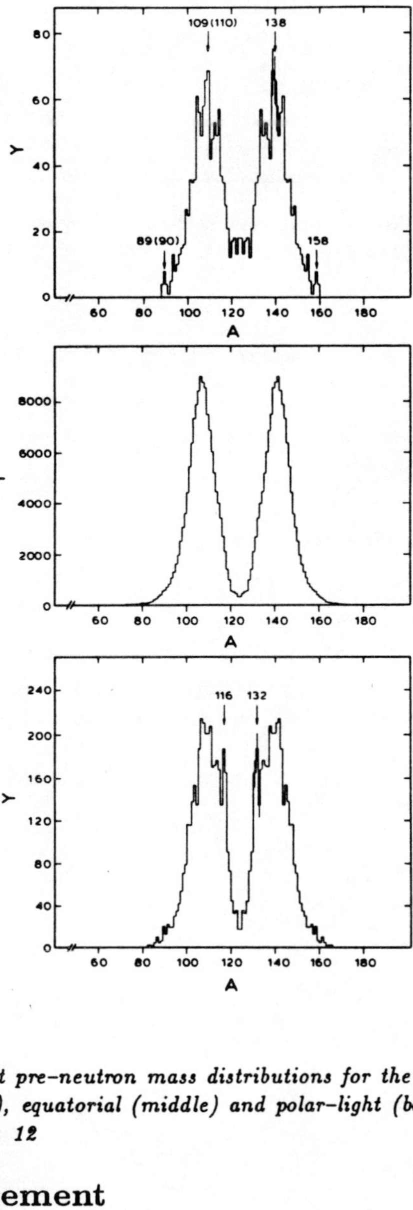

Figure 10: *Main fragment pre-neutron mass distributions for the three alpha particle emission regimes polar–heavy (top), equatorial (middle) and polar–light (bottom), corrected for neutron emission according to ref. 12*

6 Acknowledgement

This work was funded by the German Bundesminister für Forschung und Technologie (BMFT) under contract nr. 06 DA 453 I. We thank the Gesellschaft für Schwerionenforschung (GSI) Darmstadt for computer time and the Institut v. Laue–Langevin (ILL) for financial support and neutron beam time.
Instructive discussions with N. Carjan, T. Fliessbach, and B. Leroux are gratefully acknowledged.

References

[1] R.W. Hasse
these proceedings

[2] F. Gönnenwein, J.P. Bocquet, and R. Brissot
Proc. XVII Int. Symp. on Nuclear Physics, Gaussig (1987), in press

[3] P. Heeg, J. Pannicke, M. Mutterer, P. Schall, J.P. Theobald, K. Weingärtner, K.F. Hoffmann, K. Scheele, P. Zöller, G. Barreau, B. Leroux, and F. Gönnenwein
Nucl. Instr. and Meth. in preparation

[4] C.R. Guet, H.A. Nifenecker, C. Signarbieux, and M.Asghar
Proc. Symp. on Physics and Chemistry of Fission, Jülich (1979), IAEA Wien (1980) Vol. II, 247

[5] O. Tanimura and T. Fliessbach
Z. f. Phys. A 328 (1987) 475

[6] N. Carjan
PhD Thesis D 17, T.H. Darmstadt (1977)

[7] P. Schall, P. Heeg, M. Mutterer, and J.P. Theobald
Phys. Lett. B 191 (1987)339

[8] J.P. Theobald
Proc. Sem. on Fission, Castle of Pont d'Oye (Habay-la-Neuve) (1986)
C. Wagemans ed., report BLG 586, Brussels, Belgium, page 63

[9] G.A. Pik-Pichak
Sov. J. Nucl. Phys. 40(1984) 215

[10] V.A. Rubchenya and S.G. Yavshits
Z. f. Phys. A329 (1988) 217

[11] H.H. Knitter, private communication (1988)

[12] G.K. Mehta et al.
Phys. Rev. C7 (1973) 373

TERNARY FISSION: EXPERIMENT AND TRAJECTORY CALCULATIONS

M.Ya. Borkovski, Yu.I. Gusev, Yu.K. Zalite, and D.M. Seliverstov
Leningrad Nuclear Physics Institute, Gatchina, USSR

1. INTRODUCTION

Recently a number of experimental detailed studies on the fission of compound nuclei ^{236}U, ^{252}Cf accompained by alphas and other charge particles have been finished /1,2/. On the base of the obtained results in alpha accompained fission of ^{236}U the new stage in trajectory computations for the definition of scission conditions was performed in /3/ but the result seems to be not unequivocal. A quantum mechanical approach to the emission of alpha particles of fissioning nuclei sensitive to the descent time from the saddle to the scission point /4/ occurs to be very promising. But it is early to say about a good agreement between theory and experiment. The difference in the value of the prescission kinetic energy of fragments E_F^0 obtained in trajectory calculations, see for example /5,6/, still remains very large, of the order 20 MeV. In this paper we do the attempt to decrease such uncertainty in E_F^0, performing calculations and comparing them with all observable variables from thermal fission of ^{239}Pu accompained by alphas.

2. TRAJECTORY CALCULATIONS

We carried out computations in three point charge approximation by two ways. In first case as in /3/ we randomly distributed initial parameters: kinetic energies of fragments and alphas E_F^0, E_3^0, positions on X and Y axises, angles of emission Θ_{3L}^0 in wide intervals: $(0 \leq E_F^0 \leq 70)$ MeV, $(0 \leq E_3^0 \leq 10)$ MeV, $0 \leq X_0 \leq 1.0$, $(0 \leq Y \leq 3.5)$ Fm, $(0 \leq \Theta_{3L}^0 \leq 180)°$. X-axis coincides with the fragment axis, $X_0 = X/D_0$, where D_0 is the distance between centres of fragments. In each set of calculations we fixed only mass ratio R and total energy $E_t = E_F + E_3$ (± 2 MeV). Experimental energy and angular spectra as energy spectra at the fixed value of Θ_{3L} were fitted by calculations. Comparing was carried out event by event in the cells of the phase volume $\Delta E_3 = \pm 1$ MeV, $\Delta \Theta_{3L} = \pm 1°$. Usually for each R and E_t one calculated more than 10^6 trajectories. From this mode of calculations one may do the main conclusions.

1. One observes a low energy threshold in E_F^0(25÷30) MeV below which it is impossible to reproduce all experimental data at any set of other initial parameters X_0, Y, Θ_{3L}^0.

2. Alphas have the energy distribution very closed to Maxwellian

shape: $N(E_3^O) \sim E_3^O \exp(-E_3^O/T_o)$ and T_o is proportional to E_F^O.

3. The angular width (FWHM) is almost fully defined by the position of the emitted particle on X axis. The distribution of X_o values has Gaussian shape and is peaked at $X_o = 1/1+K$, where $K = (Z_H/Z_L)^{1/2}$. The value of $FWHM(X_o) = 0.27$ corresponds to experimental value 19.4°.

In the second case the experimental distribution of E_F at fixed R was taken into account. The distribution of E_3^O was used in Maxwellian shape, X_o, E_F^O, $V_c = Z_L Z_H/D_o$ in Gaussian ones. The correlation between $FWHM(E_F^O)$ and $FWHM(V_c)$ was taken such a way in order to obtain full width equal to experimental width and the best fit to measured values. Thus we had five parameters of the calculation: E_F^O, T_o, $W(E_F^O)$, $W(V_c)$ and Y_{max}.

3. COMPARING WITH EXPERIMENT

As main observables E_3, $FWHM(E_3)$ and $FWHM(\theta_{3L})$ are practically independent on $W(E_F^O)$ and $W(V_c)$ the fit of these variables was made using only two parameters E_F^O and T_o. The distribution on Y was taking in the isotropic form with $Y_{max} = (2 \div 3.5)$ Fm because only with these values of Y_{max} it is possible to satisfy to experimental results. The quality of such fit is seen in Fig.1. In computations one used the ratio of $\bar{E}_F^O/T_o$= Const. The real distribution of the total energy used in calculations reduced the uncertainty in E_F^O up to value not more than ± 5 MeV at mean value of 30 MeV. The comparison of calculations with experimental data is shown in Fig.2-5. We have chosen the next set of initial parameters - V1: $\bar{E}_F^O = 30$ MeV, $W(E_F^O) = 10$ MeV, $W(V_c) = 15.6$ MeV, $T_o = 1.4$ MeV, $W(X_o) = 0.27$, $Y_{max} = 3.5$ Fm, $(0 \leqslant \theta_{3L}^O \leqslant 180)°$. The agreement with experiment becomes essentially worse if we use the distribution on θ_{3L}^O from 0° to 360°.

The existence of a low energy tail and even separate lines in alpha spectrum due to scattering of particles on moving fragments, in particular in the range of (60-70)°, occurs to be critical to the choice of the value $\bar{E}_F^O$. The comparison of calculated and measured spectra both fitting by Gaussians shows a good agreement in Fig.2(a, b). Experiment: $\bar{E}_3 = (16.5 \pm 0.2)$ MeV, $FWHM(E_3) = (10.0 \pm 0.3)$ MeV.

Calculation: $\bar{E}_3 = (16.1 \pm 0.2)$ MeV, $FWHM(E_3) = (10.3 \pm 0.3)$ MeV.

A contribution of scattering particles in the total spectrum is shown in Fig.2(c,d). Unfortunately, an insufficient angular resolution in our experiment ($\Delta\theta_{3L} = \pm 7°$) did't permit to study more precisely the influence of E_F^O on the shape of the energy spectra at different angles.

The agreement of experiment and calculation observed in Fig.2-5 is rather satisfactory if to take into account the absence of any corre-

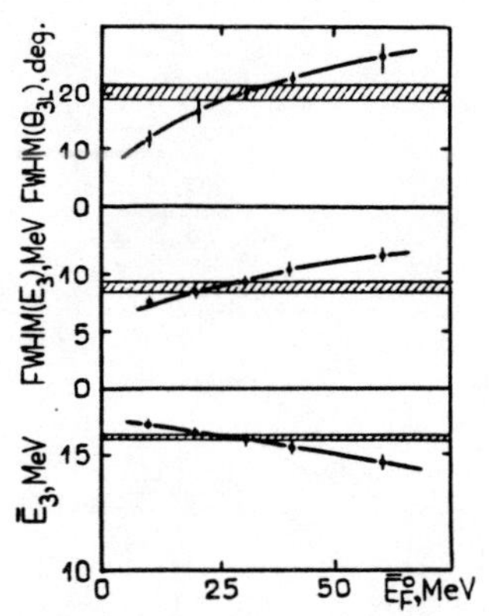

Fig.1. Dependence of average energy widths of energy and angular distributions of α-particles on the fragment kinetic energy. Calculations (V-1) have been done with $\bar{E}^{o}_{F}/T_{o}$=Const=21.4. The hatched region indicates experimental values.

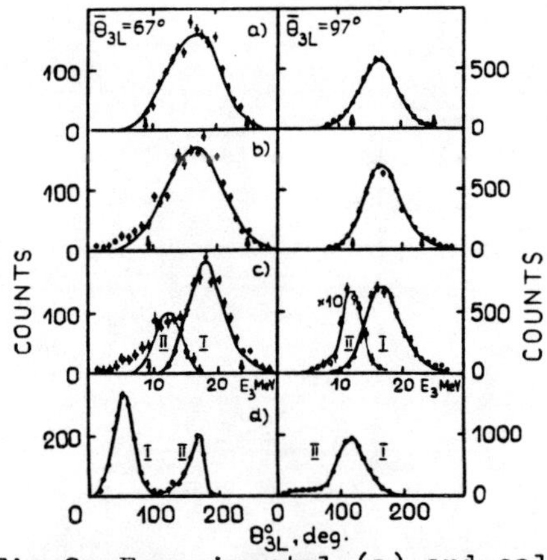

Fig.2. Experimental (a) and calculated (b,c) energy spectra for two values of Θ_{3L}: 67° and 97° at R=1.4. Curves (a,b,c) are Gaussian fits. Curve in initial angular distributions (d) is a visual fit. Arrows show limits of fitting. Regions I and II correspond to direct and scattering α-particles.

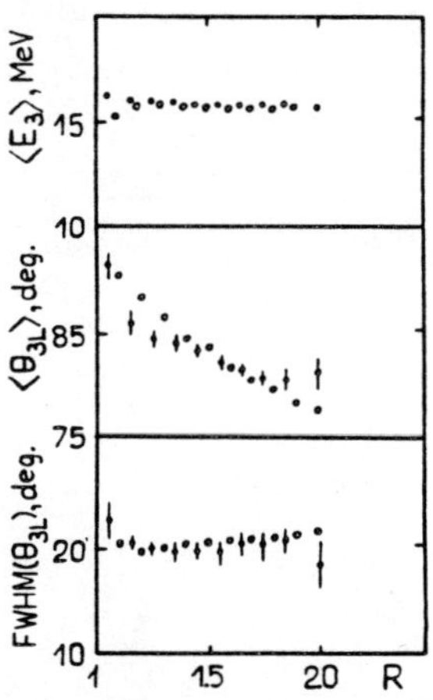

Fig.3. Parameters of energy and angular distributions of α-particles as functions of mass ratio R. $\langle E_3 \rangle$, $\langle \Theta_{3L} \rangle$ are the most probable values.
Full dots are experimental data, open ones are results of calculations (V-1).

lations between initial parameters. The sensitivity of some correlations to based parameters is seen in Fig.4,5. Increasing of $W(E^{o}_{F})$ from 10 MeV to 15.6 MeV remarkably changes the dependence of $\bar{E}_3$ on E_F, while other correlations remain the same. On the other hand decreasing of $\bar{E}^{o}_{F}$ from 30 MeV to 10 MeV essentially distorts the correlation between FWHM(Θ_{3L}) and E_3 as well as all other observables and correlations.

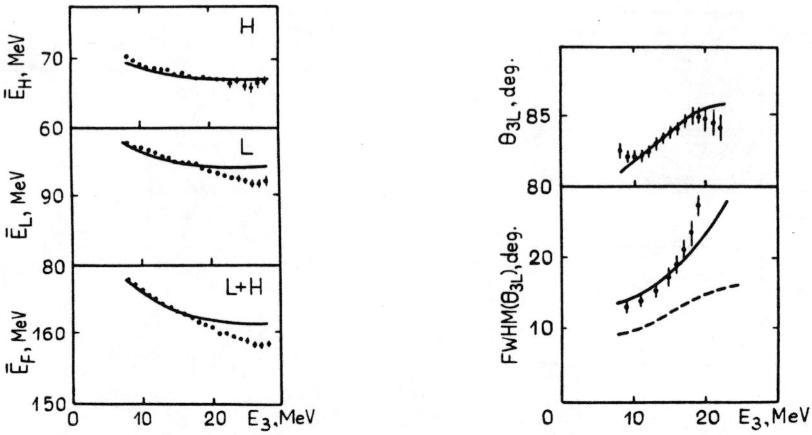

Fig.4. Average values of kinetic energy of heavy (H), light (L), the sum of two fragments (L+H) and parameters of angular distributions of α-particles as functions of α-particle energy. Points are experimental data, full lines show results of calculations (V-1), the broken line calculated values (V-1) with $\bar{E}_F^o$ = 10 MeV and T_o = 0.45 MeV.

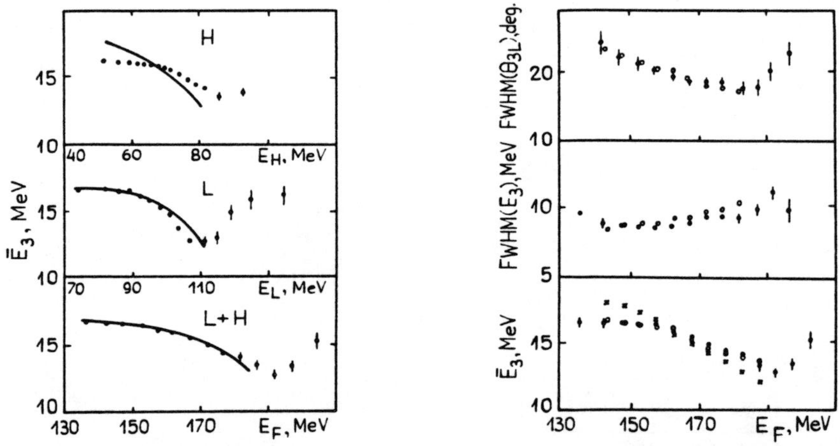

Fig.5. Parameters of energy and angular distributions of α-particles as functions of the energy of light (L), heavy (H) and the sum of fragments. Full lines and open dots represent results of calculation (V-1), crosses are results of V-1 calculation with the values of FWHM(E_F^o) = 15.6 MeV and FWHM(V_c) = = 10 MeV.

REFERENCES

/1/ J.P.Theobald. Proceedings of the 17th International Symposium on Nuclear Physics, Dresden-Gaussig, 1987, report IKDA 87/35.

/2/ V.T.Grachov, Yu.I.Gusev, D.M.Seliverstov. Sov.J.Nucl.Phys., 47, 1987, 622.

/3/ J.P.Theobald. Proceedings of the Seminar on Fission, Castle of Pont d'Oye, BL6, 1986, p.63.

/4/ O.Tamimura and T.Fliesbach. Z.Phys., A328, 1987, 475.

/5/ C.R.Guet, H.A.Nifenecker, C.Signarbieux, M.Asghar. Proceedings of the Fourth Symposium on Physics and Chemistry of Fission, IAEA, Vienna, v.2, 1980, p.247.

/6/ G.A.Pik-Pichak. Sov.J.Nucl.Phys., 40, 1984, 336.

ANGULAR DISTRIBUTION OF PROMPT γ-RAYS IN THE TERNARY FISSION OF ^{252}Cf

W. Pilz, W. Neubert
Central Institute for Nuclear Research, Rossendorf/Dresden, Germany
K. Arnold, D. Lucas, and H. Märten
Technical University of Dresden, Germany

ABSTRACT

We found a shift of the maxima of the angular distributions of the prompt γ-rays in comparison to the binary fission mode. This shift depends on the selected γ-ray energy region. It indicates that the alignment of the angular momenta of "ternary" fission fragments may be influenced by the α-particles emitted in equatorial direction with respect to the fission axis.

1. INTRODUCTION

A 0 deg./90 deg.- anisotropy of prompt γ-rays accompanying the binary fission mode has been known since 30 years /1/,/2/.
The shapes of the angular distributions are characterized mainly by dipol-and quadrupol transitions of γ-rays emitted from the rotating nuclei which are aligned perpendicularly to the fission axis /3/. Average angular momenta of fission fragments of about 7ħ were obtained by several authors /4/,/5/,/6/.
The origin of angular momenta in a fission process can be understood if one assumes that the nucleon mass distributions of the preformed fragments are asymmetrical with respect to the fission axis during the duration of the Coulomb repulsion /3/.
The emission mechanism of the light charged particles (LCP) which accompany about 0.3% of the fission processes may influence both the magnitude and the direction of the angular momenta. The γ-ray emission from fission fragments, following neutron emission, is expected to be similar in the binary and the LCP accompanied fission mode /7/.
We measured double differential emission probabilities of γ-rays from ^{252}Cf fission fragments. The fission fragments were detected in coincidence with the γ-rays and the LCPs emitted in equatorial direction.

2. EXPERIMENTAL SET-UP

A sketch of the experimental set-up is shown in fig.1. The angular distributions of the γ-rays were measured by the determination of the fragment direction for fixed γ-ray detectors. We used two large NaJ(Tl) crystals (6.5ø x 5 inches) mounted on both sides of the ^{252}Cf source at a distance of 50 cm. The direction of the emitted fission fragments was measured by using an one-dimensional position sensitive PPAC. This detector covered an in-plane angular range somewhat more than 90 deg.. The equatorial LCP were registerd by two surface-barrier silicon detectors. Both detectors were shielded by Al foils (36µm thick) in order to stop the fission fragments and the α- particles from the g.st. decay of ^{252}Cf. These detectors accepted about 25% of the full solid angle with regard to the known angular distribution of the predominant ternary α-particles.

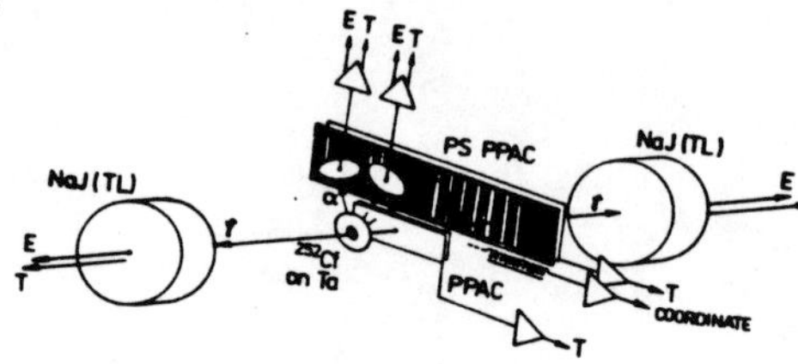

Fig.1 Detector arrangement

The strobe signal delivered a small transmission PPAC arranged in the vicinity of the ^{252}Cf source. The TOF of the fission fragments was measured between this counter and the position sensitive PPAC. A minimum distance of 6 cm between the transmission counters and the ps PPAC was sufficient for a separation of the light and the heavy groups. The fast anode signal of the photomultiplier FEU 110 coupled to the NaJ crystals was used as a stop signal for the TOF of the γ-rays and the fission neutrons. A distance of 50 cm was sufficient to separate the peak of the prompt γ-rays from the neutron distribution. The lower threshold for the spectroscopic signals from both NaJ detectors was set at 100 keV γ-response amplitude.

The activity of the ^{252}Cf source was about 10^4 fission acts per second. The ternary/binary reduction factor (including solid angles and efficiencies) for our set-up was 5.8×10^{-4}. The resulting coincidence counting rate (fission fragment x γ x LCP) amounted to 1.3 events per min.. In order to get the necessary statistics a continuous run of 41 days was required. Systematic errors were minimized by additional measurements of the angular distributions of γ-rays and neutrons in the binary fission, which were carried out under nearly the same conditions. We want to remark that the source of the γ-emission (light or heavy fragment) cannot be defined in this experiment, but we represent the angular distributions in the conventional way with respect to the direction of the light fragment.

3. EXPERIMENTAL RESULTS

The deexcitation process of the fragments by γ-rays seems to be quite similar in the binary and the ternary fission mode. This is supported by a comparison of the γ-ray spectra shown in fig.2. Fig.3 gives the probability of γ-emission under 45 deg. to the fragment direction as a function of the fragment TOF. The upper lines give raw mass scales obtained from the measured velocity distribution after corrections for the stopping power.

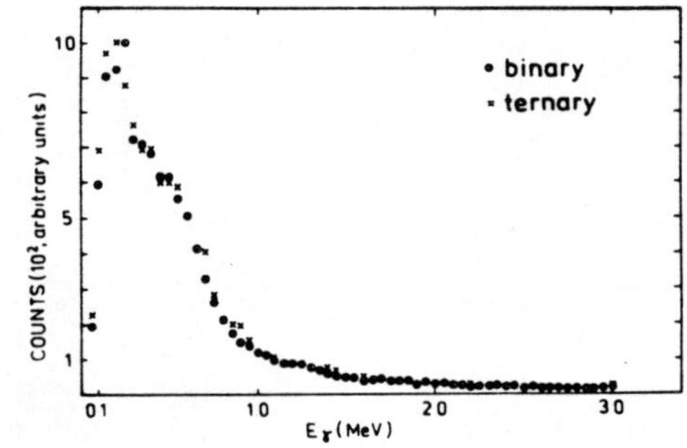

Fig.2 Binary and ternary γ-ray spectra

No striking differences could be found in the results in backward and forward geometry.

Fig.4 shows the angular distributions of prompt γ-rays for 6 bins of the γ-ray energy. The distributions are corrected for multiple scattering in the surrounding materials by using the result of a measurement with a shadow cone. Each angular interval is normalized by the geometrical efficiency of the ps PPAC measured in coincidence with the ternary α-particles. Disturbing effects as contamination and accidental coincidences were found less than 0.6% and 4%, respectively. The evaluated response matrix of the NaJ(Tl) crystals was taken into account in the shown angular distributions /6/.

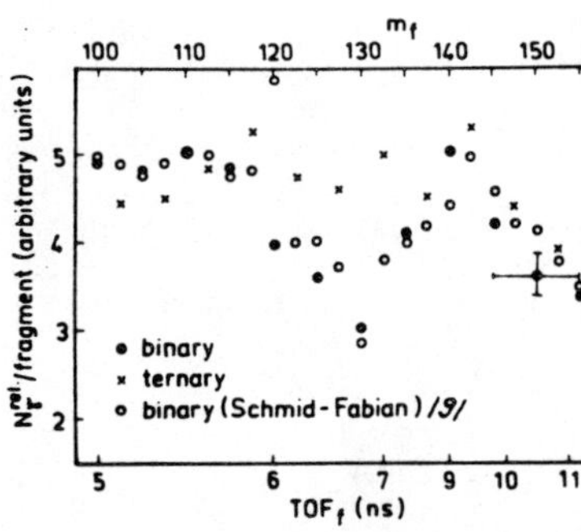

Fig.3 Relative probabilities of γ-emission

The anisotropies (maximum-minimum ratio of the angular distributions) as function of the γ-ray energy are also quite similar in the binary and ternary case /7/. The only measured 0 deg./90 deg. asymmetry amounts to be 1.015 /8/. This value is integrated over all γ-ray energies. We found a similar ratio which is 1.03 if the NaJ(Tl) detector response was not taken into account. The most interesting feature of the angular distributions of "ternary" γ-rays is the maximum at 30 deg. which shifts to about 20 deg. depending on the γ-ray energy.

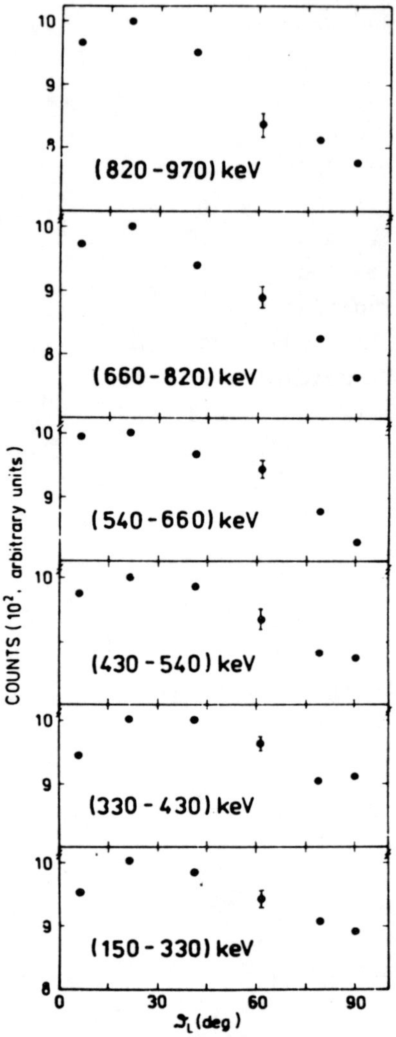

Fig.4 Angular distribution of γ-rays in ternary fission

REFERENCES

/1/ R.B. Leachman, Proc. 2. Int.Conf. on the peaceful uses of atomic energy, 15, Geneva (1958) 665

/2/ G.A. Petrov, Pisma ZhETF 6(12) (1964) 2064

/3/ V.M. Strutinsky, Pisma ZhETF 3(9) (1959) 861

/4/ A. Wolf, E. Cheifetz, Phys. Rev. C13 (1976) 1952

/5/ K. Skarsvåg, Phys. Rev. C22 (1980) 638

/6/ D.Lucas, H.Märten, W. Neubert, W. Pilz, Annual Report ZfK-621 (1986) 134

/7/ W. Pilz et al., Proc. JINR Dubna, D7-88-299 (1988) 59

/8/ O.I. Ivanov, Yad. Fiz. 15 (1972) 1121

/9/ R. Schmid-Fabian, graduation, Heidelberg (1988) 91

VI. SPONTANEOUS FISSION AND DECAY

NUCLEAR DEFORMATION, CLUSTER-STRUCTURE, FISSION AND CLUSTER RADIOACTIVITY

A UNIFYING POINT OF VIEW

R. Herrmann, K. Depta, D. Schnabel, H. Klein, W. Renner, D.N. Poenaru[1], A. Sandulescu[1], J.A. Maruhn, and W. Greiner[2]
Institut für Theoretische Physik der Goethe-Universität, D-6000 Frankfurt am Main, Germany

Induced fission [1]and spontaneous fission [2]were discovered long after α-particle decay [3]of the nucleus. The similarity between fission and α-particle decay was recognized and emphasized only a few years ago [4,5,19]. In fact, the theories of these two phenomena have been developed on essentially different grounds. Microcopic methods have been used in the theory of α-decay, but for many years a phenomenological liquid-drop-model (LDM) dominated fission theory.
The first attempt to consider both the collective nature of nucleonic as well as the single-particle effects by adding the shell corrections to the LDM-energy led to a good estimate of nuclear ground state deformations. The developement of the Two-Centre-Shell Model (TCSM) and its extension to asymmetric breakups (ATCSM) allowed within the framework of fragmentation theory a successful description of both regions of low and high mass asymmetry. The ATCSM is fundamental to all fusion and fission processes of nuclei.
The classical theoretical paper of Sandulescu, Poenaru and Greiner entitled "New type of decay of heavy nuclei intermediate between fission and α-decay" [6]initiated the field of cluster radioactivity in nuclei. Four years after this prediction experimental evidence for one of these new decay modes, namely ^{14}C radioactivity of ^{223}Ra was published by Rose and Jones [7]and later confirmed in Moscow [8], Orsay [9], Berkeley-Geneva [10]and Argonne [11]. Meanwhile Ogloblin and others have observed Ne, Si and even Mg radioactivity of various nuclei, all in agreement with theoretical predictions [6].
The discovery of bimodal fission by Hulet and his collaborators Schädel and Sümmerer and others [16]can also be naturally explained in fragmentation theory. This theory makes it possible to describe the breakup of nuclei from small to large asymmetries up to α-decay and also the inverse processes, i.e. fusion. Especially the cold fission and fusion valleys recently utilized by Münzenberg, Hofmann et al. for producing the pre-superheavy elements 106-109 have all been

[1]Permanent adress at Central Institute of Physics, R-79600 Bucharest, Romania

[2]Invited Speaker at the International Symposium on Nuclear fission, Dresden, November 1988

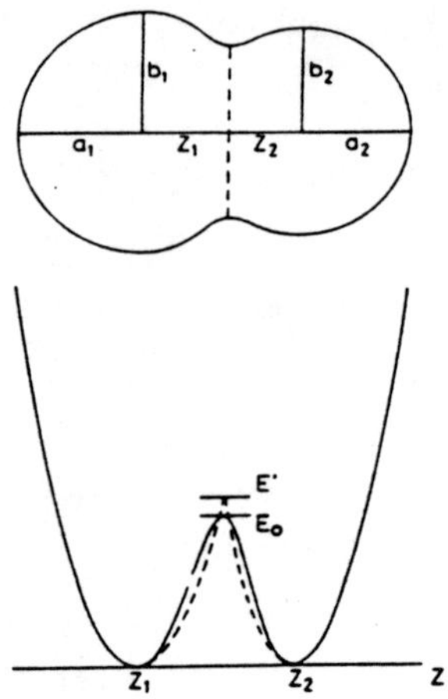

Figure 1: *Definition of the nuclear shape and the corresponding Asymmetric Two Center Shell Model (ATCSM) along the z-axis connecting the nuclear centres at* z_1 *and* z_2*. The neckparameter* $\varepsilon = \frac{E'}{E_0}$ *where* E' *is the barrier height of the two centre oscillator and* E_0 *the height of the smothed barrier, allows for a description of the necking-in of the shapes.*

predicted on the basis of fragmentation theory by Sandulescu and Greiner as early as 1976 [17].

Elements of Fragmentation Theory

The theory of fragmentation permits the treatment of two body and many body breakup channels in fission, fusion and heavy ion scattering. The unifying aspect is that the nuclear system transforms and disintegrates via collective mass transfer.

As the collective description of a nuclear system has been proven extremely successful the basic idea is the introduction of collective coordinates such as (see fig.1-3)

the mass asymmetry $\eta_A = \frac{A_1 - A_2}{A_1 + A_2}$,

the charge asymmetry $\eta_Z = \frac{Z_1 - Z_2}{Z_1 + Z_2}$,

the distance between the centers $R = z_1 + z_2$

or the total elongation $l = z_1 + z_2 + a_1 + a_2$,

the neck-parameter $\varepsilon = \frac{E'}{E_0}$

or the ellipsoidal deformations $\beta_i = \frac{a_i}{b_i}$.

For a heated nucleus in addition an excitation energy E^* or a nuclear temperature Θ is introduced.

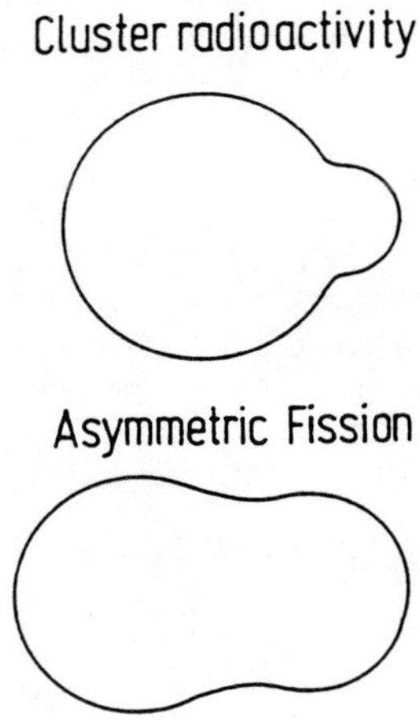

Figure 2: *Typical shapes required to describe the sticking-out of the emitted cluster in cluster radioactivity (top) and configuration for asymmetric fission processes (bottom).*

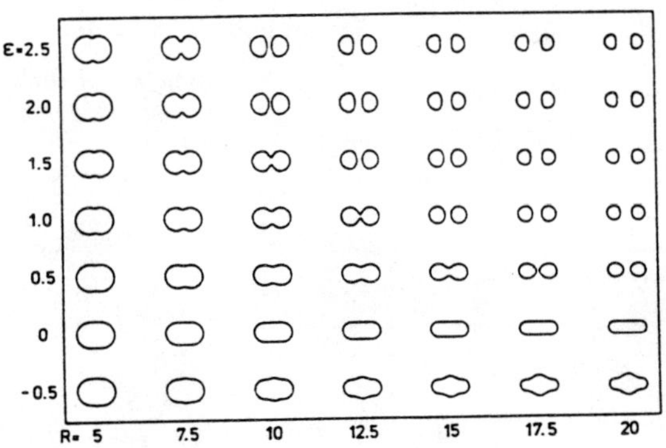

Figure 3: *Family of shapes in the R-ε-plane for the symmetric* ^{258}Fm *fission* $(\eta = 0, \beta_1 = \beta_2 = 1)$

$$E^* = \frac{A}{10}\Theta^2 \tag{1}$$

Quantization of the classical Hamiltonian

$$H = \frac{1}{2}\sum_{i,j} B_{x_i x_j}(x)\dot{x}_i\dot{x}_j + V(x) \tag{2}$$

$$x = \{R, \eta, \eta_Z, \varepsilon, \beta_i, \ldots\}$$

leads to the collective Schrödinger-equation

$$H = \left\{-\frac{\hbar^2}{2}\frac{1}{\sqrt{B}}\sum_{i,j}\partial_i\left(B_{x_i x_j}(x)\right)^{-1}\sqrt{B}\partial_j + V(x)\right\}\psi = E\psi \tag{3}$$

with the collective potential

$$V(x) = E_{macro} + (\delta U(x) + \delta P(x))e^{-\frac{\Theta^2}{\Theta_0^2}} \tag{4}$$

which consists of the macroscopic part E_{macro}, based on the LDM or more sophisticated, the droplet- or Yukawa-plus-exponential (YPE) models and a microscopic contribution based on a single particle shell model, e.g. the ATCSM. $B_{x_i x_j}$ denotes the collective mass tensor whose elements are deduced from irrotational flow models e.g. Werner-Wheeler-Method or from the single-particle levels as within the Cranking model. $\left(B_{x_i x_j}\right)^{-1}$ is the inverted mass tensor, B its determinant.

For fission processes the problem is reduced by minimizing the potential with respect to $\beta_i, \ldots$. For a certain R and taking Θ as a parameter in the adiabatic approximation one obtains wavefunctions $\psi_k^{(R)}(\eta)$ as solution of the Schrödinger-equation

$$\left\{-\frac{\hbar^2}{2\sqrt{B_{\eta\eta}}}\frac{\partial}{\partial_\eta}\frac{1}{\sqrt{B_{\eta\eta}}}\frac{\partial}{\partial_\eta} + V(R,\eta,\Theta)\right\}\psi_k^{(R)}(\eta,\Theta) = E_k\psi_k^{(R)}(\eta,\Theta) \tag{5}$$

The fission-fragment mass yield normalized to 200% is given by

$$Y(A_2) = |\Psi_k^{(R)}(\eta,\Theta)|^2\sqrt{B_{\eta\eta}}\frac{400}{A} \tag{6}$$

where

$$|\Psi_k^{(R)}(\eta,\Theta)|^2 = \frac{\sum_{k=0}^{\infty}|\psi_k^{(R)}(\eta,\Theta)|^2 e^{-\frac{E_k}{\Theta}}}{\sum_{k=0}^{\infty} e^{-\frac{E_k}{\Theta}}} \tag{7}$$

expresses the Boltzmann-like occupation of the excited states. Hence the collective (coherent) mass transfer in fission is described by wavefunctions $\psi_k^{(R)}(\eta.\Theta)$ and yields straightforwardly

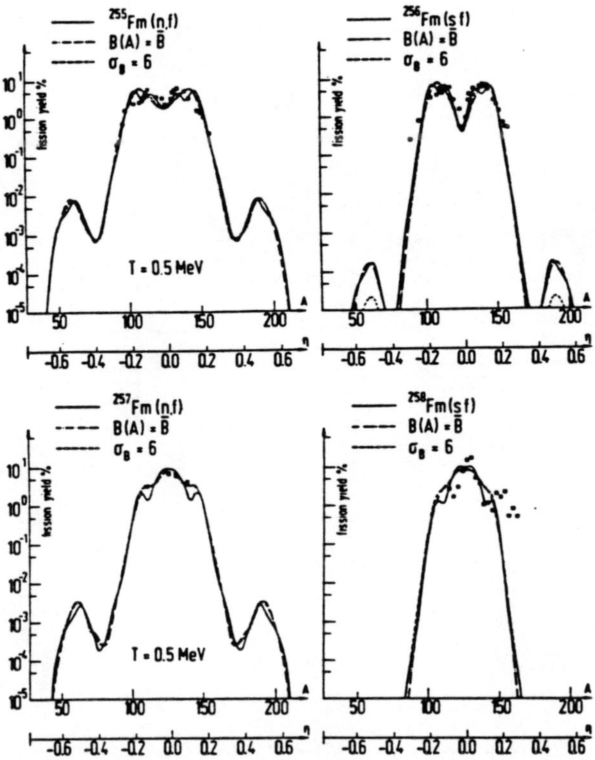

Figure 4: *Mass yields for various Fermium isotopes using* $B_{\eta\eta}(\eta)$ *(full curves), the averaged mass (B, chain curves) and the smoothed masses (broken curves). The experimental data are denoted by points (see ref. [14]).*

the mass distribution. Other distributions (e.g. charge distributions, neck vibrations, ...) may be calculated similarly.

The example of Fm-fission, see fig. 4 , shows the dominant influence of shell effects. A transition from asymmetric fission for the lighter isotopes like ^{255}Fm to pure symmetric for the heavier ones like ^{260}Fm was confirmed by the experimental data of Hoffman et al. [15]. It is important to stress, that the dominant symmetric decay of ^{258}Fm, measured by Hulet et al. [16]is caused not only by the behaviour of the LDM, but by the double magic shell closure of the $^{132}_{50}Sn_{82}$ fragment. This is a consequence of the fact that magic configurations in the potential energy surface (PES) correspond to steep valleys in the potential and therefore to enhanced fission yields at this position. The superasymmetric maxima in figure 4 near $\eta = 0.5$ correspond to ^{48}Ca or ^{208}Pb superasymmetric fission events for $Z = 20$ or $Z = 82$ respectively. This is a rather general phenomenon also predicted e.g. for the fission of No- and Cf- isotopes. We urge

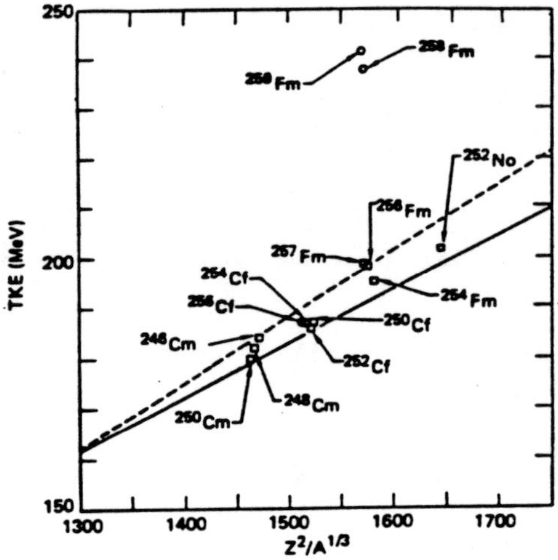

Figure 5: *Correlation of the averaged TKE with the Coulomb energy parameter $Z^2/A^{\frac{1}{3}}$ (see ref. [8]).*

to search for it!

In fact, the charge and mass distributions obtained with Lohengrin and reported by Micheldauer and Gönnewein at this conference, seem to give first indications for superasymmetric fission. For pre-actinide nuclei, the superasymmetric fission is to be expected in the region $Z = 50$, $N = 82$ which forms ${}^{132}_{50}Sn_{82}$. Experimental results of Itkis et al. [20]show indeed pronounced structure of the fission yield at the same position.

Abnormal kinetic energy of fission products

The experimental data by Hulet et al. [16]show once again the importance of shell structures in the fission process. The measured mass- and kinetic energy -distributions in the spontaneous fission of ${}^{258}Fm$, ${}^{259}Md$, ${}^{260}Md$, ${}^{258}No$ and ${}^{260}Ku$ cannot be explained within the LDM only. The already mentioned change from asymmetric to symmetric mass division near ${}^{258}Fm$ is accompanied by a sharp increase of the most probable total kinetic energy (TKE) of the fragments, see fig.5. The distributions of the TKE deviate strongly from the simple Gaussian form. In fact they can be decomposed into two Gaussian distributions, see fig. 6. These two components are peaked near either 200 or 235 MeV. The Gaussian component at higher energies corresponds to the value of ${}^{258}Fm$ in the systematics.

The results of Itkis et al. [20], see fig.7, who studied the fragment mass and energy distributions for fission in the pre-actinide region, indicate the necessity of an decomposition even up

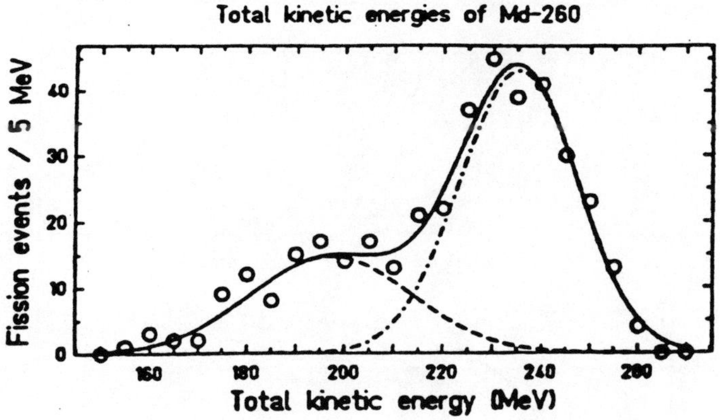

Figure 6: *Distributions of the TKE of correlated fission fragments and their suggested decomposition into two Gaussians in the case of ^{260}Mo (see ref. [16]).*

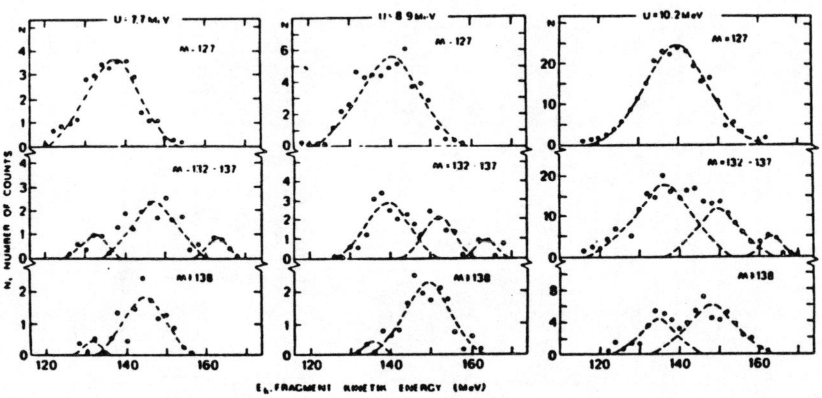

Figure 7: *The fission-fragment kinetic energy-spectra, $N(E_k)$ for ^{213}At given for various mass intervals at three values of the excitation energy U. The dashed lines show the decomposition of $N(E_k)$. (from ref [20])*

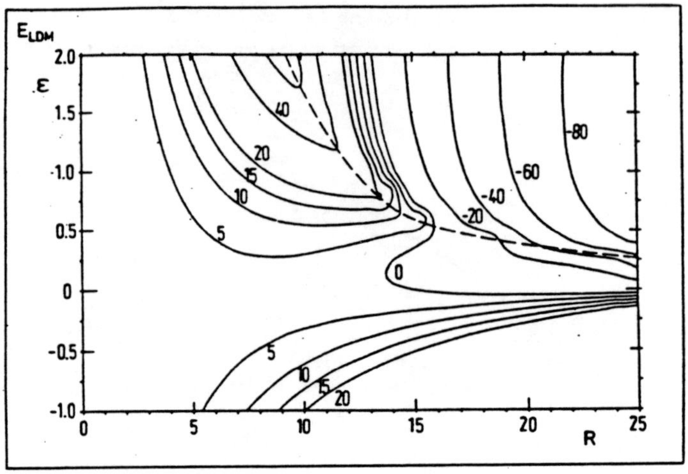

Figure 8: *The liquid drop potential energy surface E_{LDM} for the symmetric ^{258}Fm fission depending on the relative coordinate R and the neck parameter ε. The dashed curve indicates the scission line.*

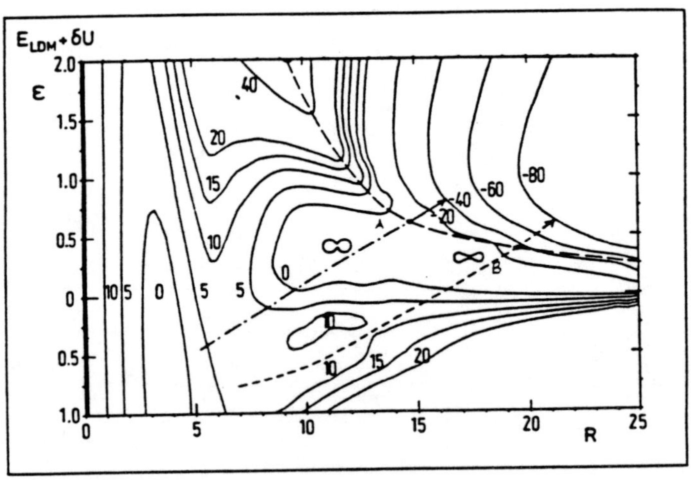

Figure 9: *The total potential energy as a sum of the liquid drop E_{LDM} and the shell corrections δU depending on R and ε for symmetric division of ^{258}Fm. The dashed curve indicates, as in fig. 8, the scission line, the two paths correspond to the possible decay channels.*

two three gaussians to describe the kinetic-energy spectra.

This decomposition can be understood as the result of structures in the PES. For ^{258}Fm the mass distribution is symmetric ($\eta = 0$). In this case, the fission problem may simply formulated in two-dimensions,e.g. the relative distance R and the neckparameter ε. The ellipsoidal deformations can be fixed, thinking of the double magic spherical Sn-nucleus.

As we can see from the family of shapes in the R-ε-plane in fig. 3, the neck parameter describes a neck-in for positive values and an additional belly in the central region of the shape for negative values. This is a direct consequence of the strong hexadecupole-dominated influence of ε on the deformation of the nucleus.

As we can read off from fig. 8, the liquid-drop part of the PES is not enough to describe a double-humped TKE-distribution. According to the liquid-drop-PES alone, the fissioning system can have just one gaussian TKE-distribution.

Adding to the liquid-drop part the corresponding shell-corrections the resulting total PES offers the system two possible decay channels divided by a ridge. Following these two different paths (sketched in fig. 9) the nucleus fissions at two different points (A,B). A corresponds to two touching spheres, while B developes a very elongated shape. Dominated by the Coulomb-part of the energy of these two different shapes, there result different kinetic energies in the exit channel, as measured by the experiment. It is worth to mention, that along the paths for intermediate states also the negative region of ε is touched. But already this simple static picture gives a natural explanation of the experimental data. We already mention here that the decay into spherical fragments might be interpreted as a radioactive decay into two big clusters.

Geometric interpretation of the second minimum

It would be helpful, to get a physical understanding of these shell-structures, which are responsible for the new observed minima in the PES near fermium.

The interpretation of fission-yields for example is based on the fact, that local minima as a function of mass asymmetry are caused by magic projectile-target combinations in the exit channel. It is essential, that the positions of these local minima, which are fixed asymptotocally for separated configurations, vary only smothly as a function of elongation even for connected fragments (see fig. 10 and 11). In other words, the asymptotic ($l \to \infty$) shell structure dominates the potential up to the top of the barrier. This is the reason for the existence of "cold" valleys, i.e. valleys in the collective potential reaching from the touching configurations inwards up to

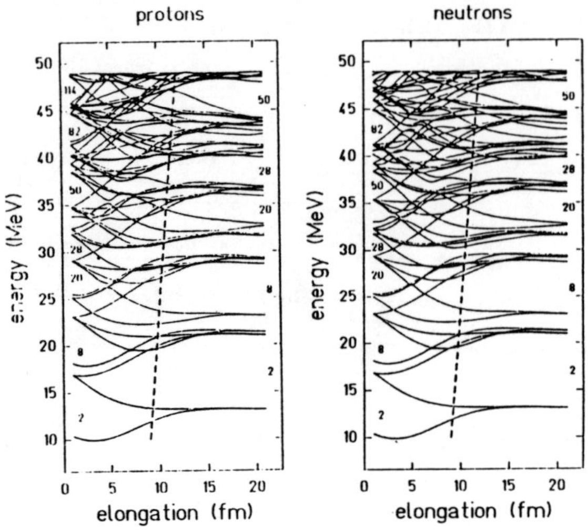

Figure 10: *Typical single particle spectrum of the ATCSM as a function of the elongation R. The dashed line indicates the onset of the asymptotic shell structure. It coincides approximately with the top of the potential barrier.*

the top of the barrier (approximately). These valleys are most important for the cold fission and cold fusion processes.

Obviously the cold valleys are already revealed by the two-centre level structure as calculated in the early 70's by Maruhn and Greiner and others of the Frankfurt School.

An important result is the prediction of proper projectile and target combinations for the production of superheavy elements. The method is based on the idea, to choose the reaction partners such that the compound system is formed with a minimum excitation energy. The theoretically predicted optimum combinations [12,17,21]correspond exactly the experimentally sythesized isotopes Z=102,104,106 with *Pb* as target nucleus (see references given in [17]) and allow for prediction of an optimum choice of reaction partners for an optimum cross section for synthesis of new isotopes $Z \geq 108$ (see fig. 12 and [22]).

We intend to apply this sucessful ideas onto the interpretation of the Fm-PES. To calculate the PES for ^{258}Fm in the R,ε-plane, the two-centre mass asymmetry was fixed ($\eta = 0$). Hence up to first order we would not exspect any structures. But there is a relation between negative/positive neck parameter ε in the Asymmetric Two Centre Shell Model and a positive/negative neck parameter ε in the Asymmetric Three Centre Shell Model [14]. Figure 13a shows a typical nuclear shape in the ATCSM with positive ε, the hatched area represents the inner region between the two centres z_1 and z_2. If we fix the total length L of the shape and

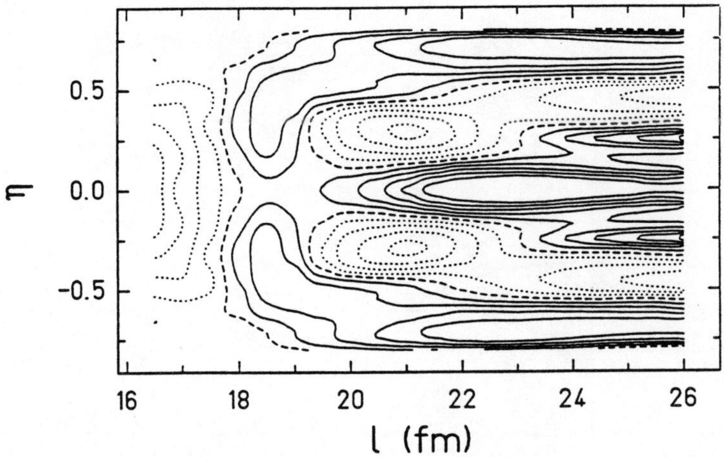

Figure 11: *Shell-corrections δU for the superheavy element ${}^{298}_{114}X$ as a function of the total length l and the mass asymmetry η. The solid lines denote $\delta U > 0$. For dashed lines the shell correction is negative $\delta U < 0$. This indicates the position of local minima in the PES for magic η. The scission configuration for $\eta = 0$ lies at $l_c \approx 24.3$ [fm]. Obviously the corrections change only smoothly for $l < l_c$ (connected configurations).*

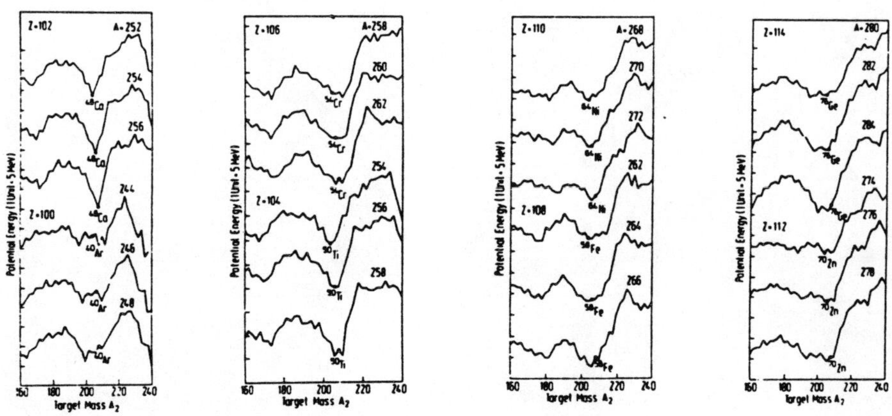

Figure 12: *Potential energy as a function of the target mass A_2 at the distance of closest approach for various compound nuclear masses (A) with a) Z=100-102, b) Z=104-106, c) Z=110 and d) Z=112-114. The projectile nuclei relevant to Pb-targets are also shown.*

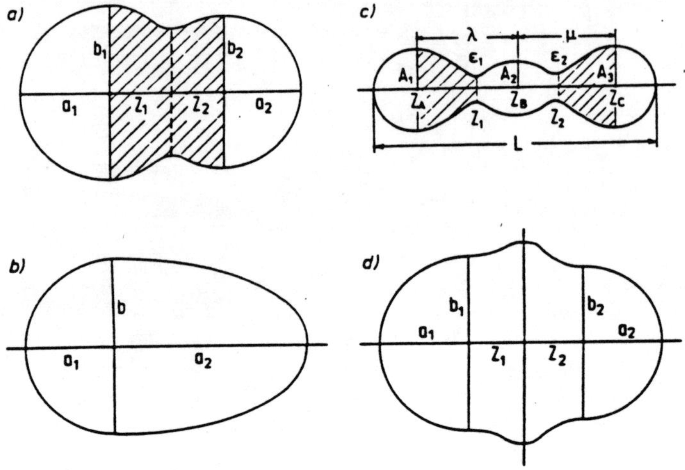

Figure 13: *Nuclear shapes in the two (a) and three (c) center model with the corresponding limiting cases of one center (b) and two center configurations with negative neck parameter (d).*

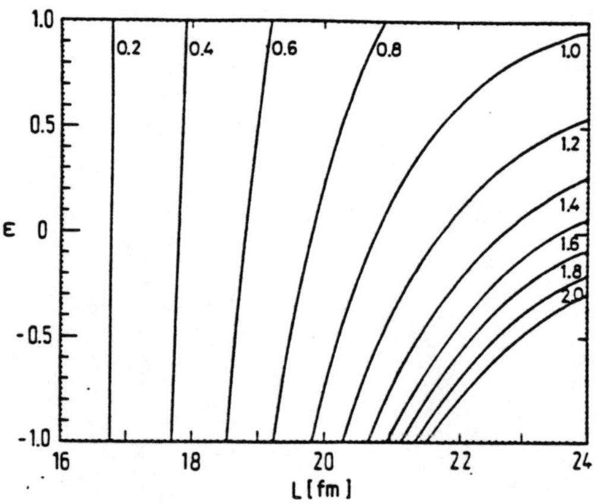

Figure 14: *Contour map of the three center mass asymmetry η_3 for symmetrical two center configurations of ${}^{264}Fm$, depending on the total elongation of the system L and the neck parameter ε. The shapes look similar to fig. 3.*

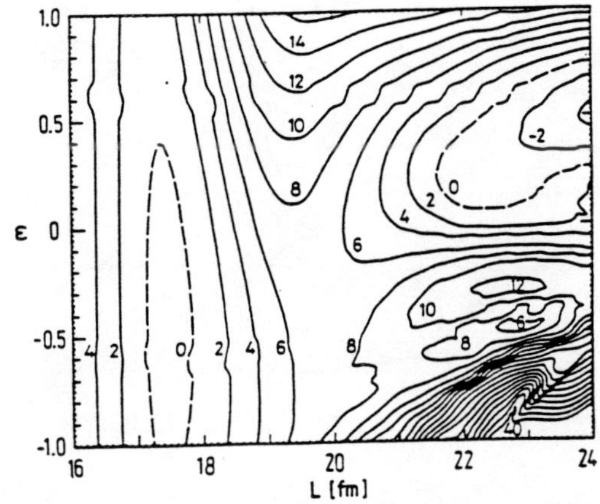

Figure 15: *Potential energy surface (PES) of the total potential of ^{264}Fm, in the L- ε-plane, same parameters as in fig. 9.*

look at the limit of vanishing centre distance $R = z_1 - z_2$, we obtain a one centre shape (see fig. 13b).

Applying the same procedure to a three centre shape with positive neck (see fig. 13c), i.e. looking for the limit of vanishing $\Delta z_A = z_A - z_1$ and $\Delta z_C = z_C - z_2$ we get a two centre shape with negative neck parameter ε (see fig. 13d). This simple description gives the linkage between two- and three-centre configurations and allows for an interpretation of a given two centre case as a three centre configuration as well.

In analogy to the two centre mass asymmetry η we can define for a symmetric two centre configuration a three centre mass asymmetry η_3 by the ratio of the central to the left mass, $\eta_3 = \frac{A_2}{A_1}$. Figure 14 shows a contour map of η_3 in the L-ε-plane for ^{264}Fm.

The corresponding family of shapes looks similar to fig.3, i.e. we have symmetric two centre configurations too. Looking now at the magic three centre configurations for ^{264}Fm given in table 1 we recognize that different three body breakups with magic proton or neutron numbers have different η_3 values in general, but in some cases the magic numbers accumulate at the some η_3 values, which causes a coherent addition of the shell correction contribution to the total potential energy. For example $Z = 20, 40$ and $N = 28, 40$ result in $\eta_3 \approx 0.5$, other magic shell closures lead to $\eta_3 \approx 1.3$, 2.0 and 6.4.

A given magic η_3 in the L-ε-plane corresponds to a magic three centre configuration and causes minima in the PES of ^{264}Fm (see fig. 15). The ridge dividing the two minima is sourrounded

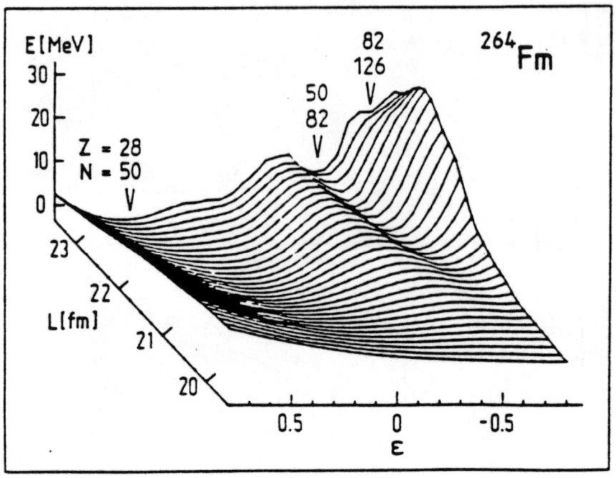

Figure 16: *Three dimensional graph of the PES from fig. 15, indicating the magic shell closures responsible for the minima.*

(see fig. 14) by the magic η_3 lines for 1.3 and 2.0. They indicate the possible fission paths similar to the ^{258}Fm case. Even the ground state properties at $\eta_3 \approx 0.5$ are described surprisingly well.

This interpretation of the minima in the PES connects binary and ternary fission, giving us a better insight to fission phenomena. Another hint is visible in fig. 16: A third minimum belonging to a three body breakup with $Z = 82$ and $N = 126, 20$ appears in addition to the already known minima. This third minimum located at a position where the strongly increasing liquid drop energy dominates, opens a different ternary channel. This suports the existence of a third gaussian in the TKE spectra or may be interpreted as a possible direct path to three centre fragmentation above this three centre saddle point. So nuclei near ^{264}Fm may not only serve as examples for binary fission, but in addition may show interesting properties in ternary fission as well.

Mass distributions for bimodal fission

Besides the kinetic energies of the fragments, the corresponding mass distributions have also been measured [16]. It was found that both distributions are strongly correlated. The total mass distribution is symmetric and can be decomposed into two parts: One part belonging to the high kinetic energy component having a sharp mass distribution, while the distribution belonging to the low energy branch is a broad one. In order to understand this phenomenon

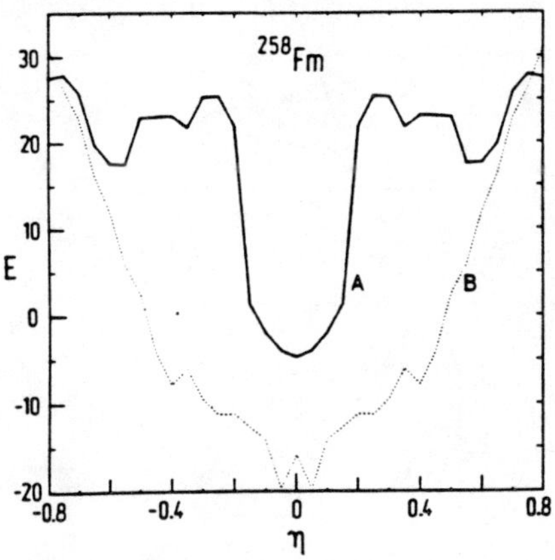

Figure 17: *Dependence of the ^{258}Fm potential on the mass asymmetry degree of freedom at the two points (A,B) on the scission line of fig.8. The curve A corresponds to two spherical fragments, while at B a long neck is developed.*

theoretically, let us consider the dependence of the potential on the mass asymmetry degree of freedom.

Looking at fig. 17, this dependence is shown at the possible scission points A and B (see fig.8). We recognize that the potential as a function of η for the light energy component A is steeper than the potential for B, belonging to the low energy component.

Since the mass distribution at A and B reflects this behaviour – thinking of our previous results – this explains semi-quantitatively also the mass distributions.

A collective description of asymmetric nuclear shapes

In order to describe the unexspected strong preformation of clusters such as ^{12}C or ^{24}Ne in heavy elements leading to cluster radioactivity, or the existence of static octupole or even dipole deformations, we have to extend the standard collective potential energy-methods, which are applied in fission theory, to very asymmetric shapes.

Fig. 18 illustrates the basic geometrical properties of the new model [18]:

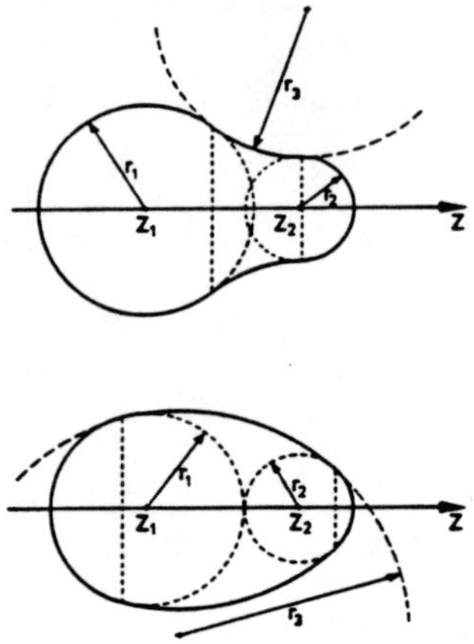

Figure 18: *Two possible shapes of the Three–Spheres–Model. The first one corresponds to a positive neck parameter* $c_3 = \frac{1}{r_3}$ *having positive curvature in the intermediate region. The lower one has negative curvature describing ellipsoidal deformations with negative neck parameter.*

Table 1: *Magic three centre configurations for* ^{264}Fm *and the according three centre mass asymmetry* η_3. *The most frequent configurations are collected in the prediction using* η_3.

Protons:			η_3
magic outer nucleus			
50	—	50	0.00
40	20	40	0.50
28	44	28	1.57
20	60	20	3.00
magic inner nucleus			
9	82	9	9.11
25	50	25	2.00
30	40	30	1.33
36	28	36	0.77
40	20	40	0.50
Neutrons:			η_3
magic outer nucleus			
82	—	82	0.00
50	64	50	1.28
40	84	40	2.10
28	108	28	3,85
20	124	20	6.20
magic inner nucleus			
19	126	19	6.63
41	82	41	2.00
57	50	57	0.87
62	40	62	0.64
68	28	68	0.41
72	20	72	0.27

prediction for strong shell effects:		
η_3	Z	N
0.5	20,40	28,40
1.3	28,40	50
2.0	50	40,82
6.4	(82)	20,126

An arbitrary nuclear shape consists of two spherical shapes at distance $\Delta z = z_2 - z_1$, smoothly joined by a neck with curvature $c_3 = \frac{1}{r_3}$, r_3 beeing the radius of a third "rolling" sphere. The shapes in this Three-Spheres-Model (TSM) are axially symmetric and the new neck parameter describes, like ϵ, the necking-in as well as ellipsoidal deformations for negative c_3.

The ATCSM, which until now has been used to describe fission processes, already contains odd-multipole moments due to the use of the mass asymmetry coordinate η. For small separations and very large asymmetries this model had to be generalized to allow for a sticking out of the small nucleus.

If we use the center-separation R and the asymmetry η as primary collective coordinates the octupole moment can be computed by use of a projection technique

$$Q_{30}(R,\eta) = \sqrt{\frac{7}{4\pi}} \int Y_{30}(\theta,\varphi) r^3 \rho(\vec{r}, R, \eta) d^3r \tag{8}$$

and similarly the dipole moment is given by

$$Q_{10}(R,\eta) = \sqrt{\frac{3}{4\pi}} \int z \rho(\vec{r}, R, \eta) d^3r \tag{9}$$

where $\rho(\vec{r}, R, \eta)$ is the charge density assumed in our model.

Consequently the PES calculated according to eq.(4) contains multipole-moments for each shape considered in the collective space. Near the ground state the lowest moments are given approximately by:

$$Q_{10} \approx 0 \tag{10}$$

$$Q_{20} \approx \left(\frac{2R}{R_0} - 1\right) z R^2 \tag{11}$$

$$Q_{30} \approx \eta \left(\frac{2R}{R_0} - 1\right) z R^3 \tag{12}$$

where $R_0 = r_0 A^{\frac{1}{3}}$.

The dipole moment vanishes, because of the assumption $\eta = \eta_Z$, i.e. equal mass/proton ratio. If this assumption is relaxed or clusters with unequal charge-mass-ratio are included, dipole moments do appear. This is what the Yale-group (M.Gai et al.) do observe.

As an example, fig. 19 shows a PES for ^{224}Ra. This is only a semirealistic calculation because as mentioned before the ATCSM does not yet describe extremely asymmetric shapes reasonable. However, it should be sufficient for pointing out some features in this collective approach. The major structure in this PES is the broad minimum near the groundstate which determines

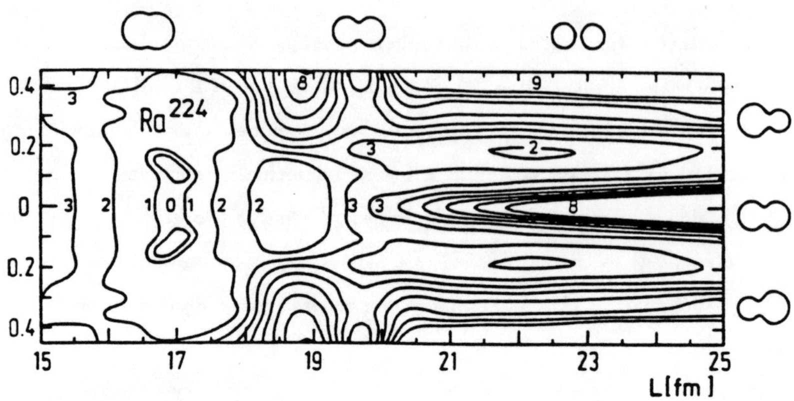

Figure 19: *Minimized PES of a typical heavy nucleus as a function of elongation L and asymmetry η. Embedded are asymmetric local minima which indicate nonvanishing static octupoledeformation, while for larger L-values the valleys crossing the fission barrier appear.*

the observable lowest odd multipole moments. Let $\psi_0(R,\eta)$ be the ground state wavefunction located in the vicinity of $R \approx R_0$ and spread out in the fragmentation degrees of freedom. The observable groundstate octupole moment will be given by an average over the local moments

$$< Q_{30} >_0 = \int \psi_0^*(R,\eta) Q_{30}(R,\eta) \psi_0(R,\eta) d\tau \tag{13}$$

with the volume element $d\tau = \sqrt{B_{RR}B_{\eta\eta} - B_{R\eta}^2} dR d\eta$.

A nonvanishing effective octupole moment $< Q_{30} >_0$ does not require an octupole deformed equilibrium deformation: Softness in the octupole direction η is completely sufficient.

Another interesting feature of the groundstate wavefunction follows from the quantity

$$P_0(\eta) = \int d\tau |\psi_0^*(R,\eta)|^2 \tag{14}$$

Obviously one might interprete $P_0(\eta)$ as the preformation probability for a cluster configuration characterized by η.

This quantity is most important in the traditional calculation of radioactive decay by letting preformed clusters tunnel through a barrier (Gamov-penetrability). In the Gamov-theory one assumes that at a certain preformation of clusters, i.e. at a certain asymmetry $\eta \approx \eta_0$ the tunneling process appears: The cluster is tunneling through the potential keeping $\eta \approx \eta_0$. This is of course, very questionable. Indeed all preformations in the groundstate (each η asymmetry) can contribute to a final fragmentation η at scission, i.e. to a final break-up. In particular the

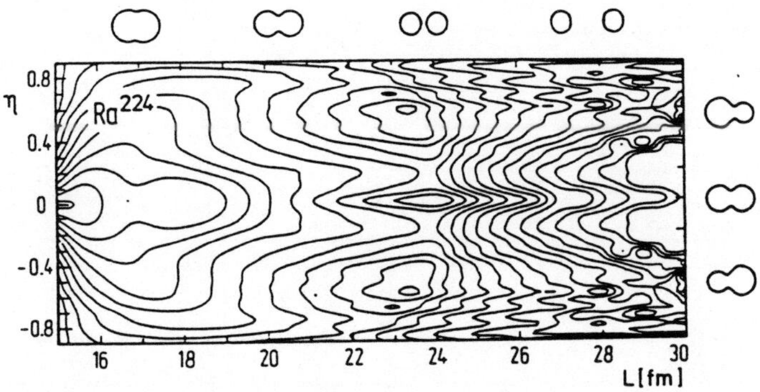

Figure 20: *PES for ^{224}Ra as a function of L and η for fixed $\varepsilon = \beta_1 = \beta_2 = 1$*

asymmetric valley has also an influence on the fission (emission) probability. Softness along a particular valley in the surface will lead to an enhanced probability density at that point and yield a larger preformation probability for this specific breakup. To show to what extent valleys in the PES determine the dynamics we show in fig. 20 a PES for ^{224}Ra for fixed $\varepsilon = \beta_i = 1$. This PES contains all information for the fissioning system from the ground state up to and beyond the scission point. Corresponding fission (emission) paths leading to various final asymmetries are plotted in fig. 21. These are determined by minimizing the WKB-integral [19]:

$$P = \int \left(\frac{2\,(V(R,\eta) - E)}{\hbar^2} \sum_{i,j} B_{ij} \frac{\partial x^i}{\partial s} \frac{\partial x^j}{\partial s} \right)^{\frac{1}{2}} ds \tag{15}$$

over all collective paths. The mass parameters B_{ij} were determined from the irrotational flow model with the Werner-Wheeler method.

Apparently the valleys in the potential cause a strong concentration of dynamic paths and the fission properties will not be given by the barrier height itself, but also by those structures in the surface closer to the ground state.

Conclusion

We have seen that various processes like ordinary fission, superasymmetric fission, bimodal fission, cold fission, cluster decay (α- and new radioactivities), groundstate preformation of clusters, groundstate multipole moments and cold fusion are all interrelated. Cold fusion and fission valleys are determined by the shell structure of the final fragments as has been emphasized already in the late 60's and early 70"s when the Two-Center-Shell-Model was developed

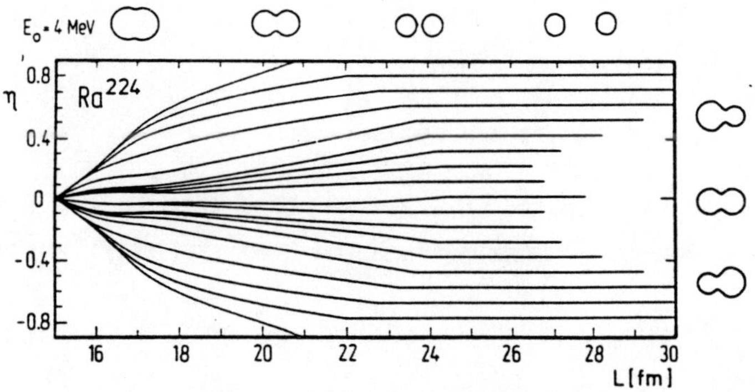

Figure 21: *Family of fission paths in the surface of fig. 20 determined by a minimization of the penetration integral eq.(15) with a fixed final asymmetry.*

by the Frankfurt school.

There is no principal difference between fission and cluster emission processes; only the final (asymptotic) valley structure of the potential energy surfaces determining these processes. Many of these features have been experimentally observed. In particular the theoretically predicted cluster readioactivity has been found and also the fusion of the pre-superheavy elements $Z = 106 - 109$ was achieved with exactly those projectile–target–combinations predicted by cold valley arguments in the mid-70's [17]. We emphasize that cold fusion, cold fission as well as cluster decay and also the high energy mode of bimodal fission proceed through "cold valleys". Fragmentation theory thus yields a unifying description of all these processes.

References

1. O. Hahn and F. Strassmann, Naturwissenschaften **27**(11), 89 (1939)
2. K. A. Petrjak and G. N. Flerov, JETP **10**, 1013 (1940)
3. H. Becquerel, Compt. Rend. 122 (1896)
4. N. Bohr and J. A. Wheeler, Phys. Rev. **56**, 426 (1939)
5. J. Frenkel, Phys. Rev. **53**, 987 (1939)

6. A. Sandulescu, D. N. Poenaru and W. Greiner, Sov. J. Part. Phys. **11**(6), 528 (1980)

7. H. J. Rose and G. A. Jones, Nature **307**, 245 (1984)

8. D. V. Alexandrov, A. F. Belyatsky, Y. A. Guhov, E. Y. Nikolsky, B. V. Novatsky, A. A. Ogoloblin and D. N. Stepanov, pis'ma JETP **40**, 152 (1984)

9. S. Gales, E. Hourani, M. Hussonnois, H. P. Shapira, L. Stab and M. Vergnes, Phys. Rev. Lett. **53**, 759 (1984)

10. P. B. Price, J. D. Stevenson, S. W. Barwick and H. L. Ravn, Phys. Rev. Lett. **54**, 297 (1985)

11. W. Kutschera, J. Ahmad, S. G. Armato III, A. M. Friedmann, J. E. Gindler, W. Henning, T. Ishii, M. Paul and K. E. Rehm, Phys. Rev. **C32**, 2036 (1985)

12. J. A. Maruhn and W. Greiner, Z. Phys. **251**, 431 (1972)

13. J. A. Maruhn and W. Greiner, Phys. Rev. **C13**, 2404 (1976)

14. H. J. Lustig, J. A. Maruhn and W. Greiner, J. Phys. G **6**, L25 (1980)

15. D. C. Hoffman et al., Phys. Rev. **C21**, 972 (1980)

16. E. K. Hulet et al., Phys. Rev. Lett. **56**, 313 (1986)

17. A. Sandulescu, R. K. Gupta, W. Scheid and W. Greiner, Phys. Lett. **60B**(3), 225 (1976)

18. K. Depta et al., Mod. Phys. Lett. **A1**, 377 (1986)

19. R. Herrmann, J. A. Maruhn and W. Greiner, J. Phys. G: Nucl. Phys. **12**, L285 (1986)

20. M.G. Itkis, V. N. Okolovich, A. Y. Rusanov and G. N. Smirenkin, Z. Phys. A **320**, 433 (1985)

21. R. K. Gupta, C. Parvulescu, A. Sandulescu and W. Greiner, Z. Phys. **A283**, 217 (1977)

22. G. Münzenberg et al., Z. Phys. **A300**, 107 (1981)

SPONTANEOUS FISSION IN A WIDE RANGE OF MASS ASYMMETRY INCLUDING HEAVY ION RADIOACTIVITIES

D.N. Poenaru[1,2], W. Greiner[2], M. Ivascu[1], and I. Ivascu[1]
[1] Central Institute of Physics, Bucharest (R-76900), P.O.B. MG6, Romania
[2] Institut für Theoretische Physik, Pf 111932, Frankfurt am Main, Germany

The results obtained within analytical superasymmetric fission model, unifying the description of heavy cluster radioactivities, alpha-decay and cold fission are resumed. Predicted halflives for C-14, Ne-24 and Mg-28 emission in the range $10^{11}-10^{25.5}$s, have been confirmed within 1.5 orders. Most probable cold fission fragments are reproduced. Action integral calculated with Werner-Wheeler inertia and Yukawa-plus-exponential potential extended for different charge densities, shows that cluster-like shapes are favored for emitted ions lighter than Si-31.

Spontaneous fission of a heavy nucleus in its ground state, predicted by N.Bohr and J.Wheeler in 1939, had been discovered by Flerov and Petrjak in 1940, shortly after induced fission (Hahn and Strassmann, 1938), but long time after alpha-decay. A real progress in the understanding of various phenomena, including fission decay of shape isomers (Polikanov et al. 1962) had been achieved in 1966, when Strutinsky introduced the shell correction method. Fragmentation theory and the asymmetric two center shell model, developed by Frankfurt school, offer new possibilities to study both fission and fusion dynamics.[1].

New kinds of natural radioactivities with mass asymmetry intermediate between fission and alpha-decay have been predicted in 1980, when the results obtained within three fission models (fragmentation theory, numerical (NSAFM) and analytical (ASAFM) superasymmetric fission models) and the traditional theory of alpha-decay extended to heavier clusters have been resumed. It was shown that even alpha emission can be considered fission process. More recently cold fission phenomena are interpreted as cluster decays. Among other emissions discussed in 1980, it was mentioned that ^{14}C should be the most probable emitted ion from $^{222,4}Ra$ and in 1984 Rose and Jones discovered ^{14}C radioactivity of ^{223}Ra. Theory and experiments have been recently reviewed [2-10]. Other groups [11-13] are sharing with us the idea of a fission process.

ASAFM, developed in 1980 and improved since then, was particularly useful in obtaining physical conclusions to guide the experiments. The metastability of every nuclide with known mass value against emission: of alpha particles, of more than 200 ions with Z<28, and of all possible light fragments with Z≥28 up to symmetrical cold fission had been studied systematically. In 1984 we had shown that the largest branching ratio relative to alpha decay ($T_\alpha/T \simeq 6\times10^{-10}$) was measured from the beginning. Highest emission rates are to be obtained for parent nuclides with Z>82 leading to daughters around ^{208}Pb, but all nuclei with Z>40, including the "stable" ones are emitters of several heavy ions. For more than 150 kinds of cluster emissions, predicted halflives of parent nuclei are shorter than 10^{23} years, and the kinetic energy of light fragments is about 2 MeV/nucleon.

Three of the new decay modes (^{14}C, ^{24}Ne and ^{28}Mg) of $^{222-4,6}$Ra, ^{230}Th, ^{231}Pa, $^{232-234}$U, have been already confirmed, by using various experimental techniques allowing to reduce the strong background of alpha particles [5-8]. For heavier clusters [9] the experiments are more difficult due to fission fragments. The measured lifetimes in the range 10^{11}-$10^{25.5}$ s, and the corresponding branching ratios (Fig. 1) agree with our early predictions within 1.5 orders of magnitude. After including an even-odd effect, the agreement was further improved. Some possible emitters of ^{32}Si and ^{34}Si are $^{238-240}$Pu nuclides (predicted branching ratios are 10^{-16} -10^{-17} [9]). Cluster radioactivities to excited states of the daughter, predicted by Martin Greiner and Werner Scheid, have so far not yet been observed.

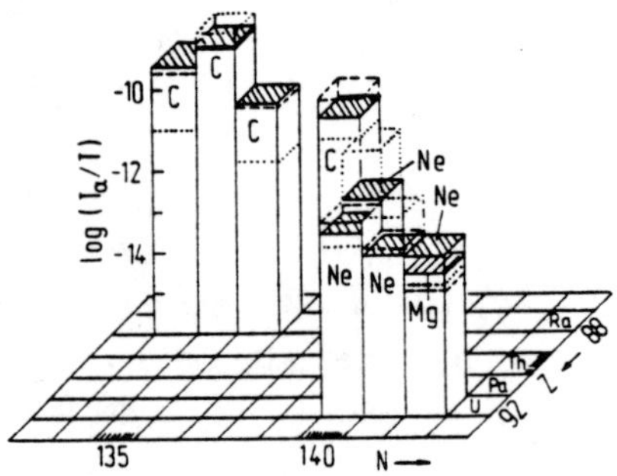

Fig. 1. Experimental branching ratios (hatched areas) compared to early ASAFM predictions (dotted lines) and after including an even-odd effect (dashed lines).

As it can be seen from Fig. 1, for the parent ^{234}U, both Ne and Mg emissions are measured [7]. This nucleus is unique, because all three groups of decay modes (alpha, clusters and cold fission) described in a unified manner within ASAFM [3], are experimentally determined. The most probable cold fission fragments observed in two regions of nuclides [14-16], are reproduced by calculations [3]. The analogue of shell stabilized nucleus ^{208}Pb, leading to highest emission rates of clusters is ^{132}Sn in cold fission with spherical shapes. Hence it is expected that this mechanism should be the strongest for ^{264}Fm. Nevertheless one has to take into account the fact that ^{208}Pb is on the line of beta-stability, but ^{132}Sn is far off on the neutron-rich side. Neither liquid drop nor shell effects of ^{132}Sn are as strong as those of ^{208}Pb. An improved description of the fission dynamics in a wide range of mass asymmetry is obtained by replacing the reduced mass with the Werner-Wheeler [17] inertia tensor, B, corrected for center of mass motion and calculating the deformation energy within Yukawa-plus-exponential model, extended by us in 1979 for different charge densities like we did within NSAFM in 1979. For two parametrizations of intersected spheres (with R_2=constant or V_2=constant) one obtains analytical formulas:

$$B(R) = B_1(R) + B_2(R) + B_c(R) \tag{1}$$

where subscripts denote the contribution of left and right fragments and the correction due to the center of mass:

$$4r_o^3B_i/3m = (z_i')^2V_i/\pi + (-1)^i 2z_i'R_iR_i'(R_i+D_i)^2 + (R_iR_i')^2[2R_i^2/H_i - 4.5R_i -$$

$$- 3.5D_i + 6R_i\ln(2R_i/H_i)] \qquad (i=1, 2) \tag{2}$$

$$4r_o^3B_c/3m = -(3/4R_o^3)I^2; \; I = I_1 + I_2 \tag{3}$$

$$I_i = z_i'V_i/\pi + (-1)^iR_iR_i'(R_i + D_i)^2 \qquad (i=1, 2) \tag{4}$$

in which r_o is the radius constant, m is the nucleon mass, R_i and z_i are the radius and center position of the sphere i, $D_1=z_s-z_1$, z_s is position of separation plane, $H_1=R_1-D_1$, $R=z_2-z_1$, $D_2=R-D_1$, $H_2=R_2-D_2$, V_i-the volume of fragment i, and the superscript prime means differentiation with respect to the shape parameter (for example R or the distance between center of mass of the fragments z_m). When the origin is in the center of mass, z_1 should be replaced by z_1-z_c and $B_c=0$; $z_1=0$ for the origin in the center of the first sphere and $z_1=-D_1$ for separation plane.

It is interesting to compare the action integrals K_j for two fission trajectories: "cluster-like" (with R_2=constant, j=1), and more compact shapes (V_2=constan j=2) leading to separated fragments

$$K_j = (2/\hbar)\int_{R_a}^{R_b}\{2B(R)[E(R) - Q]\}^{1/2}dR; \; E(R_a) = E(R_b) = Q \tag{5}$$

One can see easily that K_j is independent of the shape coordinate because $B(z)=B(R)(dR/dz)^2$. In the overlapping region the double and triple integrals [2] of E(R) are calculated by Gauss-Legendre numerical quadrature. In the range A_e=4 to 50 of all mass numbers of emitted fragments we considered the daughter ^{208}Pb, and for each A_e, the proton number Z_e giving the highest emission rate. For cold fission A_e=94 to 132 and the heavy fragment ^{132}Sn.

The results are plotted in Fig. 2, showing clearly that cluster-like shapes are preferred when the emitted cluster is relatively light (A_e<31). For heavier clusters and cold fission with smaller asymmetry the action integral along the fission trajectory with more compact shapes takes lower value.

In conclusion, we would like to stress the necessity to take into account the corrections due to the center of mass motion, when hydrodynamical masses are calculated. In spite of the strong dependence of inertia B on the shape parameter (for V_2=constant B(R) is an increasing function, but $B(z_m)$ is a decreasing one), the action integral is independent on arbitrarily chosen coordinate.

From the dynamical point of view, cluster-like shapes (R_2=constant) are favored with respect to more compact ones (V_2=constant) when the mass number of emitted ion is less than 31.

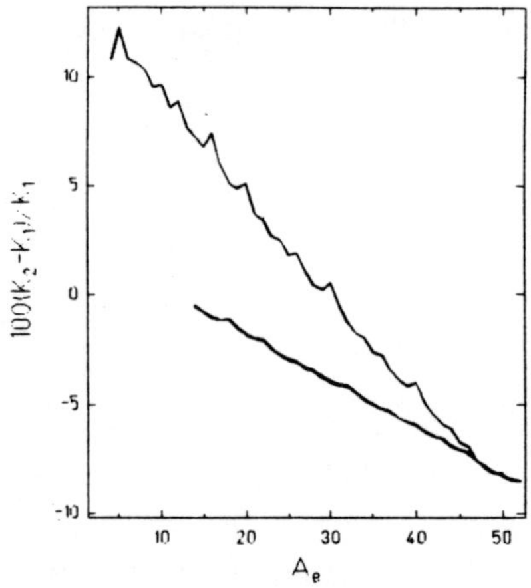

Fig. 2. Relative difference of action integrals for spheres with V_2=constant and R_2=constant. The lower curve refers to cold fission (A_e are shifted with 80 units).

References

1 GREINER, W.: invited talk at this Symposium.

2 PARTICLE EMISSION FROM NUCLEI. Vol. 1: Nuclear Deformation Energy. Vol. 2: Alpha, Proton and Heavy Ion Radioactivities. Vol. 3: Fission and Beta-Delayed Decay Modes. Poenaru, D.N. and Ivascu, M. Eds, CRC Press, Boca Raton, Florida, USA, 1988.

3 POENARU, D.N., IVASCU, M., GREINER, W.: Chapter 7, Vol. 3, Ref. 2.

4 GREINER, W., IVASCU, M., POENARU, D.N., SANDULESCU, A.: in Treatise on Heavy Ion Science, Vol. 8, D.A. Bromley Ed., Plenum Press, New York, in print.

5 HOURANI, E., HUSSONNOIS, M.: Chapter 6, Vol. 2, Ref. 2.

6 HENNING, W., KUTSCHERA, W.: Chapter 7, Vol. 2, Ref. 2.

7 PRICE, P.B., BARWICK, S.W.: Chapter 8, Vol. 2, Ref. 2.

8 HASEGAN, D., TRETYAKOVA, S.P.: Chapter 9, Vol. 2, Ref. 2.

9 POENARU, D.N., IVASCU, M., MAZILU, D., IVASCU, I., HOURANI, E., GREINER, W., invited talk at the Int. Symp. on Developments in Nucl. Cluster Dynamics, Sapporo, Japan, August 1-3, 1988.

10 POENARU, D.N., IVASCU, M.: Chapters 3-5, Vol. 2, Ref. 2, and invited talk at "Cluster'88" Int. Conf., Kyoto, Japan, July 25-29, 1988.

11 SHI, Y.J., SWIATECKI, W.: Nucl. Phys. A464 (1987) 205.

12 GREINER, M., SCHEID, W.: J. Phys. G12 (1986) L229.

13 BARRANCO, F., BROGLIA, R.A., BERTSCH, G.F.: Phys. Rev. Lett. 60 (1988) 507.

14 GÖNNENWEIN, F.: Int. Symp. on Collective Phen., Bad Honnef, 1986.

15 HULET, E. K. et al.: Phys. Rev. Lett. 56 (1986) 313.

16 HOFFMAN, D. C., SOMERVILLE, L.P.: Chapter 1, Vol. 3, Ref. 2

17 DAVIES, K.T.R., SIERK, A.J., NIX, J.R.: Phys. Rev. C13 (1976) 2385.

SPONTANEOUS FISSION PROPERTIES OF THE HEAVY ELEMENTS: BIMODAL FISSION

E.K. Hulet
University of California, Lawrence Livermore National Laboratory,
Livermore, CA 94551

ABSTRACT: We have measured the mass and kinetic-energy distributions from the spontaneous fission of ^{258}Fm, ^{258}No, ^{259}Md, ^{260}Md, ^{260}[104], and ^{262}No. All are observed to fission with a symmetrical division of mass, whereas the total-kinetic-energy (TKE) distributions strongly deviated from the Gaussian shape characteristically found in the fission of all other actinides. When the TKE distributions are resolved into two Gaussian's, the constituent peaks lie near 200 and near 233 MeV. We conclude two modes or bimodal fission is occurring in five of the six nuclides studied. Both modes are possible in the same nuclide, but one generally predominates. We also conclude the low-energy but mass-symmetrical mode is likely to extend to far heavier nuclei; while the high-energy mode will be restricted to a smaller region, a region of nuclei defined by the proximity of the fragments to the strong neutron and proton shells in ^{132}Sn.

I. INTRODUCTION

Our motivation for investigating the spontaneous fission (SF) properties of these very heavy nuclides was to explore further, a particularly interesting region beyond the heavy fermium isotopes where we and others had found a sudden onset of mass symmetry and high fragment energies in neutron-induced and spontaneous fission. Elsewhere on the nuclide chart, only the isotopes of the elements Tl through Ac fission with symmetrical mass distributions.[1] Asymmetrical (two-humped) mass distributions are a common feature in low-energy-induced fission and spontaneous fission of the actinides until ^{258}Fm is reached. The very sharply symmetrical mass distributions and TKEs approaching 240 MeV found for ^{258}Fm[2,3] and ^{259}Fm[4] are remarkable. To determine the range of this behavior, to provide critical tests of theory, and to improve our predictions for heavier and more distant nuclei, it was necessary to extend these fission studies to nuclides with greater atomic and neutron numbers.

An opportunity to study the alignment of forces between the macroscopic and microscopic components that make up the fission barrier was another factor that prompted our investigation. A major portion of the fission barrier for lighter nuclei is imparted by the liquid-drop (macroscopic) component, but its contribution diminishes rapidly with increasing Z because of the squared expansion of the Coulomb repulsive force. However, the total height of the fission barrier is the sum of the macroscopic portion and a microscopic part produced by single-particle effects from spin-orbit interactions. Upon reaching element 100 (Fm), the contribution to the barrier by the macroscopic part drops to less than a third of the total, the remainder being due to shell stabilization of the deformed ground states. Nuclei with $Z \geq 106$, particularly the predicted superheavy elements, owe their entire stability against SF to barriers derived from shell effects.[5] Hence, the nuclei we proposed to study between $Z = 100$ and $Z = 104$ were directly in the region where the effects on fission properties induced by liquid-drop potential energies should largely fade and be replaced by those from shell energies. Knowing the enormous impact of fragment shells on the SF properties of ^{258}Fm and ^{259}Fm, we might now hope to assess the changes in fission wrought by the intrinsic structure of the fissioning nucleus.

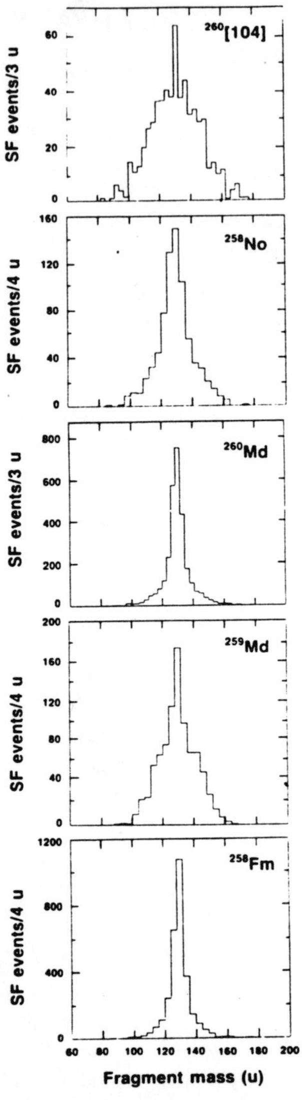

FIG. 1. Provisional mass distributions (no neutron corrections) obtained from correlated fragment energies. The mass bins have been chosen to be slightly different for each nuclide. The distributions are net after subtracting a small ^{256}Fm component.

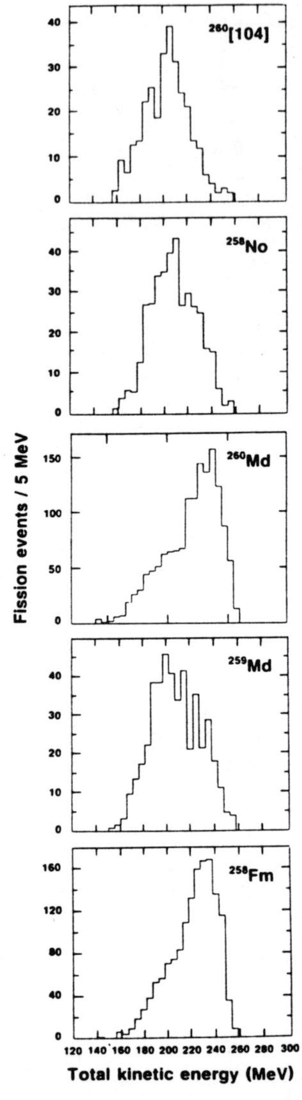

FIG. 2. Provisional total-kinetic-energy (TKE) distributions. A small contribution equivalent to the known amount of ^{256}Fm has been subtracted from all but the the ^{260}Md distribution.

All of the nuclides we have studied were selected on the basis of having neutron numbers ≥156 and atomic numbers ≥100. In addition to the reasons given above, this region was chosen because both experimental SF half-lives and theory[6,7,8] indicated the outer peak of the "double-humped" fission barrier was dropping in energy below the ground state for such nuclides. This was believed to be one factor in the sharp departure from the fission behavior of the lighter actinides. Moreover, the principal cause being advanced for the unusual fission properties of ^{258}Fm and ^{259}Fm was strong shell effects in the emerging fission-product nuclei, which were driving the reaction toward the doubly-magic ^{132}Sn nucleus.[9,10,11] Thus, the transition from asymmetrical mass division in the light fermium isotopes to symmetrical in the heavier ones would be due to fragments approaching closed proton and neutron shells (Z=50, N=82).[12] By studying heavier nuclei, we could establish the range of Z and N where fission properties were being influenced by closed-shell fragments. This would be a further test of the predictions of the two-center shell model because this model predicted a return to an asymmetrical division of mass when Z was increased beyond 100 while maintaining N essentially constant.[8]

Because it is exceedingly difficult to produce more than three or four hundred atoms of these isotopes, our measurements of their fission properties have been restricted to determining the energies of coincident fragments. The sum of these energies provides the TKE while the masses of the fragments are deduced from an inverse relationship between mass and fragment energies due to the conservation of momentum and an assumption that mass is conserved. In an earlier brief Letter,[13] we presented a portion of the results given here together with our main conclusion concerning bimodal fission.

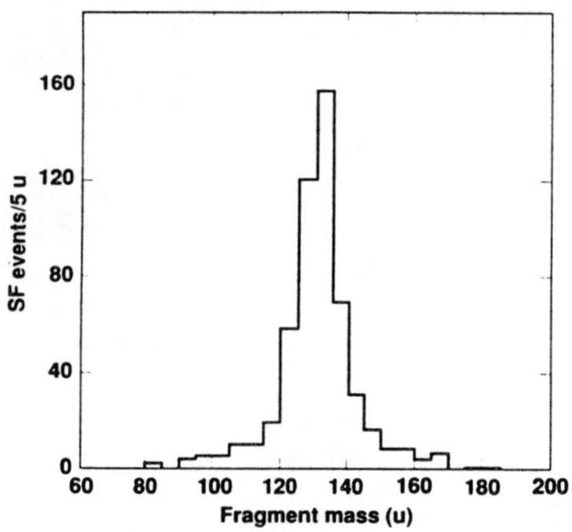

FIG. 3. Provisional mass distribution obtained for 5-ms ^{262}No.

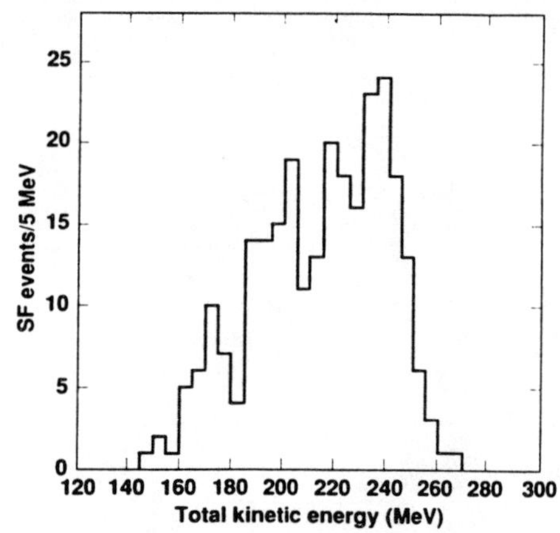

FIG. 4. Total-kinetic-energy distribution from the spontaneous fission of ^{262}No.

II. RESULTS

We present in Figs. 1 and 2 the mass and TKE distributions obtained for five of the six nuclides after subtracting background distributions contributed by small and known amounts of ^{256}Fm. No background corrections were necessary for ^{260}Md. Unlike most previous studies where ^{256}Fm was a major fission component, we found only a slight impact on any distribution from subtracting the contribution from ^{256}Fm. In Figs. 3 and 4, we show similiar distributions for 5-ms ^{262}No, which we discovered this past year.[14]

The most significant and unique feature of the TKE distributions is their pronounced deviation from a single Gaussian shape. In five of the six nuclides, decided asymmetry is imparted by conspicuous tailing in either energy direction from the central peak. This is the first observation of this phenomenon, the TKE distributions from other actinides being uniformly Gaussian with only minor divergences. Detection of this feature was made possible by reducing the interference from the SF of ^{256}Fm and improving the fragment-energy resolution over that of our earlier work. Closer inspection of these TKE distributions reveals that the peak of each distribution is not randomly located along the energy axis, but is positioned near either 200 or 233 MeV. The asymmetric tails of the TKE

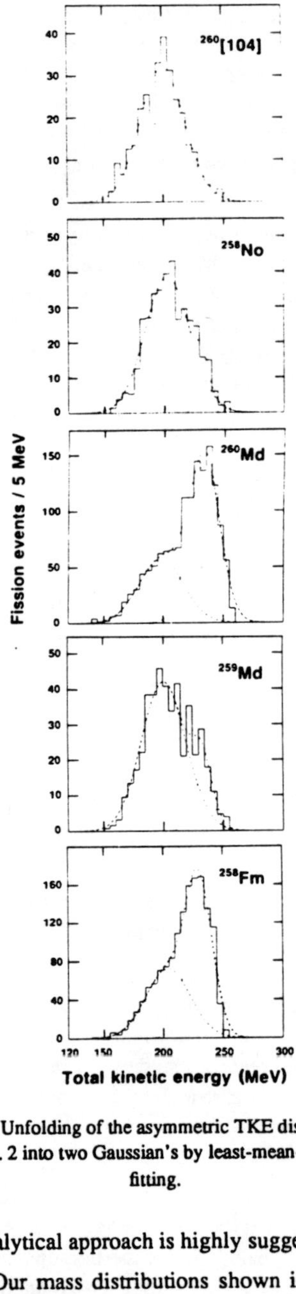

FIG. 5. Unfolding of the asymmetric TKE distributions of Fig. 2 into two Gaussian's by least-mean-squares fitting.

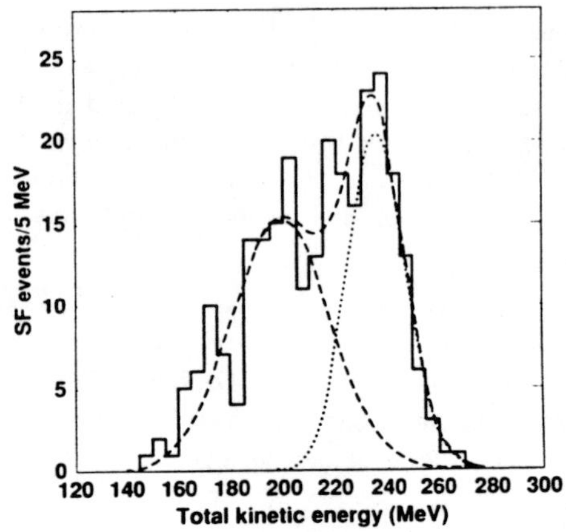

FIG. 6. Same as for Fig. 5 but for ^{262}No.

curves results in distributing an appreciable portion of the events into one or the other of these two main energy regions.

Based on these observations, we considered that the TKE curves for at least five of the nuclides were a composite of two separate energy distributions, with each most likely being Gaussian. Taking the FWHM from the TKE distribution for ^{260}[104] as a fixed parameter and model for the lower-energy Gaussian, we resolved each of the gross TKE distributions from ^{258}Fm, ^{258}No, ^{259}Md, and ^{260}Md into two Gaussian distributions by LMS fitting. The results are shown in Fig. 5 for five isotopes and Fig. 6 for ^{262}No. Reduced-χ^2 values resulting from the fitting of two Gaussian's, together with the centroids of the TKEs and percentage abundances of the low- and high-energy peaks are given for the six nuclides in Table I. Because the reduced chi-squares are near unity, the fitting of two Gaussian functions appears to be a good approximation to the parent distribution. Although the extraction of two Gaussian's by this analytical approach is highly suggestive of such a composite, it should not be construed as proof.

Our mass distributions shown in Figs. 1 and 3 are all symmetrical with the FWHM ranging from 7.9 u for ^{260}Md to 36 u for ^{260}[104]. It is apparent that nuclides having the narrowest mass distributions also have a preponderance of high-TKE events. Conversely, the very broad mass distribution found for ^{260}[104] is associated with a single TKE peak with a low energy. These obvious correlation with TKE are further reinforced when we plot the mass distributions from events sorted by their TKEs. Arbitrarily choosing 220 MeV as the dividing line between the low- and high-

TKE peaks, we show in Fig. 7 the mass distributions obtained for five of the nuclides after sorting events into bins lying above and below this energy. We find it remarkable that the high-energy mode of fission consistently produces such narrowly symmetrical mass distributions as found in these nuclides. While still symmetrical, the mass distributions for events with TKEs less than 220 MeV are much broader, spanning over twice the mass range as those from the high-TKE mode. It is these lower energy events that are responsible for the wings of the main peaks, extending far outward in mass, that we see in Fig. 1 for most of the nuclides. If we choose SF events with TKEs less than 200 MeV, the mass distributions become even broader and are nearly flat but remain symmetrical, with the exception of ^{258}Fm and ^{259}Md, which revert to asymmetrical distributions. Thus, the gross differences seen in the mass distributions from events binned by energy clearly adds another distinctive trait that separates the low-energy mode from the high-energy mode of fission.

TABLE I. Parameters obtained from least-mean-squares fitting of two Gaussians to the TKE curves. Reduced χ^2 is a measure of the quality of fit, where values from 0.5 to 1.5 indicate a reasonable probability of a good match.

Nuclide	Low-energy Peak (MeV)	Abundance (%)	High-Energy Peak (MeV)	Abundance (%)	Reduced χ^2
^{258}Fm	205	50	230	50	1.32
^{258}No	204	96	232	4	0.55
^{259}Md	202	88	234	12	0.81
^{260}Md	200	42	235	58	1.10
^{260}[104]	200	100	----	----	0.63
^{262}No	199	53	235	47	0.94

III. DISCUSSION AND CONCLUSIONS

We have investigated enough nuclei in this region to comment on trends in fission properties, the underlying causes of these uncommon modes of fission and, above all, direct observation of bimodal fission and its relevance to advancing our fundamental understanding of the fission process. Even though we can make several important qualitative connections to theory, we conclude that theory has serious weaknesses in explaining other features that we see. Based on one such connection, our results imply that the low-energy mode of fission we observe will likely extend to far heavier nuclei, well beyond the five studied so far. We can state that symmetrical mass division and high TKEs are no longer unique to just ^{258}Fm and ^{259}Fm. Our results from these six nuclides show that all fission symmetrically and five of the six have a significant component with high TKEs. The concept of this being a small islet of symmetrical fission emerging for a singular reason should now be rejected.

We suggest that two different fission processes are separately responsible for distinctive regions of the TKE distributions displayed in Figs. 2 and 4. In five of the six nuclides, we find anomalous TKE distributions, skewed in energy sufficiently as to be easily describable by two Gaussian distributions. Asymmetrical tailing from the peak energy occurs toward either higher or lower energies. Furthermore, we find that the peak in each of the gross TKE curves falls in one of two distinct positions, either near 200 or near 233 MeV. When resolved into two Gaussian distributions, the constituent peaks also lie very close to the same two energies as shown in Figs. 5 and 6. We find the division of mass to be symmetrical for every nuclide studied; however, very sharply symmetrical mass distributions are correlated with events belonging to the high-energy mode of SF. The low-energy mode is marked by broadly symmetrical mass distributions, as portrayed in the SF of ^{260}[104]. From these distinguishing features, we conclude there are two distinct modes, or bimodal fission.

Because our TKE distributions are wholly unlike those for lighter nuclei, we believe it necessary to offer an explanation. Our analysis of the TKE distributions tells us they are composed of two distributions with very different Coulomb repulsion energies. By necessity, the high-energy mode is compact and spherical at the scission point, while the low-energy mode must be highly deformed and elongated when the fragments separate. We can account for the high-energy mode on the basis of fragment shells that are emerging between the saddle and scission point. As noted in the Introduction, fragment shells near the doubly magic ^{132}Sn lower the potential-energy path and, thereby, favor the mass division into spherical Sn isotopes near the 82-neutron closed shell. As N decreases below 158 neutrons and Z of the fissioning species increases beyond 100, the opportunity to divide into two fragments near these magic proton and neutron numbers diminishes. Thus, we observe a trend away from spontaneous fission characterized by unusually high TKEs and toward the low-energy mode represented by ^{260}[104].

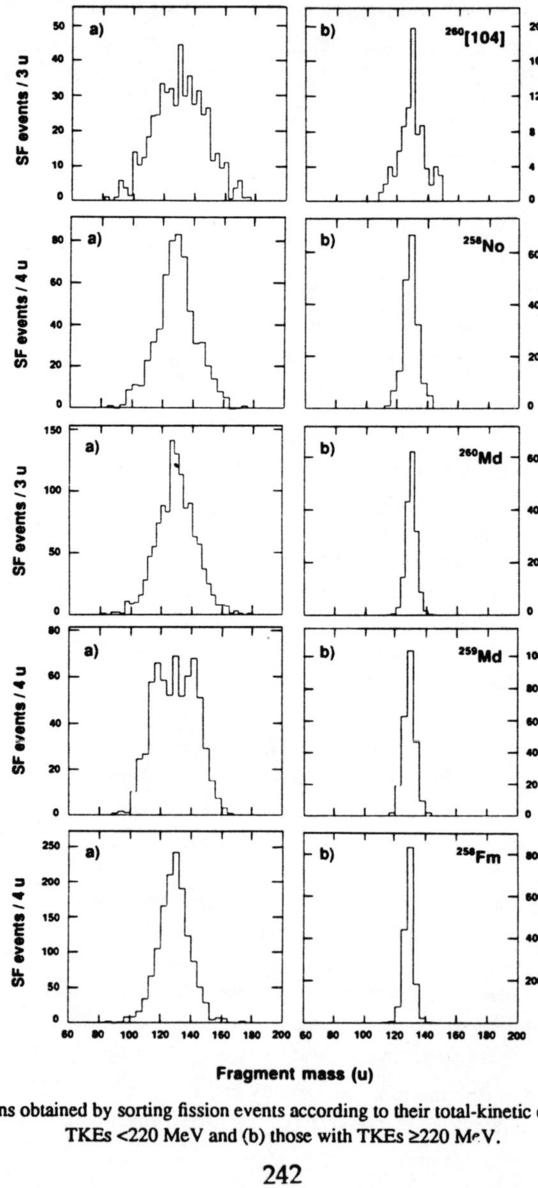

FIG. 7. Mass distributions obtained by sorting fission events according to their total-kinetic energies: (a) for events with TKEs <220 MeV and (b) those with TKEs ≥220 MeV.

An equally satisfying explanation for the low-energy mode is less apparent. Lacking a credible alternative, we are persuaded to believe its appearance is associated with the dropping of the second fission barrier below the ground state. This event is predicted to commence in the same Z and N space as for fragmentation into near-magic nuclei. Theorists have determined that the second or outer fission barrier is lowered by 0.5 to 2 MeV when shapes from asymmetric deformations are included in their calculations of potential-energy surfaces (PES).[15] This lower energy path on the PES may be responsible for the asymmetrical mass distributions found in all but the heaviest actinides.[15] On the other hand, the inner barrier favors reflection-symmetric shapes, being stiff toward any asymmetrical deformations. Upon the disappearance of the second barrier below the ground state, passage through the remaining inner barrier should qualitatively yield symmetrical mass distributions and TKE values that conform to those of liquid-drop fission. For the low-energy mode, our average TKE of 200 MeV and the broadly symmetrical mass and TKE distributions are entirely consistent with those expected from liquid-drop fission.

It would not be too surprising to find this mode of fission if the barrier were due solely to liquid-drop potential energies. It is a well known property of the liquid-drop model to produce a single barrier with a strong preference for reflection-symmetric shapes. However, the barriers for our nuclides are largely built from single-particle couplings rather than from liquid-drop energies.[5] Even so, mass-symmetric shapes before the second saddle are clearly preferred throughout this region of heavy nuclei.[8] Therefore, no matter how the first barrier is constituted, we would expect the resultant fission behavior to be roughly the same. Essentially, our observation of liquid-drop fission properties for $^{260}[104]$, in which the liquid-drop portion of the barrier is about 15%,[5] indicates experimental verification of this hypothesis. To produce the broad spread observed in these mass distributions, we presume that in the descent from the first saddle to scission, factors associated with collective motions will cause a sizable redistribution of mass between the two emerging fragments.

The explanations we offer for each mode of bimodal fission are based upon very general features previously established by PES calculations for static deformations. Each mode is derived from the effects of shell structure: one in the parent fissioning nucleus and the other from single-particle couplings in the fragments. How each can coexist and occur with near equal probability in the same nucleus presents a challenging problem that, so far, remains unsolved. Originally, we had mentioned that the mapping of two distinct paths on the PES, separated by a ridge to prevent re-equilibration during the descent to scission, was necessary to accommodate our results.[13] Theorists responded broadly and generously to this suggestion and have, indeed, found appropriate paths on the PES.[16,17] We refer the reader to the cited references for views of these new valleys in the PES landscape. Still, all is not well with the current picture because a ridge between the paths is missing from the new PES, and more importantly there has been no physical grounds advanced that would allow near-equal populations to traverse each path. We would liken the requirement to having a traffic policeman stand at the juncture of the two routes and directing about equal numbers along each. In one attempt, Möller has estimated that the inertial masses for compact shapes are considerably less than those in the trajectory taken by the elongated mode.[17] The consequence is a very much shorter lifetime for nuclei traveling the less-deformed route, hence, increasing the probability for taking this path enormously. Of course, this clearly violates our measured ratios for the relative populations. We strongly suspect that dynamical aspects govern the choice of routes, but the capability to solve the dynamics when combined with microscopic features does not exist yet.

The concept of two fission modes, characterized by their mass distributions, appeared many years ago when Turkevich and Niday suggested that there are two fundamentally different modes by which fission may proceed.[18] One mode supposedly predominates at low excitation energies and the other at high energies with the relative proportions changing with excitation energy. In their purest form, the two modes led either to a symmetrical or asymmetrical mass distribution. Over the intervening years this suggestion has resurfaced[19,20] only to be discounted because of the lack of sufficiently clear evidence. However, the two modes we report are unrelated to the ones proposed by Turkevich and Niday because ours are likely due to an accidental alignment of favorable shell structures in the fragments <u>and</u> the fissioning nucleus and within the same select group of nuclei. Excitation energy is a curse to the modes we suggest since shell structure is destroyed as internal excitation increases.

In contrast to the lighter actinides where the differences in fission properties from one isotope to the next are subtle and nearly imperceptible, we find the addition of a single nucleon results in abrupt and striking changes in the TKE and mass distributions. Adding a proton to ^{258}Fm causes the high-TKE mode to recede sharply in ^{259}Md and the addition of a neutron to the latter nuclide brings about a sudden return of this mode in ^{260}Md. On the other hand, there seems to be little change in extent of the high-energy mode in the transition from ^{259}Fm to ^{260}Md. Within this group of four nuclides having values of $N \geq 158$, the SF properties of ^{259}Md appear to be at odds with those of its three neighbors for reasons that we assume can be delegated to its Nilsson structure. Assigning a predominant role to fragment properties in the case of ^{259}Md is difficult because there should be little difference in the fragments from this nuclide compared to those from the SF of its neighboring nuclei, ^{258}Fm, ^{259}Fm, and ^{260}Md. Such sudden changes in the end results of the fission process appear to reflect a strong coupling of the collective motions with the intrinsic internal structure of the nucleus during deformation. Furthermore, dynamical aspects of fission, *e.g.* inertial mass, can be strongly affected by bunching and debunching of the Nilsson levels as the nucleus deforms. These notions require the entire fission process to be adiabatic.

Because macroscopic forces are expected to vary smoothly with Z and N, we are compelled to conclude that the sharp changes in fission properties from nuclide to nuclide are due to shell effects. Clearly, single-particle couplings in the fissioning nucleus and its fragments play an even larger role in the fission process than we previously expected. However, all static models of fission tend to average the contribution from single-particle couplings as a way of easing the calculations and are, therefore, incapable of reproducing the abrupt changes in half-lives and fission properties that we observe experimentally. In any theoretical description of fission, we believe it is essential to include explicitly these parameters, which depend so strongly on Z, N, and deformation, rather than smoothly varying them.

The North and East boundaries marking the Z and N region for which the high-energy mode of fission exists have not been determined experimentally. Exploration of the very neutron-rich region at the Eastern limit may never be possible because the combination of targets and projectiles necessary to reach this area are unavailable. With respect to the Southern and Western borders, our current knowledge indicates that Fig. 1 in Ref. 8 provides a reasonable estimate of their location, provided a preponderance of the high-energy mode is required for their definition. Farther to the West, we observe a trend signaled by fading of the high-TKE mode together with broadening mass distributions when $Z \geq 102$ and $N \leq 156$. This conforms to our expectations in that there is a diminishing opportunity to divide into two Sn fragments when Z of the fissioning species increases beyond 100. For example, the mass distribution for ^{260}[104] is not nearly as sharply focused around mass 130 as it is for nuclides closer to Fm. However, if N increases, as in the Md isotopes, and approaches 164, the influence of the 82-neutron closed-shell in the fragments again tends to enhance the high-energy mode of fission. The upward extent of Z in which the proximity to 164 neutrons influences fission properties is not yet defined, but it is seen from our fission studies of ^{262}No that it persists into the heaviest No isotopes to nearly the same degree as we found for the neutron-rich Md and Fm isotopes. This finding is consistent with the forecast cited above in Ref. 8. We hesitate to speculate about the

Northern and Eastern limits to this mode of fission but would suggest that it will disappear as suddenly as it appeared, which is when strong fragment shells are no longer available. However, we believe it safe to venture that the combination of magic neutron and proton numbers in the fragments from the fission of ^{264}Fm could lead to totally "cold fission" with exceedingly sharp TKE and mass distributions. By "cold fission", we mean that virtually all of the energy available for the fission reaction (Q value) is removed by the kinetic energy of the fragments, leaving almost nothing for their internal excitation.

We presume that symmetrical mass division will be typical of SF throughout a region of the most massive nuclei that borders and includes those still undiscovered. We make this projection on the grounds that all PES calculations show the second barrier not only dropping below the first barrier but completely vanishing for all nuclei with $Z \geq 106$, thereby providing what we believe to be the necessary basis for low-energy symmetrical fission to prevail. This new region parallels that of the preactinides, where mass-symmetrical fission is observed whenever passage to scission is through or over a single, but primarily liquid-drop barrier.[1] Mass-asymmetry reported as probable for the SF of ^{262}[105] seemingly challenges our prophecy.[21] However, the evidence for mass asymmetry from the fission of this nuclide is not strong when one considers that only 181 ± 51 fission events out of a total of 950 were due to ^{262}[105], the rest coming from the SF of ^{256}Fm. It seems somewhat questionable to us as to whether or not a meaningful mass distribution can be isolated from an overbearing asymmetrical one arising from ^{256}Fm. In support of our expectation of a low-energy fission mode for this nuclide, the authors did not find any high TKE events near 250 MeV, which would have been the signature of our high-energy mode.

In conclusion, we wish to note that fission theory appears to be lagging well behind experimental findings, whereas we would greatly prefer the reverse to be true. Most of the results and conclusions we report were unanticipated, even though much of the theoretical framework necessary for their prediction was in place. Half-life estimates for SF, with their 5 to 9 orders of magnitude uncertainty, offer almost no guidance to the landscape ahead in the search for new and heavier nuclides. Our foremost concern is this lack of predictive powers within the current static models of fission. Rather than illuminating the way, theoretical studies are reacting to what has been already found. After much discussion with theorists concerning the limitations of fission models based on the Strutinsky-Nilsson formulation, we have arrived at the not so unpopular conclusion that these models should be abandoned in favor of a new approach. The reasoning behind this view is the near exhaustion of the posibilities of introducing the necessary physical parameters that are needed in these models to predict detailed fission properties. Now that the necessary computational power is available, we have been persuaded that the time has arrived to apply the Hartree-Fock and, eventually, the time-dependent Hartree-Fock method to the fission process.

This research was performed under the auspices of the U. S. Department of Energy by Lawrence Livermore National Laboratory under Contract No. W-7405-Eng-48.

REFERENCES

1. *e.g.* M. G. Itkis, V. N. Okolovich, A. Ya. Rusanov and G. N. Smirenkin, Z. Phys. A **320**, 433 (1985).
2. W. John, E. K. Hulet, R. W. Lougheed, and J. J. Wesolowski, Phys. Rev. Lett. **27**, 45 (1971).
3. D. C. Hoffman, J. B. Wilhelmy, J. Weber, W. R. Daniels, E. K. Hulet, R. W. Lougheed, J. H. Landrum, J. F. Wild, and R. J. Dupzyk, Phys. Rev. C **21**, 972 (1980).
4. E. K. Hulet, R. W. Lougheed, J. H. Landrum, J. F. Wild, D. C. Hoffman, J. Weber, and J. B. Wilhelmy, Phys. Rev. C **21**, 966 (1980).
5. G. Münzenberg, Rep. Prog. Phys. **51**, 57 (1988).
6. J. Randrup, S. E. Larsson, P. Möller, S. G. Nilsson, K. Pomorski, and A. Sobiczewski, Phys. Rev. C **13**, 229 (1976).

7. A. Baran, K. Pomorski, A. Lukasiak, and A. Sobiczewski, Nucl Phys. **A361**, 83 (1981); H. C. Pauli and T. Ledergerber, in Proceedings of the Symposium on the Physics and Chemistry of Fission, Rochester, New York, 1973 (International Atomic Energy Agency, Vienna, Austria, 1974), Vol. I, p. 463.
8. M. G. Mustafa and R. L. Ferguson, Phys. Rev. C **18**, 301 (1978).
9. J. R. Nix, Annu. Rev. Nucl. Sci. **22**, 341 (1972).
10. U. Mosel, J. Maruhn, and W. Greiner, Phys. Lett. **34B**, 587 (1971).
11. U. Mosel and H. W. Schmitt, Phys. Rev. C **4**, 2185 (1971).
12. M. G. Mustafa, U. Mosel, and H. W. Schmitt, Phys. Rev. Lett. **28**, 1536 (1972).
13. E. K. Hulet, J. F. Wild, R. J. Dougan, R. W. Lougheed, J. H. Landrum. A. D. Dougan, M. Schädel, R. L. Hahn, P. A. Baisden, C. M. Henderson, R. J. Dupzyk, K Sümmerer, and G. R. Bethune, Phys. Rev. Lett. **56**, 313 (1986).
14. R. W. Lougheed, E. K. Hulet, F. F. Wild, K. J. Moody, R. J. Dougan, C. M. Gannett, R. A. Henderson, D. C. Hoffman, and D. M. Lee, Nucl.Chemistry Division Ann. Rept., Lawrence Livermore National Laboratory, Report No. UCAR 10062/88, 1988 (unpublished), p. 135.
15. P. Möller and S. G. Nilsson, Phys. Lett. **31B**, 283 (1970); H. C. Pauli, T. Ledergerber, and M. Brack, Phys. Lett. **34B**, 264 (1971); G. Gustafsson, P Möller, and S. G. Nilsson, Phys. Lett. **34B**, 349 (1971).
16. U. Brosa, S. Grossmann, and A. Müller, Z. Phys. A **325**, 242 (1986); V. V. Pashkevich and A. Sandulescu, Rapid Communications No. 16-86, Joint Institute for Nuclear Research, Dubna, USSR (1986), p. 19-23; K. Depta, J. A. Maruhn, W. Greiner, W. Scheid, and A. Sandulescu, Mod. Phys. Lett. A **1**, 377 (1986); V. V. Pashkevich, Nucl. Phys. **A477**, 1 (1988); S. Cwiok, P. Rozmej, and A. Sobiczewski, in AIP Conf. Proc. No. 164, 5th Int. Conf. on Nuclei far from Stability, Lake Rosseau, Canada, 1987, edited by I. S. Towner (Am. Inst. Physics, New York, 1988), p.821; S. Cwiok, P. Rozmej, A. Sobiczewski, and Z. Patyk, submitted to Nucl. Phys. A (1988).
17. P. Möller, J. R. Nix, and W. J. Swiatecki, Nucl Phys. **A469**, 1 (1987); P. Möller, J. R. Nix, and W. J. Swiatecki, submitted to Nucl. Phys. A (1988).
18. A. Turkevich and J. B. Niday, Phys Rev 84, 52 (1951).
19. H. B. Levy, H. G. Hicks, W. E. Nervik, P. C. Stevenson, J. B. Niday and J. C. Armstrong, Jr., Phys. Rev. 124, 544 (1961).
20. R. C. Ragaini, E. K. Hulet, R. W. Lougheed, and J. Wild, Phys. Rev. C 9, 399 (1974).
21. C. E. Bemis, Jr., R. L. Ferguson, F. Plasil, R. J. Silva, G. D. O'Kelley, M. L. Kiefer, R. L. Hahn, C. C. Hensley, E. K. Hulet, and R. W. Lougheed, Phys. Rev. Lett. **39**, 1246 (1977).

PARTICLE DECAY AND FISSION*

I. Rotter
Zentralinstitut für Kernforschung Rossendorf, Dresden 8051, Germany

Abstract

Consideration of the coupling between discrete and continuous states of the nuclear system in a straightforward manner entails a unified description of particle decay and fission. The particle width consists of two terms one of which corresponds to the standard term of R-matrix theory while the other one originates from the feedback of the continuum to the nucleus. The standard term dominates in particle decay while the feedback term, containing the coherent effects, dominates in fission. The corrections from the feedback term to the particle decay widths and from the standard term to the fission process are discussed. The fission dynamics of hot nuclei and the fission process of cold nuclei are dominated by the same feedback term. They show some similarities.

1. Introduction

Recently, the unified description of alpha decay, heavy ion radioactivities and cold fission has been illustrated[1] for ^{234}U. The cold fission is a small fraction of fission events without excitation of the fragments. It differs from particle decay modes only by the mass ratio of the fragments.

Thus, cold fission is, on the one hand, a special decay mode as shown on the basis of numerical calculations[1] and, on the other hand, a special fission mode which is the result of the large amplitude collective motion implied in the fission process. By this, a unified description of fission and decay modes is suggested, where both modes appear as borderline cases in the process of interaction of the nucleus with its decay products.

In order to find out what is common of decay and fission let us ask the question: What is characteristic of a finite nucleus and what is proposed in the standard description of nuclear states? The answer is the following. The standard nuclear structure description proceeds from single-particle wavefunctions bound in a potential. The residual interaction between the bound nucleons as well as their collective motion are considered but the coupling to the surrounding continuum is not taken into account. The Hamiltonian is proposed to be Hermitian, the eigenvalues are real, providing the energies but not the widths of the states. Using R-matrix theory, the decay of the nuclear states is approximately described by means of the product of a pentration factor and a spectroscopic factor. The feedback from the decay channels is not considered.

In reality, every existing nucleus has excited states most of which decay by the emission of particles. These states are coupled to the surrounding continuum of break-up channels or, in other words, they are embedded in the continuum. Consequently,

* XVIIIth International Symposium on Nuclear Physics: Physics and Chemistry of Fission, Gaussig/GDR, Nov. 21-25, 1988

the Hamilton operator of a nuclear structure calculation has to be a non-Hermitian operator, providing the energies and widths of the states, on the condition that is passes into a Hermitian operator, allowing for the vanishing widths of low-lying stable states of the many-body nuclear system. Only such an extended description is complete and allows to draw conclusions on the life-times of the different modes in a straightforward manner.

Recently, a method has been developed for the description of the nuclear system by taking into account the coupling to the surrounding continuum from the very beginning. This method[2], considering the nucleus as an open quantum mechanical system[3], may be regarded as an attempt towards a more complete description of the many-body nuclear system. Numerical results are obtained, up to now, only by restricting to channels with one nucleon in relative motion to the residual nucleus consisting of A-1 bound nucleons. The results show, despite of these restrictions, characteristic features of the interplay between long-living and short-living states. In the expression for the partial widths, the coupling to the continuum leads to an additional term involving the feedback of the continuum to the spectroscopic properties. The interplay between the standard expression of R-matrix theory and the additional feedback term can be seen immediately. In fission, the feedback term is dominant while in decay, the dominant part is the standard term.

In Sect. 2, the formalism for describing the open quantum mechanical system is sketched and the formula for the spectroscopic values, including the corrections from the coupling to the continuum, are given. The interplay between the standard and the feedback term is illustrated in Sect. 3 by means of numerical results obtained for nucleon and alpha decay in light nuclei. Some consequences for fission are discussed in Sects. 4 and 5 while some conclusions are drawn in Sect. 6.

2. The corrections to the standard spectroscopic values

Characteristic of the open quantum mechanical nuclear system is the appearance of the constraint operator

$$Z = Q + G_P^{(+)} H_{PQ} . \tag{1}$$

Here, Q is the projection operator onto the function space of the system consisting of products of bound single-particle wavefuntions while P projects onto the function space of break-up channels. The Green function $G_P^{(+)}$ describes the motion of a particle in the P subspace and $H_{PQ} \equiv PHQ$ describes the coupling between nucleus and continuum. The formalism to handle numerically channels with one nucleon in the continuum (continuum shell model) is described in detail in refs.[2,3].

The constraint operator (1) is a sum of two terms. The operator Q contains the stationary part of the process which can be described by a nuclear structure model since the Q subspace is identical with the function space used in the standard nuclear structure calculations for discrete states[2,3]. The second part of Z corresponds to the dynamical part of the process in which the feedback from the continuum (P subspace) is involved. By means of the operator Z, the corrections to the spectroscopic values which are caused by the coupling to the continuum, can be formulated.

In the continuum shell model, the Hamilton operator reads[3]

$$H_{QQ}^{eff} = H_{QQ} + H_{QP}G_P^{(+)} H_{PQ} = QHZ \tag{2}$$

while the wavefunction of a state R is given by

$$\tilde{\Omega}_R = \tilde{\Phi}_R + \tilde{\omega}_R = Z\tilde{\Phi}_R . \tag{3}$$

Here, $H_{QQ} \equiv QHQ$ is the Hamilton operator of the standard nuclear structure calculation performed in the Q subspace. Further, H_{QQ}^{eff} is non-Hermitian and has complex eigenvalues $\tilde{E}_R - \frac{i}{2}\tilde{\Gamma}_R$ and complex eigenfunctions $\tilde{\Phi}_R$. For states below particle decay thresholds, it is Hermitian with $\tilde{\Gamma}_R = 0$. Thus, the widths $\tilde{\Gamma}_R$ of the nuclear states follow immediately, without additional assumptions, within the model. The wavefunction $\tilde{\Omega}_R$ of the state R has the true asymptotic behaviour.

The partial width $\tilde{\gamma}_R^c$ describes the coupling of the state R to the channel c. It is, by definition,

$$\tilde{\gamma}_R^c = (2\pi)^{1/2}\langle \chi_E^{c(-)} |V| \tilde{\Omega}_R^{(+)}\rangle \tag{4}$$

where $\chi_E^c = u_{\varepsilon\tau lj} \mid t\rangle$ describes the final state observed and $u_{\varepsilon\tau lj}$ is the scattering wavefunction of the particle with energy ε and quantum numbers τlj in relative motion to the target state $|t\rangle$. It is

$$\tilde{\gamma}_R^c = \gamma_{1,R}^c + \gamma_{2,R}^c = Z\gamma_{1,R}^c \tag{5}$$

where

$$\gamma_{1,R}^c = (2\pi)^{1/2}\langle \chi_E^{c(-)} |V| \tilde{\Phi}_R^{(+)}\rangle \tag{6}$$

and

$$\gamma_{2,R}^c = \sum_{c'} \int_{\varepsilon_{c'}}^{\infty} dE' \langle \chi_E^{c(-)} |V| \xi_{E'}^{c'(-)}\rangle \frac{1}{E^{(+)}-E'} \tilde{\gamma}_R^{c'} . \tag{7}$$

The first term (6) can be factorized into a spectroscopic amplitude and a penetration amplitude[4]. It corresponds to the standard expression of a partial width in R-matrix theory. The second term (7) cannot be factorized. In $\gamma_{2,R}^c$, the coherent action of all nucleons is involved according to the sum over all (open and closed) channels, the integration over energy, the explicit appearance of the threshold energies ε_c, which are characteristic of the whole nucleus and the explicit dependence on all $\tilde{\gamma}_R^{c'}$.

The total width $\tilde{\Gamma}_R$ is determined by the eigenvalue of the operator (2). For isolated states, it follows

$$\tilde{\Gamma}_R = \sum_c (\tilde{\gamma}_{Rc})^2 \; . \tag{8}$$

In the standard R-matrix theory, the width $\tilde{\Gamma}_R$ is calculated by using relation (8). In the continuum shell model, it has been shown numerically that relation (8) is, indeed, fulfilled to a good approximation as long as the states R are isolated.

Thus, the difference between the standard nuclear theory and the description of the nucleus as an open quantum mechanical system consists in the appearance of the dynamical terms to the Hamilton operator, eq. (2), to the wavefunction of a nuclear state, eq. (3), and to the partial width, eqs. (5). The dynamical terms are all of the same type and expressed by means of the constraint operator Z. The additional terms to the Hamilton operator and to the wavefunction of a state take into account immediately the well known fact that most nuclear states decay. They represent therefore obvious corrections to the standard nuclear structure theory. The relation (5) for the partial width appears as a consequence of (3) and does not contain any new assumptions. The feedbeck term $\gamma^c_{2,R}$ does not appear in the standard theory.

It depends on the special case considered which of the two terms in (5) is dominant. While (6) describes the direct connection between the individual state R and the channel c, the term (7) involves the connection between R and c due to the coherent action of all the nucleons in the nucleus and in the channels. The term (6) is dominant generally for the separation of nucleons and of small nucleon groups like alpha particles, from a light parent nucleus. In the case of fission, however, the coherence effects involved in the dynamical term (7) are important so that it dominates.

An estimation shows that the standard term term $\gamma^c_{1,R}$ of the decay width is small for particle groups separated from heavy nuclei. The dominant part is $\gamma^c_{2,R}$ which involves all $\tilde{\gamma}^{c'}_R$ for open as well as for closed channels. Closed channels are important only, if the corresponding $\tilde{\gamma}^{c'}_R$ are large due to the energy denominator in (7). By this, shell effects are involved also in $\gamma^c_{2,R}$ if the excitation energy of the state R is not too high.

The interplay between the two terms can be seen best in decays which are suppressed or forbidden due to the nuclear structure or quantum numbers involved, and in cold and ternary fission which occur also with a relatively small probability. Both processes will be discussed in the following.

3. Suppressed and forbidden decays

In the standard R-matrix theory, the partial width against decay of the state R into the channel c is given by an expression similar to $\gamma^c_{1,R}$. As it is very well known, the standard theory provides values which are mostly in good agreement with the experimental data for the decay into nucleon channels. Nevertheless, the existence of the second term $\gamma^{c}_{2,R}$ can be seen even in nucleon decay. An example are the isospin forbidden transitions in light nuclei the large matrix elements of which are hardly to explain in the framework of the standard theory.

An example of these transitions is the excitation of the isospin-forbidden analogue resonance $J^{\pi} = 3/2^{-}$, $T = 3/2$ at 15.1 MeV in the mirror nuclei ^{13}N and ^{13}C. The results of calculations in the continuum shell model, given in table 1, show that not only the coupling to open decay channels leads to an enlargement of $\tilde{\Gamma}_R$ but also the coupling to closed channels has to be taken into account[5]. Agreement with the large width $\tilde{\Gamma}_R \approx 1$ keV observed experimentally can be obtained if the channel coupling involved in $\gamma^{c}_{2,R}$ is taken into account.

Table 1
Width $\tilde{\Gamma}_R$ of the $3/2^{-}$, 3/2 level at 15.1 MeV in ^{13}N calculated in the continuum shell model

Number of channels	States of ^{12}C with $T \approx 0$ (open channels)*	States of ^{12}C with $T \approx 1$ (closed channels)**	$\tilde{\Gamma}_R$/keV
1	0^+_1		0.19
2	0^+_1, 0^+_2		0.37
3	0^+_1, 0^+_2	0^+	0.38
3	0^+_1, 0^+_2, 2^+		0.42
4	0^+_1, 0^+_2, 2^+	0^+	1.01
5	0^+_1, 0^+_2, 2^+	2^+_1, 2^+_2	4.69
6	0^+_1, 0^+_2, 2^+	0^+, 2^+_1, 2^+_2	6.44

* 0^+_1 at 0 MeV (g.s.), 0^+_2 at 7.7 MeV, 2^+ at 4.4 MeV

** 0^+ at 17.8 MeV, 2^+_1 at 16.1 MeV, 2^+_2 at 18.7 MeV

The existence of the term $\gamma^{c}_{2,R}$ in the reduced widths can be seen also in alpha decay. The alpha transitions in light nuclei can be classified in the following manner[6].

(i) Structure-allowed transitions. The state has alpha-particle structure, i.e. the main component of the shell-model wavefunction is characterized by a Young tableau with maximal symmetry. An example is the first 2^+ level at 4.4 MeV in ^{12}C.

(ii) Structure-suppressed transitions. The main component of the shell-model wavefunction is characterized by a broken Young tableau. An example is the 0^+ level at 10.1 MeV in ^{12}C.

(iii) Structure-forbidden transitions. None of the components of the shell-model wavefunction is compatible with an alpha representation. An example is the 1^+ level at 12.7 MeV in ^{12}C.

This classification is based on the symmetry properties of the wavefunction and is not directly connected with the isospin. Isospin forbidden alpha transitions may belong to the structure-suppressed or structure-forbidden transitions in dependence on the quantum numbers of the corresponding states. An analogy to the structure-forbidden alpha transitions (iii) does not exist in nucleon transitions.

It is very well known that the alpha widths calculated by means of $\gamma^{c}_{1,R}$ in the standard theory for structure-allowed transitions in light nuclei are in good agreement with the experimental data due to their large spectroscopic factors. Calculations for structure-suppressed and structure-forbidden alpha transitions in ^{12}C, using the same method, showed the following general regularity[6] (table 2):

Table 2

Alpha widths of structure suppressed (S) and structure-forbidden (F) alpha transitions in ^{12}C calculated in the shell model with R-matrix theory

Levels of ^{12}C J^{π} T	E/MeV	Levels of ^{8}Be J^{π} T	E/MeV	Type of the transition	S_{α} (exp)*	S_{α} (theor)
0^+0	10.3	0^+0	0	S	$8\cdot10^{-1}$	$5\cdot10^{-2}$
1^+0	12.7	2^+0	2.9	F	$3.5\cdot10^{-3}$	$1\cdot10^{-3}$
1^+1	15.1	2^+0	2.9	F	$2\cdot10^{-6}$	$2\cdot10^{-6}$
2^+1	16.1	0^+0	0	S	$4\cdot10^{-5}$	$3\cdot10^{-6}$
2^+1	16.1	2^+0	2.9	S	$1.5\cdot10^{-3}$	$2\cdot10^{-5}$
0^+1	17.8	0^+0	0	S	$2\cdot10^{-3}$	$7\cdot10^{-7}$
0^+1	17.8	2^+0	2.9	S	$1.5\cdot10^{-2}$	$6\cdot10^{-7}$
2^+1	18.8	0^+0	0	S	$3\cdot10^{-3}$	$1\cdot10^{-5}$
2^+1	18.8	2^+0	2.9	S	$3.5\cdot10^{-2}$	$2\cdot10^{-5}$

* The experimental data are from F. Ajzenberg-Selove, Nucl. Phys. A 433 (1985) 1

In the structure-forbidden transitions, the theoretical results agree well with the experimental data. The differences between calculated and experimental values in the case of structure-suppressed transitions amount, however, at least one order of magnitude. They are much greater than may be explained with an uncertainty of, e.g., the channel radius involved in the standard theory (for details see ref.[6]). The reason for the differences between experimental and calculated values should not be sought therefore in an inaccuracy of the calculation within the standard method but in the absence of $\gamma^{c}_{2,R}$ in the standard theory. Because the absolute values of the alpha widths for structure-suppressed transitions are small, the term $\gamma^{c}_{2,R}$ may change the value $\tilde{\gamma}^{c}_{R}$ radically. In the case of structure-allowed transitions, $\gamma^{c}_{2,R}$ cannot influence the results since $\gamma^{c}_{1,R}$ itself is large in absolute value. In the structure-forbidden transitions from the 1^{+} levels in ^{12}C, there is no enhancement from $\gamma^{c}_{2,R}$ because no corresponding alpha channels exist. Thus, the successful calculation of the probabilities of structure-forbidden alpha transitions by means of $\gamma^{c}_{1,R}$ in the standard theory is not accidental. It only emphasizes the absence of alpha channels (and corresponding low-lying deformed alpha states[9] in ^{12}C) with quantum numbers 1^{+}. Calculations on the basis of the continuum shell model for the structure-suppressed and structure-forbidden transitions in ^{12}C do not exist up to now.

4. Cold and ternary fission of cold nuclei

By means of numerical calculations, it has been shown[1] recently that cold fission can be described in a unified manner together with alpha decay and heavy-ion radioactivities for ^{234}U which is, up to now, the only nucleus, where all three groups have been observed experimentally. The possibility to describe the three processes in a unified manner corresponds to the fact that the coherent part $\gamma^{c}_{2,R}$ is important also in alpha decay of heavy nuclei. This result is, of course, in contrast to the standard assumption in R-matrix theory that the decay width can be described by the spectroscopic term $\gamma^{c}_{1,R}$ alone.

The expression (7) shows that cold fission gives a non-negligible contribution to low-energy nuclear fission. This is due to the energy denominator, since most channels c' are closed, as well as to the fact that, generally, the low-lying collective strongly deformed states have large $\tilde{\gamma}^{c'}_{R}$ for cluster channels c' without intrinsic excitation of the nuclei. This result is in agreement with experimental data showing pronounced substructure effects[7] and with the results of calculations on the basis of different models which give, due to the Pauli principle, relatively small internal excitation energies[8].

Fission is accompanied sometimes by the emission of a third particle, mostly an alpha particle. Until now, very little is known about the creation mechanism of the third particle. Recently, Tanimura and Fliessbach[10] treated the emission process during fission for the first time quantum mechanically in a dynamical model by solving numerically the one-body Schrödinger equation for the motion of an alpha particle in the average time dependent potential of a fissioning nucleus. As a result, the major decay mechanism is not the tunneling[10] involved in the standard term $\gamma^{c}_{1,R}$, but the alpha particles gain sufficient energy to surpass the Coulomb barrier due to the rapid changes of the shape of the fissioning nucleus.

Experimental investigations showed a cooling down of the fissioning system by the emission of the alpha particle[11]. That means, the alpha particles cannot be considered as spectators but they turn out to be active participants in the dynamics of the fission process.

Formula (5) for the width $\tilde{\gamma}_R^c$ of a nuclear state suggests the assumption that ternary fission occurs via very short-living two-particle channels. Then, there are two possibilities to create the third particle which is assumed in the following to be an alpha particle in order to simplify the discussion.

(i) The alpha particle originates from alpha decay of the nucleus A before scission starts:

$$A \rightarrow \alpha + A_{12} \rightarrow \alpha + A_1 + A_2 \; . \qquad (9)$$

In this case, an increasing energy of the alpha particle corresponds to a decreasing excitation energy of the fissioning nucleus A_{12}, i.e. a decreasing number of open channels and a decreasing level density. As a consequence[9], the coherent term $\gamma_{2,R}^c$ decreases so that the ratio $\gamma_{1,R}^c/\gamma_{2,R}^c$ increases. Furthermore, shell effects in $\gamma_{2,R}^c$ itself are the more pronounced the lower the excitation energy of the decaying nucleus A_{12} is because closed channels c' contribute to $\gamma_{2,R}^c$ only if $\tilde{\gamma}_R^{c'}$ is large due to the energy denominator in eq. (7). As a result, the peak-to-valley ratio increases with increasing energy of the alpha particles.

(ii) The alpha particle originates from alpha decay of one of the fragments

$$A \rightarrow A_{1\alpha} + A_2 \rightarrow \alpha + A_1 + A_2 \; . \qquad (10)$$

In this case, the process of fission of the nucleus A is not directly influenced by the emisssion of the alpha particle. A correlation between the energy of the alpha particle and the peak-to-valley ratio is therefore not to be expected.

Since the experimental results point to the active role of the alpha particle in the fission process[11], at least some part of them must be created via the mechanism (9). As a matter of fact, the two processes (i) and (ii) are not very different from one another for short-lived intermediate states A_{12} and $A_{1\alpha}$ respectively. The description of the alpha emission in a potential, changing due to fission[10], seems therefore to be a good approximation to the description of the average properties.

The emission of neutrons and protons takes place mainly from the fully accelerated fragments[12]. This different behaviour of neutrons and alpha particles is surely caused by the differences in the threshold energies for the break-up of the fissioning nucleus, eq. (7).

5. Fission dynamics of hot nuclei

The properties of nuclei are described microscopically by the Hamilton operator H_{QQ}^{eff}, eq. (2). In hot nuclei, the second term $H_{QP}G_P^{(+)}H_{PQ}$ is dominant[9]. Long-lived compound nucleus modes, described mainly by H_{QQ}, and short-lived break-up modes exist simultaneously. This coexistence of modes with different time scales at high level density creates incomplete fusion of two colliding nuclei at low and intermediate energy[9].

The second term in H_{QQ}^{eff} and the second term in $\tilde{\gamma}_R^c$ arise both from the second part of the Z-operator. It is caused by the non-vanishing coupling H_{PQ} between the nucleus and the continuum of break-up channels. At high level density, the properties of the nuclear system are determined mainly by this second term in H_{QQ}^{eff} [9]. The properties of fission of a cold nucleus at low level density are determined mainly by $\gamma_{2,R}^c$. Since both terms are of the same type, the two processes are expected to show some similarities.

The emission of alpha particles and protons accompanying the fission of ^{161}Ho and ^{169}Ta at about 100 and 200 MeV excitation energy, has been investigated experimentally by Lindl et al.[13]. It could be shown that a significant amount of alpha emission takes place at times where the fragments are in close proximity. The time scale for separating the fragments is comparable to the time scale for alpha-particle emission. The emission of the protons occurs, however, mainly from the fully accelerated fragments. The mechanism of near-scission emission of alpha particles as well as the absence of a near-scission component in the proton spectra are, indeed, similar to the results observed in the fission process of cold nuclei. The light particles are emitted, obviously, by a mechanism of the type (9) and (10), respectively, in both cases.

The conclusions drawn here for the fission dynamics of hot and cold nuclei follow from the analytic form of the Z-operator which enters into the Hamilton operator of the system, the wavefunction of a nuclear state and the partial width. The Z-operator appears in the theory necessarily, if one takes into account, from the very beginning, the fact that most excited states of a nucleus decay. The Z-operator involves the interplay between short-lived and long-lived modes. It would be highly interesting to determine experimentally the time scale for separating the fragments at different excitation energies in a model-independent manner.

6. Conclusions

It has been shown in this paper that the widths for the emission of particles and for fission can be described in a unified manner. If one takes into account the coupling between discrete and continuous states in a straightforward manner, the formula for the partial width of an isolated state consists of two parts. One part corresponds to the standard R-matrix term which can be factorized into a spectroscopic factor and a penetration factor. It contains the shell-model properties. The other term originates from the feedback of the continuum to the nucleus. It cannot be factorized into a spectroscopic and a penetration factor but is characterized by the coherent effects which form the nucleus as a whole. While the first part is used, generally, in describing the particle decay of nuclei, the coherent properties of the second part are basic to the description of fission. But the corrections from the second part to particle decay as well as the shell-model corrections to fission cannot be neglected, in general. In the present paper, both types of corrections are discussed. They prove the unified description of decay and fission.

The existence of two terms in the partial width originates from the existence of two terms in the wavefunction. The statement that channel coupling is involved in the wavefunction of a nuclear state and that both the standard and the feedback term enter into the expression for the partial width are therefore equivalent to one another. Consequently, experimental results which state that channel coupling is a property of the nuclear state itself, proof immediately the unified description of decay and fission proposed in the present paper. Well-aimed experimental investigations are necessary in order to prove convincingly this property of nuclear states.

Let me conclude with a quotation taken from the forward in ref.[14]

> "A well-known parable recounts the conflicting and distorted descriptions of an elephant provided by several blindfolded men each of whom has only felt a leg, a trunk, a tusk, or a tail."

Descriptions of the nucleus are of the same type. We know the many truths of the nucleus each of which is observed and valid under certain conditions. The nucleus is, indeed, elephantine in size. We know the leg, the trunk, the tusk, the tail ... Now, we have to understand the elephant in its entirety.

References

1. D.N. Poenaru et al., Z. f. Phys. A 328 (1987) 309
2. H.W. Barz et al., Nucl. Phys . A 275 (1977) 111
3. I. Rotter, Fiz. Elem. Chastits At. Yadra 15 (1984) 762 and 19 (1988) 274
4. I. Rotter, Phys. Rev. C 27 (1983) 2261; C 29 (1984) 1119
5. I. Rotter, J. Phys. G: Nucl. Phys. 5 (1979) 75
6. V.V. Balashov and I. Rotter, Nucl. Phys. 61 (1965) 138
7. P. Koczon et al., Phys. Letters B 191 (1987) 249
8. H. Walliser et al., Z. f. Phys. A 329 (1988) 209
9. I. Rotter, J. Phys. G: Nucl. Phys. 14 (1988) 857, and to be published
10. O. Tanimura and T. Fliessbach, Z. f. Phys. A 328 (1987) 475
11. J.P. Theobald, Preprint TH Darmstadt IKDA 87/35 (1987); R. Kotte et al., Z. f. Phys. A 328 (1987) 495 and A 330 (1988) 189; W. Pilz et al., Proc. XVIIth Intern. Symp. Nucl. Phys., Gaussig/GDR, 1987
12. H. Märten, Dissertation B, TU Dresden; H. Märten et al., Contribution to this Symposium
13. B. Lindl et al., Z. f. Phys. A 328 (1987) 85
14. W. Ebeling and Yu.L. Klimontovich: Selforganization and Turbulence in Liquids. Teubner-Texte zur Physik 2 (1984)

VII. INDUCED FISSION IN UNUSUAL REGIONS

HEAVY MUONIC ATOMS AND THE DYNAMICS OF NUCLEAR FISSION

presented by **P. David**

J.F.M. d'Achard van Enschut[1], W. Bertl[2], P. David[3], R. Engfer[4], H. Folger[5], J. Fuger[6], H. Hänscheid[3], J. Hartfiel[3], E.A. Hermes[4], H. Janszen[3], T. Johansson[7], J. Konijn[8], T. Krogulski[9], C.T.A.M. de Laat[8], T. Mayer-Kuckuk[3], W. Müller[10], R. von Mutius[3], J. Pauwels[10], C. Petitjean[2], S.M. Polikanov[5], H.S. Pruys[4], H.W. Reist[2], F. Risse[3], Ch. F.G. Rösel[3], L.A. Schaller[11], L. Schellenberg[11], W. Schrieder[3], A.K. Sinha[3], L.M. Simons[2], A. Taal[8], J.P. Theobald[12], G. Tibell[13], N. Trautmann[14], D. Vermeulen[4]

[1] Philips I and E Laboratories, NL-5600 MD Eindhoven, The Netherlands
[2] Paul Scherrer Institut (PSI, formerly SIN) CH-5234 Villigen, Switzerland
[3] Institut f. Strahlen- u. Kernphysik, Universität Bonn, D-5300 Bonn, Germany
[4] Physik Institut der Universität Zürich, CH-8001 Zürich, Switzerland
[5] Gesellschaft f. Schwerionenforschung, GSI, D-6100 Darmstadt, Germany
[6] European Institute f. Transuranium Elements, JRC Karlsruhe, D-7500 Karlsruhe, Germany
[7] The Svedberg Laboratory, S-75121 Uppsala, Sweden
[8] NIKHEF-K, Amsterdam, Postbus 41882, NL 1009 DB Amsterdam, The Netherlands
[9] Department of Mathematics and Nature, University of Warsaw, Branch in Bialystok, 41 ul. Lipowa, PL 15424 Bialystok, Poland
[10] CBNM, Euratom, B-2440 Geel, Belgium
[11] Institut de Physique, Universite de Fribourg, CH-1700 Fribourg, Switzerland
[12] Institut f. Kernphysik, Technische Hochschule, D-6100 Darmstadt, Germany
[13] Department of Radiation Sciences, S-75121 Uppsala, Sweden
[14] Institut f. Kernchemie, Universität Mainz, D-6500 Mainz, Germany

Abstract

A study of heavy muonic atoms has been performed with the aim to investigate dynamical processes in prompt muon induced fission. For this purpose nuclear ground state shapes (spectroscopic quadrupole moments), probabilities of radiationless (r.l.) transitions, fission probabilities of r.l. transitions, prompt and delayed fission yields per muon stop and the spectroscopy of fission fragments have been investigated. The influence of the muon as a probe on the fission process has been revealed to be confined to an augmentation of the outer fission barrier by 0.6 Mev and to a lowering of the total kinetic energy by about 1.5 MeV. Considering the fate of the muon after a prompt fission procoess as being indicative of dynamics in fission, the probability of muon attachment to the light fragment has been determined for some fragment masses.

1) Introduction

A negative muon which cascades down in an atom of a heavy element may induce fission by two distinctively different processes.

In a prompt process, non radiative (n.r.) transitions between the lower muonic levels can lead to prompt fission by direct electromagnetic excitation of the nucleus (internal conversion). For heavy nuclei the energy of the muonic (2p→1s)-transition is of the same magnitude as the height of the fission barrier, and the energy of the (3d→1s)-transition with about 9.5 MeV is much higher. Already in 1948 Wheeler /1/ pointed out that non radiative transitions can induce fission.

In a delayed process the muon in the atomic ground state may be captured into the nucleus by weak interaction, thus exciting the nucleus with the possible consequence of delayed fission. In fact, this process may be considered as muon capture on one of the bound protons in the weak process $\mu^- + p \rightarrow n + \nu_\mu$ resulting in a broad spectrum of excitation energies in the nucleus.

Dynamics of large amplitude collective nuclear motion, as occuring in heavy ion reactions and in nuclear fission is difficult to observe.

It is the aim of this contribution to present an experimental access to this problem in prompt muon induced fission. In the picture of a potential energy surface of a heavy fissioning nucleus especially the transition from saddle to scission is of interest. In the interaction of the collective nuclear motion with the single particle degrees of freedom by non conservative forces, i.e. forces which may be treated in the picture of the nuclear viscous fluid by a phenomenological viscosity constant, energy and/or angular momentum may be transferred and the characteristic time is determined by the energy dissipation i.e. the friction process. From TDHF calculations (Negele et al. /2/) and from fission calculations applying the Fokker-Planck equation (Weidenmüller et al. /3/) the time scale involved for this process is determined to be several 10^{-21}s.

The muon has been selected as a probe, since it behaves as a spectator in the case of prompt fission and it may serve as an indicator of the process. Each probe disturbs the system it is used for to explore. Because of this the quality of the muon as a spectator has to be proven by revealing possible distorting influences on the promptly fissioning nuclear system.

Measurements on heavy muonic atoms have been performed to obtain consistent information concerning the following topics:

1. Probabilities of radiationless transistions
2. Fission probabilities of radiationless transitions
3. Augmentation of the fission barrier due to the presence of the muon
4. Fission fragment spectroscopy
5. Probability of muon attachment to the light fission fragment in a prompt fission process

Fig. 1 shows three regions of interest. In the first region the muonic atom is formed. Radiative transitions occur and feed the subsequent levels. The nuclear groundstate shape of ^{237}Np has been determined. Radiationless transition probabilities being of interest especially for the depopulation of the 3d and 2p fine structure complexes have been measured for muonic ^{208}Pb, ^{232}Th, ^{238}U and ^{237}Np. In the next step the fission probabilities of these radiationless transitions have been determined for the nuclei ^{238}U and ^{237}Np in $(\mu^-,\gamma f)$ coincidence measurements and the augmentation of the nuclear fission barrier caused by the presence of the muon has been evaluated.

In the second region the development of nascent fragments may be influenced by the dynamics of the system.

In the third region well separated fragments are developed. Results from fission fragment spectroscopy of prompt muon induced fission have been compared with those obtained in hadron induced fission.

Having revealed the distortions of the nuclear system by the muon a measurement has been designed to investigate fission dynamics by considering the fate of the muon after a prompt fission process as being indicative of the process.

2) Nuclear groundstate shapes

The muonic X-ray spectrum of ^{237}Np provided a model independent nuclear structure information regarding B(E2)↓-values and the nuclear spectroscopic quadrupole moment $Q=3.886\pm0.006$ b, which could be determined from the hfs of the muonic 5g→4f-complex to an accuracy of about 0.2%. The intrinsic quadrupole moments of the nuclear groundstates show a smooth systematic increase with increasing atomic mass in the actinides. The present value for the intrinsic quadrupole moment $Q_0=10.88\pm0.02$ b fits well into this regular trend (fig. 2) /4, 5/ and is also in good agreement with theoretical calculations. For more details see ref. /6/.

3) The probability of nonradiative decay in heavy muonic nuclei

The role of the 2p→1s and 3d→1s radiationless (r.l.) transitions in nuclear excitation leading to prompt processes like γ-decay, neutron emission or fission has been discussed in several papers /7-15/(see tab. 1). According to Teller and Weiss /12/ the probability for a radiationless 3d→1s transition ($P_{r.l.}$(3d→1s)), calculated for muonic ^{238}U to be about 15%, should be quite similar for muonic atoms of neighbouring nuclei such as ^{232}Th and ^{237}Np. For ^{238}U /13/ and ^{237}Np /15/ the probability is found to agree with this prediction. To provide more systematical data on the probability of the radiationless muonic 3d→1s transition further measurements have been performed with the target nuclei ^{208}Pb, ^{232}Th and, repeatedly, ^{238}U.

This experiment was performed by applying two Ge-detectors and a large volume CsF-scintillator in a similar way as described in ref /13,15/. From a comparison of the muonic X-ray single spectra with the measured coincidences with the 2p→1s transition detected in the CsF-crystal the fraction of missing 3d→2p muonic X-rays is obtained, giving the total decay probability of the 3d-level not populating the 2p-level. Correcting this value for the radiative 3d→1s transition (Γ(3d→1s)/Γ(3d→2p)=5%) one obtaines the probability for the nonradiative decay of the 3d-level.

In table 2 $P_{r.l.}$(3d→1s) and $P_{r.l.}$(2p→1s) are listed for the measured atoms. $P_{r.l.}$(2p→1s) is determined under the assumption that the 2p→1s transition in ^{208}Pb is 100% radiative. The 2p→1s radiationless strengths show large differences for the various nuclei. Table 2 clearly shows the threshold effect of the (γ,n) reaction influencing the total photo absorption cross section. The highest probabilities for a radiationless 2p→1s transition occurs in ^{235}U and ^{239}Pu. In both nuclei the energies of the $2p_{3/2}$→$1s_{1/2}$ and $2p_{1/2}$→$1s_{1/2}$ transitions are larger than the neutron binding energy. In ^{232}Th and ^{237}Np less nonradiative strength is observed for energies below the neutron thresholds. The nucleus ^{238}U takes a medium position, since the neutron separation energy S_n is between the energies of the $2p_{1/2} \rightarrow 1s_{1/2}$ and $2p_{3/2} \rightarrow 1s_{1/2}$ transitions.

The giant resonance concept of Teller and Weiss /12/ does not explain these large differences of the probabilities for radiationless transitions, because the energy and the width of the giant dipole resonance vary only little from nucleus to nucleus.

By comparing the single to coincidence ratio of the subcomplexes (S) of the 3d→2p transitions with the ratio of the whole complex (C) it is possible to determine the nonradiative transition probabilities of the fine structure lines $2p_{3/2} \rightarrow 1s_{1/2}$ and $2p_{1/2} \rightarrow 1s_{1/2}$ by using the values for $P_{r.l.}(2p \rightarrow 1s)$ of table 2 for the whole complex. The errors given in table 3 contain only the experimental uncertainties of this work. The errors of $P_{r.l.}(2p \rightarrow 1s)$ of table 2 must be treated as systematical errors, which do not influence the differences of the nonradiative transition probabilities of the fine structure components. The deviation $R = (P^{2\rightarrow 1}_{rad}(S) - P^{2\rightarrow 1}_{rad}(C))/P^{2\rightarrow 1}_{rad}(C)$ of the relative radiative transition probabilities is independent of $P_{r.l.}(2p \rightarrow 1s)$ of the whole complex.

There is no significant difference between $R(2p_{3/2} \rightarrow 1s_{1/2})$ and $R(2p_{1/2} \rightarrow 1s_{1/2})$ for lead, which supports the assumption of $P_{r.l.}(2p \rightarrow 1s) = 0$ (as suggested above).

The values $P^{2\rightarrow 1}_{r.l.}(S)$ for ^{232}Th show a decrease for increasing transition energies. This surprising fact may be explained by a resonant absorption of the $2p_{1/2} \rightarrow 1s_{1/2}$ transition.

The large difference between the nonradiative transition probabilities of the subcomplexes in ^{238}U exceeds the expected value according to the giant resonance model of Teller and Weiss /12/. This can be explained by the opening of the neutron channel. By regarding some hyperfine parts within the $2p_{3/2} \rightarrow 1s_{1/2}$ transition leading to different mean excitation energies of 6435 keV and 6533 keV this explanation is confirmed because they show no difference in the nonradiative transition probabilities.

4) The fission probability of non radiative transitions measured with muonic X-rays in coincidence with prompt and delayed fission

Experimentally there is a systematic and rapid increase in the probability of prompt fission with the value of the fissility parameter. The prompt fission yield of ^{237}Np is about 10 times higher than the corresponding figure for ^{238}U (see table 6). Prompt fission may be induced by non radiative 2p→1s and 3d→1s transitions. The energetic relation between the nuclear fission barrier heights and the transition energies of the $2p_{3/2} \rightarrow 1s_{1/2}$ and $2p_{1/2} \rightarrow 1s_{1/2}$ fine structure complexes have been investigated determining the fission probabilities per non radiative transition.

In the case of ^{237}Np these values were measured for the first time, using a parallel plate avalanche counter (PPAC) in a ($\mu^-,\gamma f$)-experiment. The detector consisted of 32 titanium foils with a total loading of 19.5 mg/cm^2 ^{237}Np in the form of electrolytically deposited oxide. To detect the γ-rays we made use of a germanium counter with anti-compton system (30% relative efficiency). Aluminum plates shielded the Ge-detector from the natural activity of the target. The detection efficiency of the PPAC was determined directly using the measured coincidences of muonic X-rays with fission and the single rates yielding $\epsilon_{fiss} = 0.83 \pm 0.22$ (to be published).

The prompt fission probabilities or muonic ^{238}U, first investigated in a ($\mu^-,\gamma f$)-experiment by Johansson et al. /16/ have been measured again. In this experiment we used a PPAC with 26 target foils of depleted Uranium rolled to a thickness of 6mg/cm^2 (4g total amount). Each target was situated between two thin aluminized mylar foils which formed the electrodes of the 1.5mm thick PPAC's. The experimental setup consisted of this fission chamber, two Ge-anti-compton systems, and a CsF-detector. With this arrangement the fission probabilities (P_f) and the probabilities of non radiative transitions were measured in the same experiment for ^{238}U.

The values obtained for the radiationless transition probabilities have been given in chapter 3. The corresponding probabilities for prompt fission are listed in table 4. In contrast to ^{237}Np for ^{238}U the fission probability of the 2p→1s transition is strongly dependent on the energy. Detailed calculations of P_f in the frame of the double humped barrier were performed for these nuclides /17/. With the computer code FISALL the fission barrier parameters given in table 5 were determined from hadronic fission probabilities measured by Back et al. /18/ (figure 3). For Neptunium the parameters are essentially the same as those published in ref. /18/. For ^{238}U the values are similar to those obtained by Bhandari /19/ which satisfactorily reproduce observed photofission cross sections. The dependence of the fission probability on the barrier augmentation is shown in figure 3 too, assuming the following changes of the barrier parameters: ΔE_a = 0.1 MeV, ΔE_{II} = 0.3 MeV, ΔE_b = 0.5-1.0 MeV, $\Delta h\nu_a = \Delta h\nu_{II} = \Delta h\nu_b = 0$ MeV. The measured fission probabilities of the 2p→1s transitions indicate a barrier augmentation of ΔE_b = (0.6 ±0.1) MeV.

The fission probabilities after the excitation by the radiationless 3d→1s transitions are P_f(3d→1s) = (5.7 ±1.9)% for ^{238}U and P_f(3d→1s) = (41 ±22)% for ^{237}Np. The decrease of the P_f from the corresponding values of 18% /20/ and 60% /21/ respectively as measured at the same excitation energy in hadronic reactions is another unambiguous indication of the barrier augmentation. Calculations applying the model of Kuprianov et al. /22/ and using the parameters given in table 5 result in fission probabilities for hadronic reactions of $P_f(^{238}U)$ = 16% and $P_f(^{237}Np)$ = 66%. From the measured values for the P_f(3d→1s) the increase of the outer barrier can be calculated to be $\Delta E_b = (0.6^{+0.22}_{-0.16})$ MeV for ^{238}U and ΔE_b = (0.7 ±0.6) MeV for ^{237}Np.

5) Relative total fission probabilities per muon stop for $^{233,234,235,236,238}U$ and ^{237}Np

With the same experimental setup as used in the lifetime experiment (see chapter 6) a measurement of the relative probabilities of muon induced fission was accomplished comparing the ratios of the fission countrates in the PPAC. The differences of the beam intensity at the different target positions were measured with a set of ^{235}U targets of well known target thicknesses and could be corrected for. Mesochemical influences on the relative capture rates can be excluded because all isotopes were in the form of UF_4 or NpF_4. The relative fission probabilities were determined in destinct measurements with varied order of the targets leading to consitend results which are listed in table 6a and b.

6) Muon capture rates in $^{233,234,235,236,238}U$ and ^{237}Np

The lifetime of negative muons bound in the 1s orbit was measured for the isotopes $^{233,234,235,236,238}U$ and ^{237}Np. The detection system consited of a scintillator, identifying the incident muons and a PPAC containing all the isotopes. The target material was evaporated on thin Ni foils (200μg/cm^2) with a thickness of typically 100μg/cm^2. Each target was placed between two 1.5μm thick mylar foils vapor coated with gold. They formed the cathodes of fast PPAC's mounted on both sides of the targets. The anodes consisted of Al foils, which were thick enough (20μm) to stop the fission fragments. Both fission fragments were detected in coincidence to suppress the background from the alpha activity of the target material. The muon fission time distribution of all targets were measured with the same TAC/ADC system with a time resolution of about 0.7ns. The obtained mean lifetimes and the deduced capture rates are listed in table 7. The possibility of the presence of a second lifetime in the time distribution possibly due to the decay of fission isomers was investigated. No indication was found for any of the isotopes. The PPAC and a typical time spectrum are displayed in figs. 4 and 5.

7) The distribution of mass yield and total kinetic energy

The measured distributions of fragment masses and total kinetic energies (TKE) have been compared with results from hadron induced fission reactions /23,24/. The conclusion can be drawn that the mass distributions from different reactions agree within the error bars if the excitation energy of the compound nuclei is the same.

The TKE for prompt and delayed muon induced fission are systematically lower by about 1.5 MeV as compared to hadronic fission reactions leading to the same compound nucleus. This effect may be explained by a screening of the nuclear charge by the muon. The quality of the muon as a probe to investigate fission dynamics is neither influenced by the effect described above nor by the barrier augmentation.

8) Muon attachment to fission fragments

In a number of theoretical works /25-34/ the possibility has been discussed to get information on dynamics in prompt nuclear fission from the value of the probability with which the muon is attached to the light fission fragment. The results are controversial. The necessity to measure this probability has become obvious. In particular the measurement of the dependence of this quantity on mass number and TKE(A) would give a much more solid basis for further discussions. Belovitskij et al. /35/ have performed very illustrative work with photo emulsions and have obtained a value for P_L and results on muon conversion also. Here we give preliminary results on a first electronical measurement of P_L.

In the prompt muon induced fission the muon only interacts via the electroweak force. After the fission process it is attached to either fission fragment with a probability of more than 0.98 /36/. The ratio P_L/P_H, of the attachment probability of the light and heavy fragment, respectivly, is depending on the fission time. Because of the higher binding energy of the muon P_H is bigger than P_L and it increases with the fission time.

The muon bound to a fission fragment decays into an electron with a probability of 0.05. Because the lifetime of the bound muon is large compared to the time of flight for the fragment to be spatially separated, the corresponding fission fragment can be determined by detecting the fission fragment and the track of the decay electron.

The fission fragments are registered in large area semiconductor detectors. The decay electron is detected with the magnetic spectrometer SINDRUM /37/, which consist out of 5 multiwire proportional chambers (MWPC) and a hodoscope (fig. 6).

In the data analysis electron tracks were accepted if they intersect only one detector area for assi the electrons to the fission fragments unambiguously. By applying windows on light and heavy fragr and requiring prompt fission, the attachment probabilities are obtained from the corresponding spectra. The time spectra $t(\mu^-,e^-)$ of light and heavy fragments are shown in figure 7a and b, respect Lifetimes were fitted to the data of figure 7. Preliminary results are listed in table 7.

References

/1/ J. A. Wheeler, Rev. Mod. Phys. 21 (1949) 133

/2/ J. W. Negele et al., Phys. Rev. C17 (1978) 1098

/3/ H. A. Weidenmüller et al., Zhang Jing-Sheng, Phys. Rev. C29 (1984) 879

/4/ R. B. Shera et al., Phys. Rev. Lett. 53 (1984) 1888

/5/ J. D. Zumbro et al., Phys. Lett. B267 (1986) 383

/6/ C. T. A. M. de Laat et al.; Phys. Lett. B189 (1987) 7

/7/ Balats et al., Sov. Phys. JETP 11 (1960) 1239, Sov. Phys. JETP 12 (1961) 813, Sov. Phys. JETP 22 (1966) 4

/8/ D. F. Zaretski, V. M. Novikov, Nucl. Phys. 14 (1959) 540, Nucl. Phys. 28 (1961) Sov. Phys. JETP 14 (1962) 157, Sov. Phys. JETP 14 (1962) 198

/9/ J. A. Diaz et al., Nucl Phys. 40 (1963) 54

/10/ V. Srinivasan, M. K. Sundaresan, Can. J. of Phys. 49 (1971) 621

/11/ K. P. Lohs et al., Nucl. Phys. A236 (1974) 457

/12/ E. Teller, M. S. Weiss, Trans N. Y. Acad. Sci. 40 (1980) 222

/13/ T. Johansson et al., Phys. Lett. B116 (1982) 402

/14/ F. F. Karpeshin, V. O. Nesterenko, Comm. of the JINR, Dubna 1982, E4-82-69

/15/ P. David et al., Phys. Lett. B180 (1986) 324

/16/ T. Johansson et al., Phys. Lett. B97 (1980) 29

/17/ J. Hartfiel, Thesis, Inst. f. Strahlen- u. Kernphysik, Univ. Bonn 1985

/18/ B. B. Back et al., Phys. Rev. C9 (1974) 1924, Phys. Rev. C10 (1974) 1948

/19/ B. S. Bhandari, Phys. Rev. C22 (1980) 606

/20/ P. David et al., Z. Physik-Atomic Nuclei A326 (1987) 367 J. van der Plicht et al., Nucl. Phys. A346 (1980) 349

/21/ A. Gavron et al., Phys. Rev. C12 (1976) 23

/22/ V. M. Kupriyanov et al., Sov. J. Nucl. Phys. 39 (1984) 176

/23/ P. David et al., Phys. Lett. B124 (1983) 161

/24/ P. David et al., Z. Physik-Atomic Nuclei A328 (1987) 37

/25/ V. A. Karnaukhov, Sov. J. Nucl. Phys. 28 (1978) 621

/26/ Y. N. Demkov et al., JETP Letters 28 (1978) 287

/27/ D. F. Zaretski et al., Sov. J. Nucl. Phys. 31 (1980) 24

/28/ P. Olanders et al., Phys. Lett. B90 (1980) 193

/29/ Z. Y. Ma et al., Nucl. Phys. A348 (1980) 446, Phys. Lett. B106 (1981) 159

/30/ J. A. Maruhn et al., Phys. Rev. Lett. 44 (1980) 1576

/31/ V. E, Oberacker et al., LAMPF II, Workshop Los Alamos 1982, LA-9798-P,UC

/32/ L. Bracci et al., Nucl. Phys. A423 (1984) 429

/33/ P. Boi et al., Phys. Lett. B132 (1983) 39

/34/ F. Karpeshin et al., Sov. J. Nucl. Phys. 45 (1987) 965

/35/ G. E. Belovitskij et al., see contribution to this conference

/36/ G. E. Belovitskij et al. , Lecture notes in Physics Vol 158 (1982) 71

/37/ SINDRUM, W. Bertl. et al., SIN-Proposal R-80-06.2, 1981

/38/ R. von Mutius, Thesis, Inst. f. Strahlen- u. Kernphysik, Univ. Bonn 1985

It is a pleasure to thank Professor J. P. Blaser and his staff for their encouraging support and for excellent working conditions at PSI. This work was financially supported by Bundesministerium Forschung und Technologie des Bundesrepublik Deutschland, contract number 06BN271.

Table 1: Theoretical results for $P_{r.l.}$ [%] of ^{238}U :

Theory	2p→1s	3p→1s	3d→1s
Zaretski et al. /8/	40	78	-
Teller et al. /12/	24	50	15
Karpeshin et al. /14/	11-15	48-60	25-33
Lohs et al.			
^{209}Bi /11/	-	25	3.9

Table 2: Experimental radiationless transition probabilities $P_{r.l.}$(2p→1s) and $P_{r.l.}$(3d→1s)

	$P_{r.l.}$(2p→1s) [%]	Reference	$P_{r.l.}$(3d→1s) [%]	Reference
^{208}Pb			3.0±1.5	this work
^{232}Th	15.0±7.0	/7/	11.0±1.1	this work
^{235}U	29.0±5.0	/7/		
^{238}U	23.0±4.0	/7/	13.6±1.4	this work
			14.0±5.0	/13/
^{237}Np	15.4±2.4	/38/	9.0±4.0	/15/
^{239}Pu	41.0±6.0	/7/		

Table 3: Experimental nonradiative transition probabilities of the fine structure components (E = mean transition energy of the fine structure component, R = relative radiative transition probability, see text)

	E [keV]	$P_{r.l.}$(2p→1s) [%]	R [%]	S_n [keV]
^{208}Pb				
$2p_{1/2}$→$1s_{1/2}$	5780		+0.17±0.98	7000
$2p_{3/2}$→$1s_{1/2}$	5960		−0.01±0.52	
^{232}Th				
$2p_{1/2}$→$1s_{1/2}$	6060	16.7±1.7	−2.0±1.1	6440
$2p_{3/2}$→$1s_{1/2}$	6360	11.7±1.9	+3.9±1.5	
^{238}U				
$2p_{1/2}$→$1s_{1/2}$	6130	18.2±1.0	+6.3±1.3	6150
$2p_{3/2}$→$1s_{1/2}$	6460	26.2±0.9	−4.2±1.2	

Table 4: Fission probabilities per radiationless transition, E denotes the mean excitation energy. For P_f the absolute errors as the errors of the values relative to each other (in brackets) are given.

	^{238}U		^{237}Np	
Transition	E [keV]	P_f [%]	E [keV]	P_f [%]
$2p_{1/2} \to 1s_{1/2}$	6130	2.91±1.14(±0.22)	6218	61±19(±4)
$2p_{3/2} \to 1s_{1/2}$	6435	0.74±0.25(±0.10)	6525	57±18(±4)
$2p_{3/2} \to 1s_{1/2}$	6533	1.80±0.62(±0.21)		
3d→1s	9400	5.7 ±1.9 (±0.6)	9600	41±21(±19)

Table 5: Barrier parameters of ^{238}U and ^{237}Np. All values are in MeV.

	E_a	$h\nu_a$	E_{II}	$h\nu_{II}$	E_b	$h\nu_b$	Reference
^{238}U	5.9	1	2	0.9	6.12	0.62	/18/
	5.6	1.77	2.3	0.52	6.35	0.74	/19/
	5.55	1.6	2.3	0.52	6.35	0.74	this work
^{237}Np	5.7	0.8			5.5	0.55	/18/
	5.9	0.8			5.6	0.52	/21/
	5.8	0.8	2.45	0.5	5.5	0.55	this work

Table 6a: The total fission probabilities per muon stop (P_f) relative to that of ^{238}U ($P_f(^{238}U)$). The ratios of the prompt (p) to the delayed (d) fission probabilities per muon stop were determined from the data of the lifetime experiment.

	p/d [%]	$P_f/P_f(^{238}U)$	p/P_f [%]	$p/P_f(^{238}U)$ [%]
^{233}U	20.05±0.14	3.37±0.13	16.70±0.12	56.28±2.21
^{234}U	17.69±0.08	2.20±0.17	15.03±0.07	33.07±2.56
^{235}U	12.54±0.08	2.18±0.06	11.14±0.07	24.29±0.69
^{236}U	18.56±0.12	1.40±0.06	15.65±0.11	21.91±0.95
^{238}U	8.76±0.10	1	8.05±0.09	8.05±0.09
^{237}Np	28.14±0.08	3.82±0.23	21.96±0.05	83.89±5.05

Table 6b: Fission probabilities per stopped muon as calculated with the value for ^{237}Np obtained in the ($\mu^-,\gamma f$)-experiment (chap. 4).

	total	prompt	delayed
^{233}U	0.48±0.13	0.080±0.020	0.40±0.11
^{234}U	0.31±0.08	0.047±0.012	0.26±0.07
^{235}U	0.31±0.08	0.034±0.009	0.27±0.07
^{236}U	0.20±0.05	0.031±0.008	0.17±0.04
^{238}U	0.14±0.04	0.012±0.003	0.13±0.04
^{237}Np	0.54±0.14	0.120±0.030	0.42±0.11

Table 7: Measured mean lifetimes of muons bound in the 1s-orbit and deduced capture rates Λ.

	τ [ns]	Λ [10^5/s]
^{233}U	68.9±0.3	141.4±0.6
^{234}U	70.6±0.2	137.9±0.4
^{235}U	72.2±0.2	134.8±0.4
^{236}U	74.3±0.3	130.9±0.5
^{238}U	77.0±0.4	126.2±0.7
^{237}Np	69.8±0.2	139.6±0.4

Table 8: Preliminary results of the muon attachment probability to light fission fragments of ^{237}Np depending on mass number: $P_L = N_L/(N_L + N_H \cdot f)$ with $f = P^e_L/P^e_H$. f is the ratio of the detection efficiency for the electrons coming from the light and heavy fragments. The mean lifetimes of the muon were $\tau_L = (182 \pm 27)$ns and $\tau_H = (116 \pm 4)$ns.

Mass window		number of events			
light	heavy	N_L	N_H	f	P_L
107-117	120-130	16±5	140±12	1.15	0.090±0.027
97-107	130-140	65±11	670±26	1.41	0.064±0.011
87-97	140-150	30±7	457±22	1.70	0.037±0.009
77-87	150-160	2±2	67±8	1.99	0.015±0.015
all masses		112±14	1309±37	1.41	0.057±0.007

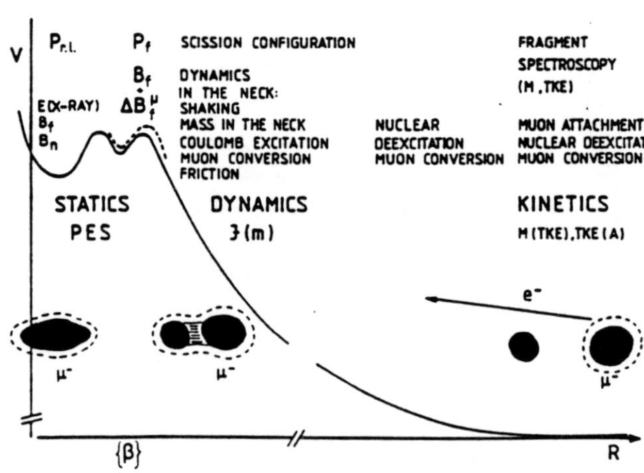

Fig. 1: Different stages studied in prompt muon induced fission

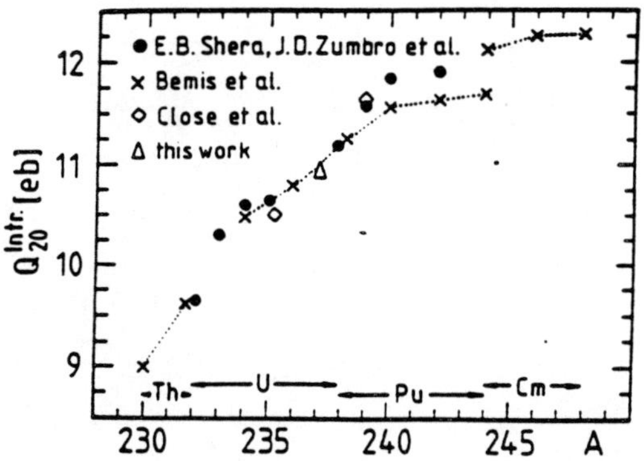

Fig. 2: Spectroscopic quadrupole moments in actinides

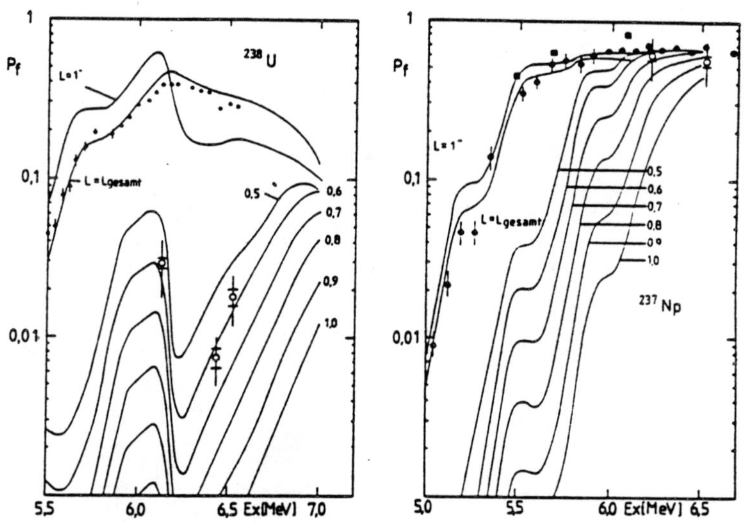

Fig. 3: Fission probabilities for different values of the barrier augmentation (0.5 - 1.0 MeV). $L=L_{gesamt}$ denotes a simulation of experimental data given by Back et al. /18/. The fission probabilities per radiationless transition $P_f(2p\rightarrow 1s)$ are shown with absolute errors (ϕ) and with errors of the values relative to each other (ϕ̄).

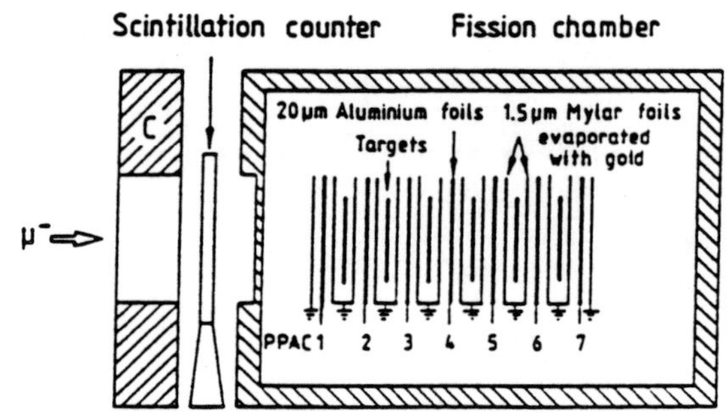

Fig. 4: Schematic drawing of the experimental setup consisting of collimator C, scintillator and fission chamber. Each target is placed between two PPACs. The aluminum foils stop the fission fragments preventing ambiguities in the target identification.

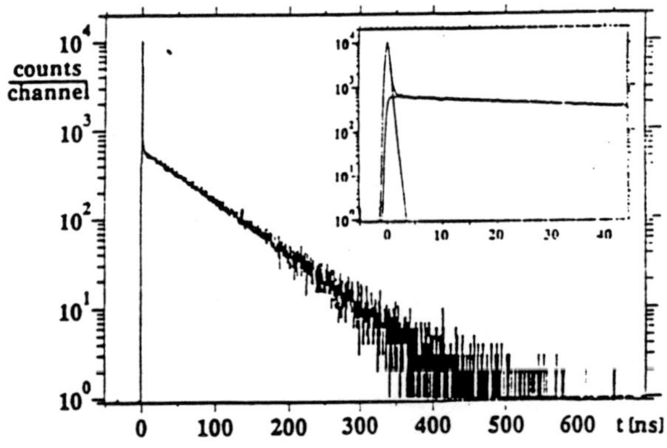

Fig. 5: Time distribution $t(\mu^-,f)$ of ^{236}U. The width of the prompt peak is 0.65 ns. A fit to the delayed part of the spectrum gives a mean lifetime of 74.3 ± 0.3 ns. The total number of events is $2.9 \cdot 10^5$. The background is 0.05 per ns.

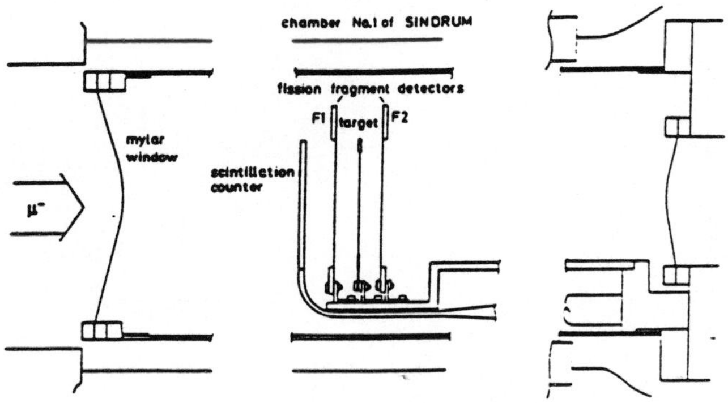

Fig. 6: Schematic view of the fission setup placed in SINDRUM.

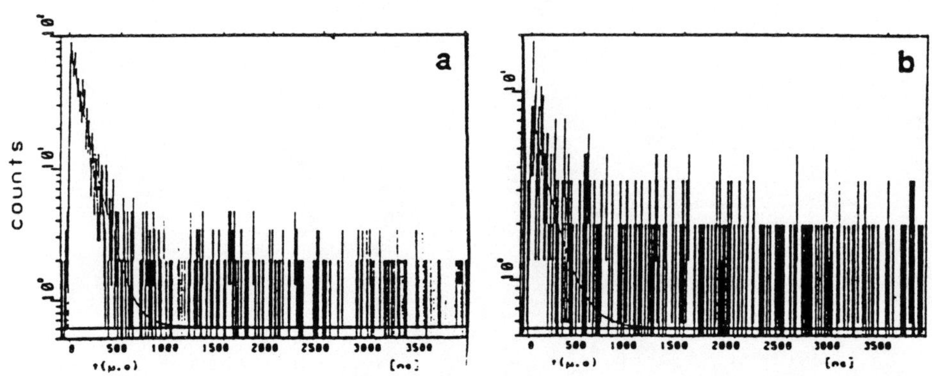

Fig. 7: (μ^-, e^-)-time spectrum of the decay muons attached to the heavy (a) and light (b) fission fragments.

EMISSION OF MUONS BY THE PROMPT FISSION OF URANIUM NUCLEI AND MULTIPOLARITY OF γ-RAY OF FRAGMENTS

G.E. Belovitzkiy, V.N. Baranov, and C. Petitjean*
Institute of Nuclear Research AN USSR

Abstract: Angular and energy distribution as well as probability of emission of conversion muons which are emitted by prompt fission of uranium nuclei are obtained. Multipolarity of γ-ray ($E\gamma$ > 3 MeV) of fragments is determined.

INTRODUCTION

The fission of nuclei can be induced by slow negative muons through two mechanisms: /a/ nuclear capture of muons, /b/ direct (radiationless) transfer to the nucleus of the energy of mesonic-atom transition. These two fission mechanisms are called "delayed" and "prompt". In prompt fission the muon is sticked by one of the fragments, and /a/ may be ejected from the excited fragment into the continuum through an internal conversion process, /b/ may undergo a μ -> e decay, or /c/ may be absorbed by the fragment with the result that particles are emitted (n, p, α).

The study of an angular and energy distribution, of the probability of the conversion of muons and the sticking muon to the light (heavy) fragments and comparison with theoretical calculation permits to determine the multipolarity of γ-ray (E_γ > 3 MeV) of the fragments. This information is very valuable for the study of the fission dynamics.

EXPERIMENT

As a target and detector of charged particles we used NIKFI-K and R emulsions of ~ 150 mm thickness which registered the muons with E_μ < 6 MeV (K) and all energies (R). The method for introduction of uranium salts into the emulsion, exposition to a beam of muons (μE4 beam at the PSI, Switzerland) and scanning of the emulsion under a microscope was described in ref. 1.

About 5.10^5 fissions were analyzed. We chose the events where the particles with Z =1 and range > 15μ are emitted from the fission point (within a time less than 10^{-13} sec) and are stopped in the emulsion.

*)Switzerland, SIN-5234, Villigen

The ranges of light and heavy fragments and a particle, space angle between them and ionization on the tracks were measured. The kinds of the particle were known from the ionization and range. The particles with Z =1 might be muons and protons (p, d, t) which were emitted before or during the fission. The exception of recoil protons was made by using the kinematics of the elastic collisions. On the R-emulsions muons were identified by $\mu \rightarrow e$ decay. We detected 252 conversation muons with energy of $E_\mu > 0.5$ MeV. The determined accuracy of energies was (2 - 10)%.

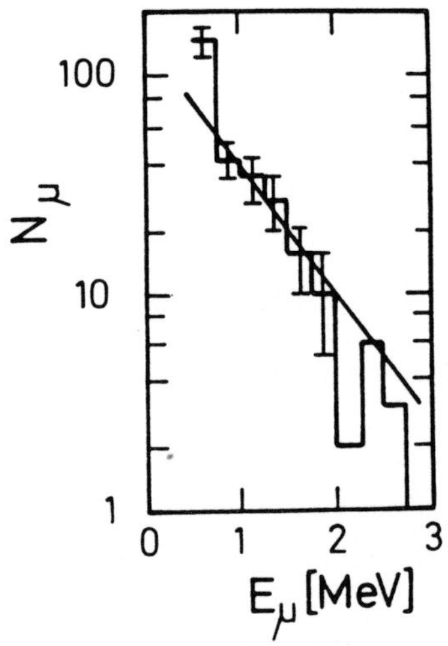

Fig. 1 Measured energy distribution of conversion muons

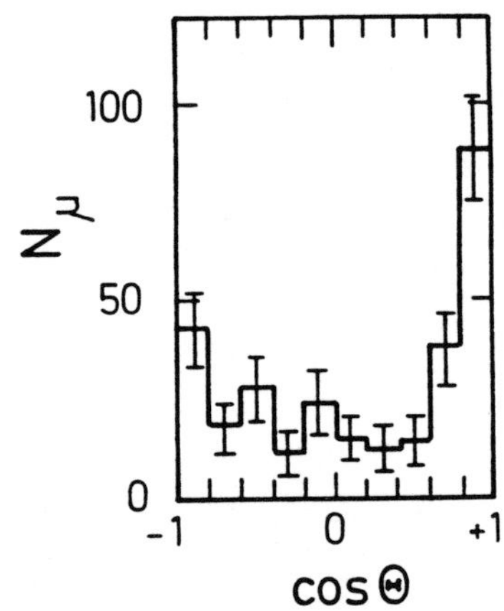

Fig 2 Angular distribution of conversion muons

ENERGY DISTRIBUTION OF CONVERSION MUONS

On the Fig.1 you can see the experimental (corrected) energy distribution of the conversion muons. On the same figure you can see the theoretical curve, which was calculated in refs. 2,3 by using γ-spectrum of ^{235}U (n_{th},f) fragments and KMK (E1 and E2), which agree with the experiment. There are, however, data indicating that the latter is somewhat softer than theoretical one. The probability of conversion of muons per one act of prompt fission is calculated by using the experimental energy spectrum: P_c (E_μ > 0.5 MeV) = (0.8 ± 0.1) . 10^{-2}. This value is in satisfactory agreement with the data obtained in ref.4. The theoretical spectrum indicates that the portion of muons with E_μ > 0.5 MeV is 50% of the total spectrum. Thus P_c (E_μ > 0) = (1.6 ± 0.2) . 10^{-2}.

ANGULAR DISTRIBUTION OF CONVERSION MUONS

It is shown in fig.2. The ratio of muon number emitted with angle less and greater than 90° between muon and heavy fragment directions of motion is used further in order to characterize the angular distribution $N(\theta < 90^\circ) / N(\theta > 90^\circ) = (1.3 \pm 0.2)$. From comparison of theoretical[3)] and experimental angular distributions it is possible to obtain very important information about contribution of light and heavy fragments in muon conversion. Two angular distributions within (l.s.) are calculated on assumption that all muons are emitted only from light or only from heavy fragments. Summing these distributions with the different weights one can obtain a good agreement with the experiment. As a result, the light fragments (P_L) give the contribution in muon conversion of about 30-35 %, and heavy (P_H) give of 65-70 %: $(P_H/P_L)_{ex} \approx 2$. The full experimental probability of conversion is $P = 1.6 \cdot 10^{-2}$. Hence, $(P)_{ex} = 1.05 \cdot 10^{-2}$; $(P)_{ex} = 0.55 \cdot 10^{-2}$.

W_L DETERMINATION

$W_L < 0.1$ was determined by the electronic method[4)]. We see that W is not certain. If $W_L = 0.09$ then all the conversion is from light fragments, if $W_L = 0.04$ - all the conversion is from heavy one. Thus we determined W_L by means of measuring the amount of the charged particles (p, α) emitted from the ends of the light and heavy fragments ($N_L = 10$, $N_H = 113$).

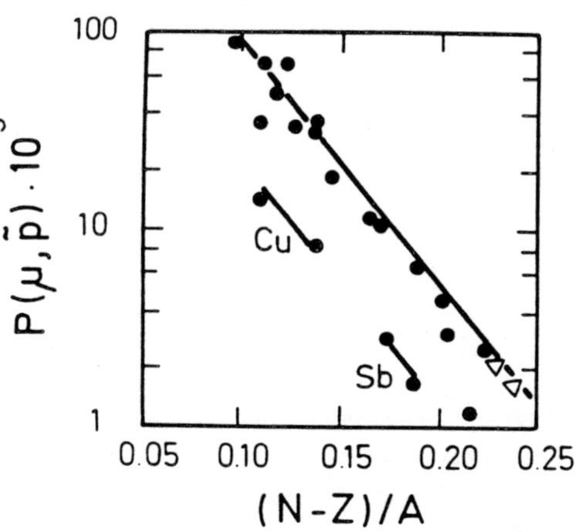

Fig. 3 Muon conversion probability versus neutron excess

It is easy to show that $W_L = K\, N_L/(N_L + N_H)$ where $K = {}_H\bar{P}_{\mu p}/{}_H\bar{P}_{\mu p}$, and ${}_H\bar{P}_{\mu p}$, ${}_L\bar{P}_{\mu p}$ are the probabilities of emission of charged particles when muon is absorbed by heavy and light fragments. In ref. 7 we use all experimental data about the $P_{\mu p}$ for 17 nuclei with $13 < Z < 83$. We obtain correlation $P_{\mu p} \approx \exp(-\alpha(N-Z)/A)$, where α = const., and A, N, Z are number of nucleons, neutrons and protons (fig.3). Using this correlation we obtain K = 0.77 and W = 0.06 ± 0.02.

γ-MULTIPOLARITY

In order to determine the multipolarity of γ-ray of fragments it is necessary to compare experimental data with the theoretical values of P_H and P_L. The total probability of muon conversion P_H, P_L - per one prompt fission for γ-ray multipolarity (EJ) - was calculated using experimental data about the exit of fragments with given Z_i[5], the probability for muon to be sticked to a fragment Z_i[6], and probability of conversion of muon from fragment Z_i for γ-ray multipolarity (EJ)[3]. The results of the calculation of P_H, P_L and their sum are presented in the table for three values of W_L.

From the table one can see that for the E1 transition the calculated values are less and for the E2 transition they are greater than experimental data for all W_L. If we take the probability of conversions for E1 and E2 transition with corresponding weights one can obtain (P_H/P_L) theoretical equal to experimental. In addition $(P_H + P_L)$ must be equal to P_c-experimental.

Full conversion probability x 10^{-2}	W_L = 0.05 E1	E2	W_L = 0.06 E1	E2	W_L = 0.07 E1	E2	Exp. data
P_H	0.42	6.1	0.42	6.0	0.41	5.9	1.05
P_L	0.55	2.0	0.70	2.5	0.87	3.1	0.55
P_H+P_L	0.97	8.1	1.12	8.5	1.28	9.0	1.60

Multipolarity of γ-rays of fragments may be obtained by the decision equation: $(P_i)_{ex} = \alpha_i\, P_i(E1) + \alpha_\lambda\, P_i(E2)$; $(\alpha_1 + \alpha_2) = 1$. When $W_L = 0.5$, the theoretical data agree with the experiment. In this case the multipolarity of γ-rays of the heavy fragments (E_γ > 5 MeV) is (0.9 E1 + 0.1 E2), the light fragments (E_γ > 3 MeV) it is - E1. Although the intensity of γ-rays (E2) is small, but since KMK (E2/E1) is ~10, the conversion of muon on > 50% connected with E2 transition.

In conclusion we show the preliminary results of measuring the conversion muon amount with energy (0.25 - 0.5) MeV. This point lies much higher than the theoretical curve. We suppose that it may be connected with the γ-spectrum of the prompt fission fragments which may be softer than γ-spectrum of $^{235}U(n_{th,f})$.

REFERENCES

1. G.E. Belovitzkiy et al. Yad. Fiz. **43**, 1057 (1986).
2. I.Y. Barit et al., Preprint P-0058 INR AN USSR (1977).
3. F.F. Karpeschin et al., Yad. Fiz. **32**, 55 (1980); **40**, 643 (1984); **45**, 1556 (1987).
4. D. Ganzorig et al., Phys. Lett. **77B**, 257 (1978).
5. P. David et al., Phys. Lett. **124B**, 161 (1983).
6. E. Ma et al., Nucl.Phys. **A348**, 446, (1980).
7. G.E. Belovitzkiy et al., Yad. Fiz. **46**, 1142 (1987).

ANTIPROTON INDUCED FISSION

T. von Egidy
Physik-Department, Technische Universität München,
D8046 Garching, Germany

Abstract: The annihilation of antiprotons with nuclei results in a very localized huge energy deposition mainly on the nuclear surface. This new excitation mode induces fission in heavy nuclei. Special reactions are involved in these processes. New results of fission experiments at LEAR/CERN are presented. Future studies at LEAR are discussed.

1. Antiproton-nucleus interactions

Nearly ten years before the discovery of the fission process and shortly before the first observation of the neutron, Paul Dirac suggested the existence of antiparticles |1|, because his relativistic quantum mechanical equation for the electron had solutions with negative kinetic energies. In 1932 positrons, 1955 antiprotons and 1956 antineutrons were discovered. The existence of antiparticles of all fermions is a bacic fact in elementary particle physics which has no equivalent in classical physics. Therefore, experiments with antiprotons open new fields and allow the study of many interesting processes. A review of antiproton-nucleus interaction is given in ref. 2.

Five years ago the most intense and purest source of low energy antiprotons was commissioned at CERN, Geneva: the Low Energy Antiproton Ring LEAR. Using a fascinating combination of several accelerators and storage rings, LEAR finally delivers for experiments a beam of about $5 \cdot 10^5$ antiprotons per s. The beam has a diameter of a few mm and an energy precision of about 10^{-3} in an energy range between 5 and 1000 MeV. The beam can be split so that several experiments can simultaneously run at LEAR.

Antiprotons are stopped in matter similar to protons. They form antiprotonic atoms and cascade down to lower orbits emitting antiprotonic x-rays and Auger electrons. If the antiproton reaches in heavier atoms an orbit with $\ell \leq 8$ and therefore $n \leq 9$ (ℓ and n are the angular momentum and principal quantum numbers, respectively), it annihilates with a nucleon at the nuclear surface, usually already where the nuclear density is only about 10% of the central density. The annihilation implies an enormous concentration of energy, 1880 MeV in the volume of one or two nucleons. Consequently it has been speculated that special states of nuclear matter, such as quark-gluon plasma, are created. In most cases various combinations of pions, on average five, are produced either directly or by decay of mesonic resonances. Some of the pions which have average energies of 230 MeV enter the nucleus and start with a multipion interaction an intranuclear cascade. In a fast phase many particles (π,p,n,d,t,^{3}He,^{4}He) are emitted |3,4|. A hot nucleus is produced with energies up to several hundred MeV |5| which has low linear and angular momentum in contrast and complementary to other excitation mechanisms. This excitation energy is removed in a slower phase by neutron or proton evaporation or by fission or even fragmentation for heavy nuclei. Finally, residual nuclei are left which have lost many nucleons and are frequently radioactive so that they can be identified by their γ-radiation |6,7|. These processes can be described by intranuclear cascade calculations |5,8-10|.

2. Fission and fragmentation induced by antiprotons

The annihilation of antiprotons at the surface of heavy nuclei can result in very unusual and extreme nuclear matter states. In this context glueballs, large quark bags, quark gluon plasma, shock waves, density waves, fireballs, hotspots, hot nuclear gas and phase transitions of nuclear matter have been discussed. However, it is not clear which signatures can be used to identify these states. The detailed study of fission and fragmentation can be a method to investigate special processes. The standard process after $\bar{p}$-annihilation in very heavy nuclei is supposed to be fast emission of about 5 to 30 nucleons and heating of the nucleus to a few hundred MeV. Since the reactions of several pions with the nucleus can result in a rather homogeneous distribution of much energy with small linear and angular momentum, total excitation energies up to 800 MeV can be expected |5|. Highly excited very heavy nuclei undergo fission. The excitation energy stays essentially in the fission fragments which evaporate consecutively neutrons and also some protons |3|. The distribution of residual fission fragments gives information on these processes |11|. Residual nuclei in the mass region $10<A<50$ would be an indication of nuclear fragmentation |7|. It has been speculated that in highly excited nuclear matter ($E>500$ MeV) a phase transition from liquid to gas states may occur which can be observed by multiple production of nuclear fragments from an explosion of the very hot nucleus.

3. Fission experiments with antiprotons at LEAR

The first experiments on antiproton induced fission were performed at Brookhaven with 2.5 GeV/c antiprotons and mica track detectors |12|. Binary fission cross sections of 2.17 b, 0.29 b and 0.21 b for U, Bi and Au targets, respectively, were measured.

A special fission experiment with antiprotons was done at LEAR in order to identify hypernuclei |13|. The principle of the experimental method is shown in fig. 1. The recoil-distance technique was applied to measure delayed fission. The annihilation is in some cases accompanied by the production of K-mesons which produce Λ-particles in the nucleus and thus hypernuclei. Prompt fission occurs in the target after about 10^{-18} s. Hypernuclei, however, have a lifetime of about 10^{-10} s and may leave the target due to the recoil from pions or prompt particles. These fission events can be observed in the upstream direction shadowed by the target. Such delayed fission events were seen with ^{238}U and ^{209}Bi targets (fig.1). The extracted lifetimes for delayed fission were attributed to the lifetimes of hypernuclei and found to be $(1.0^{+0.8}_{-0.4})\cdot 10^{-10}$ s for U and $(2.5^{+2.5}_{-1.0})\cdot 10^{-10}$ s for Bi. Prompt fission events were also observed |13|. The opening angle of the fission fragments which deviates from 180° is related to the momentum of the fissioning nucleus.

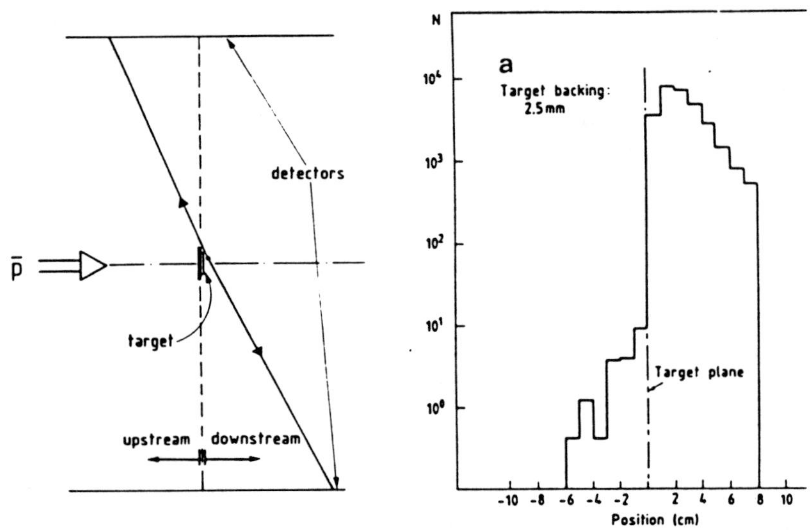

Fig. 1 The left part shows the recoil-distance method for delayed fission. The right part shows the position distribution of fission fragments. Negative positions are in the shadow region.

Fig. 2 shows the deduced momentum transfer for the Bi target together with calculations by Iljinov et al. |5|. Relative fission probabilities were determined for Au, Pb, Bi and U |14|; the results are displayed in fig. 3. It demonstrates an increase from 2% in Au to 100% in U.

Another experiment at LEAR measured neutron and gamma-ray emission from antiproton annihilation at rest in uranium |15,16|. Neutrons were measured with plastic neutron detectors in coincidence with charged particles. The neutron energy or momentum was determined by time-of-flight (fig. 4). Three components in the spectrum were identified originating from direct emission (2.0 per $\bar{p}$), evaporation (1.27 per $\bar{p}$) and fission (2.5 per $\bar{p}$). The neutron temperature of the fission component is 2.5 MeV to be compared with 1.5 MeV from thermal neutron induced fission.

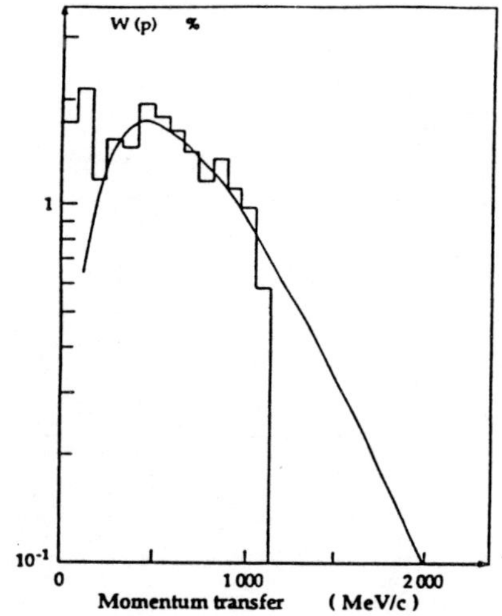

Fig. 2 Momentum distribution of prompt fissioning nuclei from experiments (full line) and cascade calculations for ^{208}Pb (histogram) |5|.

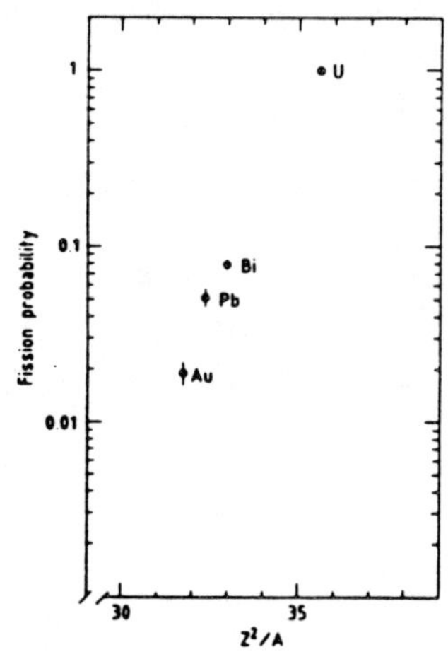

Fig. 3 Relative fission probabilities after $\bar{p}$ annihilation.

The prompt signals in the plastic detectors were assigned to γ-rays which have a rate of 2.2 per $\bar{p}$ with a temperature of 6.0 MeV. This high energy of the γ-spectrum indicates the large excitation energy of of the fission fragments. These results are compared with intranuclear cascade calculations by Jasselette et al. |9|.

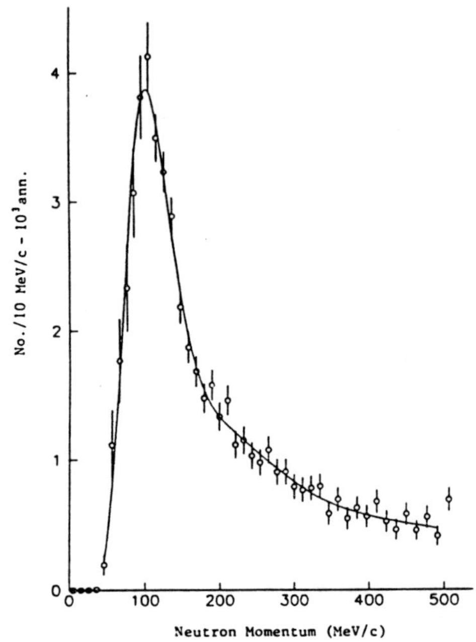

Fig. 4 Observed neutron momentum spectrum after $\bar{p}$-induced fission.

The spectra of p, d, t, ^{3}He, ^{4}He, ^{6}He, ^{8}He and Li have been measured after stopped antiproton annihilation in ^{238}U |3,4| and other targets. Most of these spectra do not show special features for U, because the prompt particle emission takes place before fission. The energy cut-off of the proton spectrum of U at 5 MeV, far below the Coulomb barrier near 15 MeV, indicates proton evaporation from fission fragments. These results and the distribution of fission fragments after stopped antiproton annihilation |11| will be discussed in a special contribution to this symposium.

A new fission experiment is planned at LEAR which will start in December 1988 |17|. Mass and energy distributions and angular correlations of fission fragments and light fragments will be measured for a series of targets with arrays of semiconductor detectors and time-of-flight techniques. New information on unexplored fission modes, multifragmentation, disassembly of nuclei, special nuclear matter states and intranuclear cascades is expected.

I wish to thank H. Daniel, P. David, F. Gönnenwein and F.J. Hartmann for fruitful discussions.

References

|1| P. Dirac, Proc. Royal Soc. A 133 (1931) 60

|2| T. von Egidy, Nature 328 (1987) 773

|3| W. Markiel, H. Daniel, T. von Egidy, F.J. Hartmann, P. Hofmann, W. Kanert, H.S. Plendl, K. Ziock, R. Marshall, H. Machner, G. Riepe, J.J. Reidy, Nucl. Phys. A 485 (1988) 445

|4| P. Hofmann, F.J. Hartmann, H. Daniel, T. von Egidy, W. Kanert, W. Markiel, H.S. Plendl, H. Machner, G. Riepe, D. Protić, K. Ziock, R. Marshall, J.J. Reidy, Nucl. Phys. A to be submitted

|5| A.S. Iljinov, V.I. Nazaruk, S.E. Chigrinov, Nucl. Phys. A 382 (1982) 378

|6| E.F. Moser, H. Daniel, T. von Egidy, F.J. Hartmann, W. Kanert, G. Schmidt, M. Nicholas, J.J. Reidy, Phys. Lett. B 179 (1986)25

|7| E.F. Moser, H. Daniel, T. von Egidy, F.J. Hartmann, W. Kanert, G. Schmidt, Ye. S. Golubeva, A.S. Iljinov, M. Nicholas, J.J. Reidy, Z. Physik A, submitted

|8| J. Cugnon, P. Jasselette, J. Vandermeulen, Nucl. Phys. A 470 (1987) 558

|9| P. Jasselette, J. Cugnon, J. Vandermeulen, Nucl. Phys. A 484 (1988) 542

|10| P.L. Mc Gaughey, M.R. Clover, N.J. DiGiacomo, Phys. Lett. 166 B (1986) 264

|11| H. Machner, Sa Jun, G. Riepe, D. Protić, H. Daniel, T.von Egidy, F.J. Hartmann, P. Hofmann, W. Kanert, W. Markiel, H.S. Plendl, K. Ziock, R. Marshall, J.J. Reidy, to be submitted to Phys. Rev. Lett., and these Conference Proceedings

|12| L. Husain, S. Katcoff, Phys. Rev. C 4 (1971) 263

|13| J.P. Bocquet, M. Epherre-Rey Campagnolle, G. Ericsson, T.Johansson, J. Konijn, T. Krogulski, M. Maurel, E. Monnand, J. Mougey, H. Nifenecker, P. Perrin, S. Polikanov, C. Ristori, G. Tibell, Phys. Lett. B 182 (1986) 146 and B 192 (1987) 312

|14| M. Berrada, J.P. Bocquet, M. Epherre, G. Ericsson, T. Johannson, J. Julien, K. Kilian, J. Konijn, T. Krogulski, M. Maurel, E. Monnand, J. Mougey, H. Nifenecker, P. Perrin, S. Polikanov, C. Ristori, G. Tibell, in "Physics with antiprotons at LEAR in the ACOL era", ed. U. Gastaldi et al., Edition Frontieres, Gif-sur-Yvette, 1985, p. 627

|15| A. Angelopoulos, A. Apostolakis, T.A. Armstrong, B. Bassalleck, G. Bueche, M. Fero, M. Gee, N. Graf, H. Koch, R.A. Lewis, M. Mandelkern, P. Papaelias, H. Poth, H. Rozaki, L. Sakelliou, J. Schultz, J. Schwertel, G.A. Smith, T. Usher, D.M. Wolfe, Phys. Lett. B 205 (1988) 590

|16| T.A. Armstrong, R. Bishop, V. Harris, R.A. Lewis, E. Minor, G.A. Smith, Preprint PSU LEPS 88/11, Pennsylvania State University (1988)

|17| H. Daniel, T. von Egidy, F.J. Hartmann, P. Hofmann, Y.S. Kim, E.F. Moser, P. Weißgerber, P. David, H. Machner, G. Riepe, H.S. Plendl, E.P. Gavathas, J.J. Reidy, K. Ziock, Experiment PS 203 at LEAR/CERN, 1988

FISSION FRAGMENT DISTRIBUTION FOLLOWING ANTIPROTON ABSORPTION AT REST ON ^{238}U

H. Machner, Sa Jun, G. Riepe, D. Protić
Institut für Kernphysik, KFA Jülich, Postfach 1913, D-5170 Jülich, Germany
H. Daniel, T. von Egidy, F.J. Hartmann, P. Hofmann, W. Kanert, W. Markiel, H.S. Plendl
Physik-Department, Technische Universität München, D-8046 Garching,Germany
K. Ziock, R. Marshall
University of Virginia, Charlottesville, Virginia 22901, USA
J.J. Reidy
University of Mississippi, University, Mississippi 38677, USA

ABSTRACT

We have studied the interaction of atomic nuclei with antimatter by measuring the charged particle spectra and the yield distribution of residual nuclei following the absorption of antiprotons by uranium nuclei. The data indicate that every antiproton absorption leads to fission of an highly excited nucleus after emission of energetic particles. The distribution of residual nuclei centers around mass 106.

1. Introduction

In order to study the properties of nuclear matter, it is useful to investigate the nucleus under special conditions of the system. Great interest deserve very high energy densities with eventual "deconfinement" of the quarks from the individual nucleons. Such a phase is called a quark-gluon plasma. In such a phase the nucleons have lost their identity. There is hope to reach this new phase of matter if the energy density considerably exceeds 1 GeV/fm^3. A large series of heavy ion experiments at relativistic energies is devoted to achieve at this goal. There are conjectures of Rafelski[1] and Strottman[2] which suggest to produce this new form of matter by applying antiproton annihilation in matter. To study this possibility as well as the reaction mechanisms in the matter-antimatter interaction we investigated the interaction of antiprotons at rest with atomic nuclei. In this contribution we will concentrate on the uranium case. Next we will describe the experiments, then present our results and finally conclude by comparing these results with model calculations.

2. Experiments

The experiments have been performed at the low energy antiproton storage ring LEAR at CERN. The antiproton beam had a momentum of 201 MeV/c. This beam was degraded in such a way that the number of stopped antiprotons in the target was maximized. During the irradiation times energy spectra of charged particles were measured. Spectra of hydrogen ions[3] were measured with a solid state counter telescope which consisted of stacked HPGe counters[4]. Spectra of helium isotopes were measured with a threefold silicon surface barrier counter telescope. The latter are published in Ref.[5].

In order to measure a distribution of residual nuclei a $32 mg/cm^2$ thick target of enriched ^{238}U was exposed to the beam in three different time intervals for altogether 4.7×10^8 antiprotons. At the end of each individual exposure

γ-ray spectra following the radioactive decays were measured with well shielded Ge(Li)-detectors. An example of such spectra is given in Fig. 1. The lowest

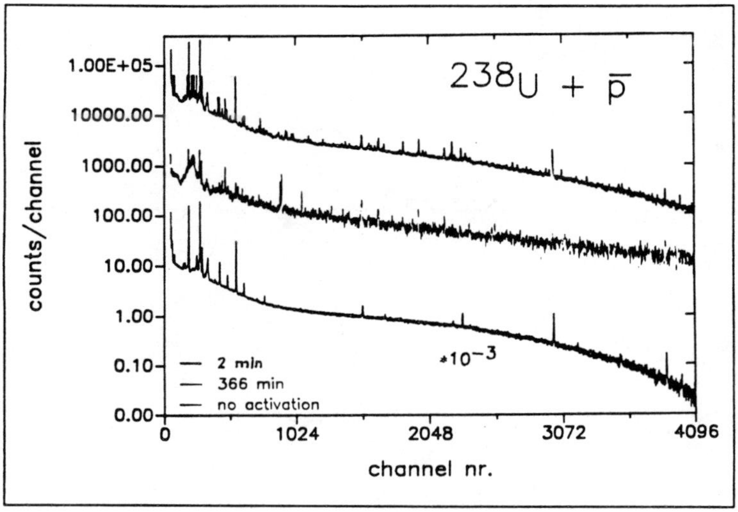

Fig. 1: Spectra of γ-rays emitted from the uranium target before the first irradiation and after two different cooling times

spectrum shows the γ-rays being emitted from the uranium target before any irradiation. Nearly all low energy γ-lines stem from the decay of ^{235}U while the higher energy γ-lines are from the decay of ^{238}U. The spectrum in the middle of the figure corresponds to 2 minutes cooling time and the upper one to 366 minutes cooling time. There are especially in the upper spectrum γ-lines visible which were induced by the absorption of antiprotons.

These photopeaks were evaluated and converted into yield per incident antiproton with the help of known decay schemes. It should be mentioned that most of the identified reaction products were cummulative yields, i.e. they include not only the yield of the specific isotope but also those of the β-precursors.

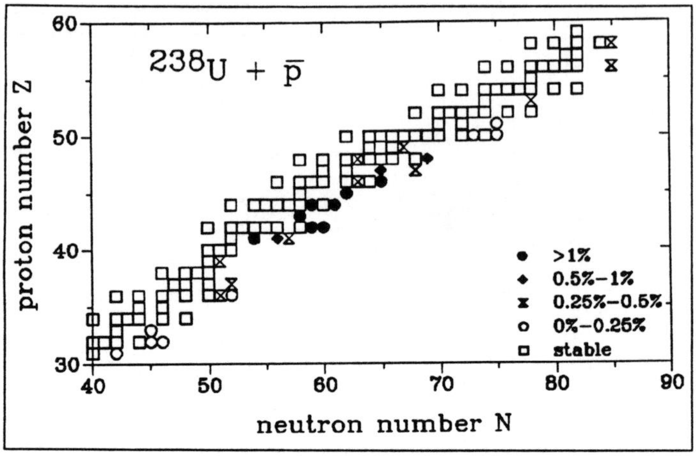

Fig. 2: A part of the chart of nuclides together with the identified isotopes and their deduced yields

3. Results and Discussion

The deduced intensities are shown in Fig. 2. The identified isotopes are close to the valley of β-stability. The yield is large for proton numbers 40–50 and neutron numbers 60–70. To account for the yields in some more quantitative way we have fitted a bi-Gaussian to the data:

$$Y(A,Z) = Y_0 \frac{4\ln 2}{\pi a z} \exp\left(-(4\ln 2)\left(\frac{(Z-Z_0)^2}{z^2} + \frac{(A-A_0)^2}{a^2}\right)\right)$$

with

$$Z_0 = z_0 + z_1 A^{1/2} + z_2 A.$$

The values obtained in the fit for z_1 and z_2 were very close to those defining the valley of β-stability. Therefore, we have used just these values and no longer adjusted them in the fit procedure. The fit performed with these constraints yielded a most probable residual system with mass number A=106.3 and Z=44.62. The overall yield is 0.82 per incident antiproton. If we assume fission to occur to a hundred percent, we can deduce the number of stopped antiprotons to 41 per 100 incident antiprotons. This agrees within errorbars with the numbers obtained from the degrader calibration[5]. The presently deduced distribution has only one maximum which is a sign for high energy fission. The distribution is rather wide with respect to the mass number when compared to distributions obtained with a few hundreds of MeV protons. However, such a wide distribution was obtained in pion induced fission[6] . This result seems to be quite natural since the antiproton–proton annihilation is known to result into approximately five pions[7].

Under the assumption of only binary fission we obtain a total number of emitted charge numbers of approximately two. This agrees with our results from direct measuring the charged particles[5,3] . The number of removed neutrons is approximately 25. This is slightly larger than the number which can be extrapolated from the work of Iljinov et al.[8] but much larger than the prediction of Jasselette et al.[9].

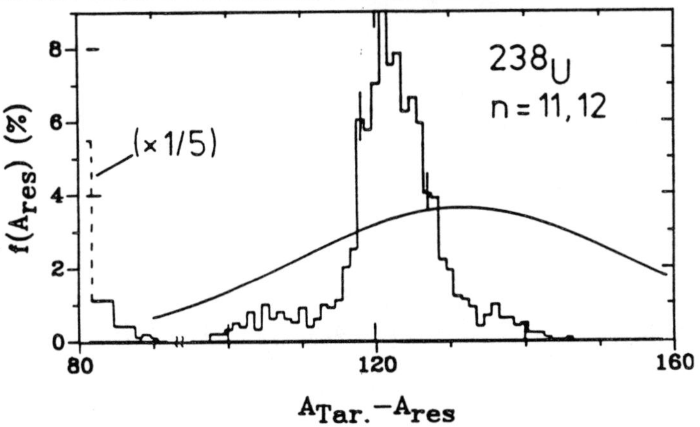

Fig. 3: Comparison of the presently deduced residual mass distribution with a prediction[9]

In Fig. 3 we compare the presently deduced mass distribution with the one predicted by Ref.[9] . The two distributions agree which each other in that they show a symmetric maximum. The calculated ones are, however, much smaller than the presently deduced ones. The shift in the mass number may be caused by the emission of neutrons from the fission products. Such processes are not included in the calculations. However, we observed such processes by detecting protons with energies far below the Coulomb barrier for heavy nuclei[5]. In a recent study of neutrons following antiproton absorption on uranium[10] the number of fission neutrons was found to correspond to approximately 0.5 of all neutrons.

In summary we have for the first time measured the distribution of residual nuclei after antiproton absorption at rest in uranium. The distribution is symmetric around mass 106.3. No target like residue was found. We, therefore conclude, that binary fission has taken place and that every antiproton absorption leads to fission. The fission fragments are highly excited and emit protons and neutrons. To gain more insights in the process and the underlying reaction mechanisms we plan to directly measure fission fragments and their kinetic energies.

REFERENCES

1. J. Rafelski, *Phys. Lett.* **72 B** (1980) 13
2. D. Strottman and W. R. Gibbs, *Phys. Lett.* **149 B** (1984) 288
3. P. Hofmann et al., to be published
4. H. Machner et al., *Phys. Lett.* **138 B** (1984) 39
5. W. Markiel et al., *Nucl. Phys.* **A 485** (1988) 445
6. K. H. Hicks et al., *Phys. Rev.* **C 31** (1985) 1323
7. T. von Egidy, contribution to this conference
8. A. S. Ilinov et al., *Nucl. Phys.* **A 282** (1982) 378
9. P. Jasselette, J. Cugnon, and J. Vandermeulen, *Nucl. Phys.* **A 484** (1988) 542
10. G. Büche et al, Physics of LEAR with low energy antiprotons, ed C. Amsler et al. (Harwood, Chur,1988) p.797

FISSION INDUCED BY 1 GEV PROTONS

L.N. Andronenko, A.A. Kotov, L.A. Vaishnene,
M.M. Nesterov, N.A. Tarasov
Leningradf Nuclear Physics Institute, Gatchina, USSR
W. Neubert
Central Institute for Nuclear Research, Rossendorf/Dresden, Germany

ABSTRACT
This paper gives a review on fission induced by 1 GeV protons. During the last 8 years we studied systematically a wide range of nuclei from uranium to nickel in order to test experimentally the predictions of the liquid drop model. Fissilities, mass and velocity distributions of the fragments, as well as the longitudinal momentum transfer indicate a significant change of the decay process in the vicinity of $Z^2/A \approx 20$ which is related to the onset of the Businaro-Gallone instabilities and/or multifragmentation competition.

1. INTRODUCTION

Fission is a process which is not restricted to heavy nuclei. The liquid drop model (LDM) permits also fission of medium-weight and lighter nuclei. However, the LDM predicts increasing fission barriers with decreasing nuclear masses reaching maximum barriers of about 50 MeV for nuclei with $A \simeq 100$ u. Consequently, energies sufficiently above this predicted barrier are required that these nuclei may undergo fission. We have found that the necessary excitation energy can be transferred in intermediate-energy proton-nucleus interactions whereby the transferred angular momentum is very low.
The region around $A \approx 100$u is of special interest because the LDM predicts also some pecularities in the fission of the regarded nuclei. With decreasing total mass the potential energy surface undergoes a topological change when the fissility parameter crosses the so-called Businaro-Gallone point. Below this point there is no longer a traditional fission saddle point and the mass distribution should show two evaporation like wings extending as far as symmetry where a minimum should be observed /1/.
Therefore, we made efforts to study experimentally fission cross sections, mass distributions and kinetic energies of the fragments as well as excitation energy deposit in a wide range of target nuclei from uranium to nickel.

2. CROSS SECTIONS AND FISSILITIES

Contrary to the traditional method utilizing solid state detectors our measurements are based on the detection of binary events in coincidence whereby the detector arrangement guarantees a nearly 4π solid angle acceptance. Thin samples of ^{238}U, ^{232}Th, ^{209}Bi and ^{208}Pb were inserted between two silicon semiconductor detectors. The proton beam passed directly through this sandwich and coincident fission events and their kinetic energies were measured. This method fails for target nuclei lighter than gold. We have found a loophole by using a double-gap parallel plate avalanche counter (PPAC) with a central electrode made from Ni which supports the target material. While the proton beam passed this counter we measured as a function of the anode voltage the coincidence rate of signals the amplitude of which corresponds to fission fragments. The observed plateau indicates full registration efficiency of the fragments. This method and the necessary corrections are described in the original papers /2/ /3/. The measured fission cross sections are given in table 1. The corresponding fissilities are shown in fig.1 together with two theoretical calculations. The experimental values level off in the region $Z^2/A \leqslant 25$. The minimum predicted by the LDM could not be found in our measurements although it is indicated in photofission data. The shape of the fissility as function of Z^2/A cannot be reproduced by the simple model of Nix and Sassi /4/(solid line) which implies a constant excitation energy. Better agreement gives the cascade-evaporation model of Iljinov et al. /5/ (dashed line). In the framework of this model, the fissility is governed by the ratio Γ_f/Γ_n which depends on the level densities and the fission barriers. A sensible parameter set for a_f/a between 1.02 and 1.11 fit the data. The fission barrier is mainly determined by the LDM part because shell effects vanish at such excitation energies realized in these disassembling nuclei.

Table 1

Target	Z^2/A	σ_f, mb	σ_{in}, mb	σ_f/σ_{in}
^{238}U*	35.56	1,480 ±60	1,720	0.865 ±0.036
^{232}Th*	34.91	940 ±47	1,688	0.557 ±0.028
^{209}Bi*	32.96	183 ± 9	1,568	0.116 ±0.006
^{209}Bi	32.96	180 ± 9	1,568	0.115 ±0.006
^{208}Pb*	32.33	132 ±13	1,563	0.084 ±0.008
^{208}Pb	32.33	142 ±14	1,563	0.091 ±0.009
^{197}Au	31.68	71 ± 7	1,504	(0.472 ±0.047) 10^{-1}
^{nat}Yb	28.32	9.7 ± 1.5	1,372	(0.707 ±0.109) 10^{-2}
^{nat}Ho	27.21	9.8 ± 1.5	1,327	(0.73 ±0.11) 10^{-2}
^{nat}Tb	26.57	9.0 ± 1.5	1,293	(0.69 ±0.13) 10^{-2}
^{nat}Sm	25.63	13.1 ± 2.0	1,240	(1.06 ±0.16) 10^{-2}
^{nat}Te	21.12	8.9 ± 1.8	1,108	(0.80 ±0.16) 10^{-2}
^{nat}Ag	20.45	6.6 ± 2.0	981	(0.67 ±0.20) 10^{-2}
^{nat}Ni	13.52	4.9 ± 1.5	627	(0.78 ±0.24) 10^{-2}

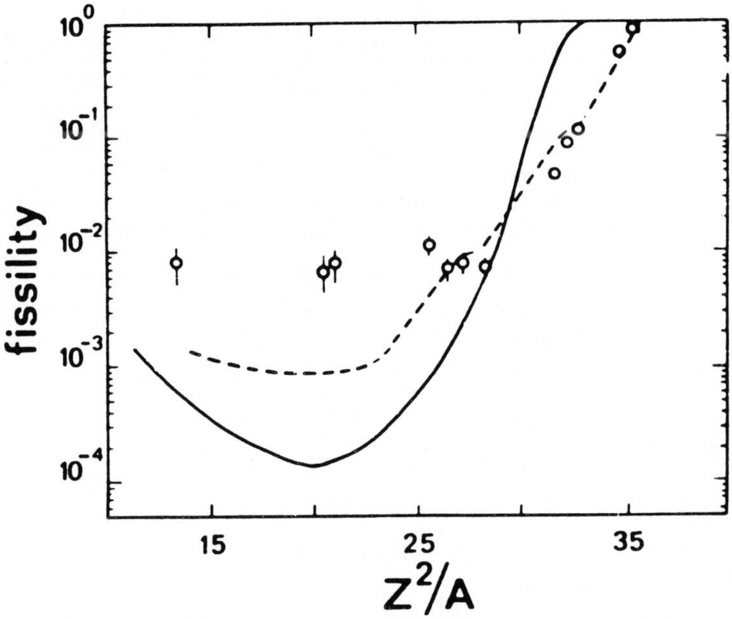

Fig.1

However, a reasonable agreement with the experimental points can be achieved only with LDM barriers lowered by a factor of 0.6...0.7. Probably, the conception of a barrier in the frame work of the LDM becomes invalidated in this region. In reply to this question we have to measure other quantities as outlined in the following sections.

3. HOW CAN WE SEPARATE FISSION FROM OTHER REACTIONS?

In fig.2 (taken from ref. /6/) we show the formation cross sections of fragments emerging from a Au target bombarded with protons from 0.2 GeV to 11.5 GeV. 1 GeV protons are available in our experiments. This energy is distinguished by the possibility of the necessary energy transfer and a still sufficient separation of the fission fragment distribution from the fragmentation and spallation contributions. Apart from these favourable conditions in heavy target nuclei, the fission cross section drops more than two orders of magnitude from uranium to the nickel region. Therefore, effective methods are required to separate the fission fragments in demand if medium-weight and light nuclei are investigated. This problem

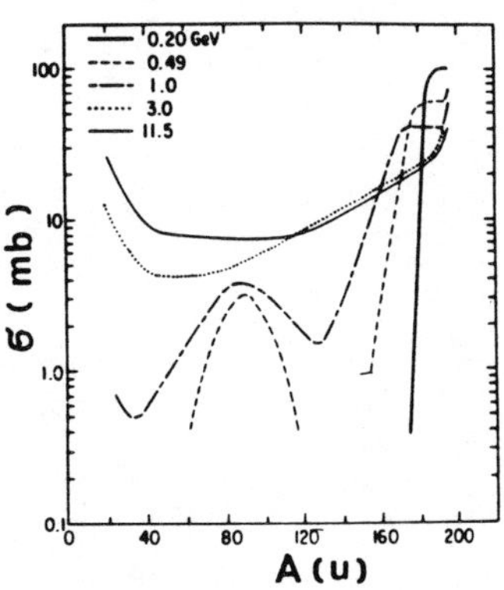

Fig.2

was solved by selective surpression of the predominant background of light particles in the PPACs and by using coincidence criteria considering the underlying kinematics. A two-body final state (like binary fission) is completely determined by the measurement of three independent variables such as two emission angles and the energy of one reaction product. However, with increasing emission of additional light particles the observables become more and more disturbed. In this case, only the mean values of the energy, momentum and relative emission angle of the fragments are conserved. Especially the angular correlations of complementary fragments disperse over a wide angular interval both in-plane and out-of-plan (see section 7).

4. EXPERIMENTAL SET UP

The experimental set up consists of a double-arm spectrometer /7/ (fig.3) which has regard to the mentioned conditions. The start detector is a double-grid avalanche counter (DGAC) which minimizes the energy-loss of the transmitting fragments /3/. This counter is operated quite near to the target. The high background level of light particles is effectively surpressed by a properly chosen gas amplification. The fired Si detector in the right-hand arm of the spectrometer

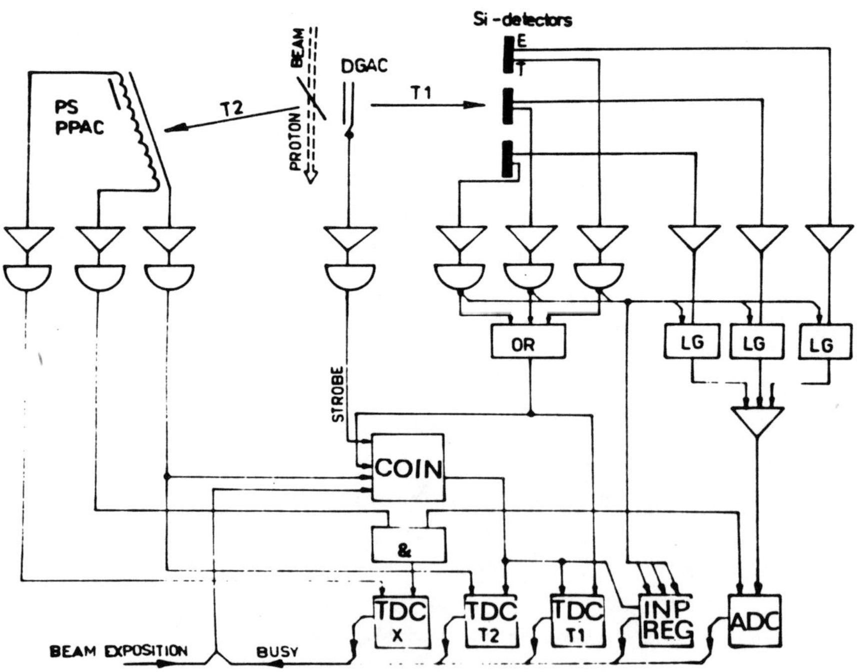

Fig.3

delivers the stop and energy signals of the detected fragment whereas the TOF T2 and the in-plane coordinate are measured with a position-sensitive PPAC with delay-line read out. An angular range of 48 deg. was covered in the reaction plane and an azimuthal angle of $\phi=\pm4$ deg. out-of-plane has been accepted. Therefore, also a considerable part of the non-coplanar events should have been registered.
If a true coincidence event hits both arms, five CAMAC modules are read out: the input register (number of the fired Si detector), the ADC (the corresponding energy), and three TDCs which give the TOFs T1, T2 and the in-plane coordinate X. We emphasize that the masses of the fragments have been determined directly from the measured TOFs and fragment energies.

5. MASS AND VELOCITY DISTRIBUTIONS

Fig. 4 shows the evolution of the mass distributions with decreasing target mass. Heavy target nuclei (Bi, Au, W) delivered symmetric fragment-mass distributions whereas a pronounced asymmetric shape was observed for fragments from Sb, Ag and Ni targets corresponding to a range $21.3 \geqslant Z^2/A \geqslant 13.5$. The measured mass distributions are not only determined by the properties of the scission point configuration but they may be affected also by (i) the involved distributions of the disassembling nuclei, (ii) the distribution of these nuclei over the excitation energy and (iii) by light particle emission from the excited fragments. Calculations carried out in the frame work of the cascade -evaporation model could role out an important influence of (i). However, a drastical change of the mass distribution shapes with excitation energy was found. The calculations show that the mass distributions for Ag undergoing fission becomes highly asymmetric with decreasing excitation energy. Therefore, the expected asymmetry effect at the Businaro-Gallone limit should not be blocked up by (ii). Indeed, multiple particle evaporation becomes a dominant process in this region. Fig.5 shows calculated mass distributions before (dashed histogram) and after light particle emission (solid histogram) for 1 GeV proton induced fission in Au and Ag. The

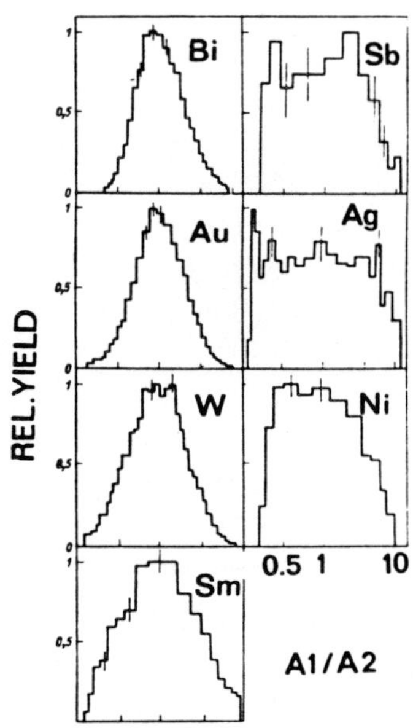

Fig. 4

evaporation has a minor influence for the Au target, but it is decisively for the Ag target. The asymmetric shape at the scission point is transformed into a step-like shape as shown by the solid histogram in fig.5. Consequently, the observation of the genuine asymmetry of the fragment mass distribution arising in the vicinity of the Businaro-Gallone limit is difficult. In order to observe this effect, it is necessary either to select fragments emitted from nuclei with rather low excitation energy or to find such fragment characteristics which are less disturbed by the accompanying light particle emission. The bulk of the light particles should be emitted isotropically in the c.m. system of the moving fragment. Therefore, the velocity vectors of the fragments are expected to be less affected. In fig.6 we plotted the measured velocity distributions of the complementary fragments. In this representation, we can distinguish clearly between the cases of symmetric and asymmetric fission (Bi and Ag targets, respectively). The asymmetric distribution could be confirmed by calculations in the framework of the cascade-evaporation model for fissioning nuclei of $A \simeq 100$ with excitation energies below 400 MeV. In our opinion, we have found an obvious indication of the predicted instability along the mass asymmetry coordinate.

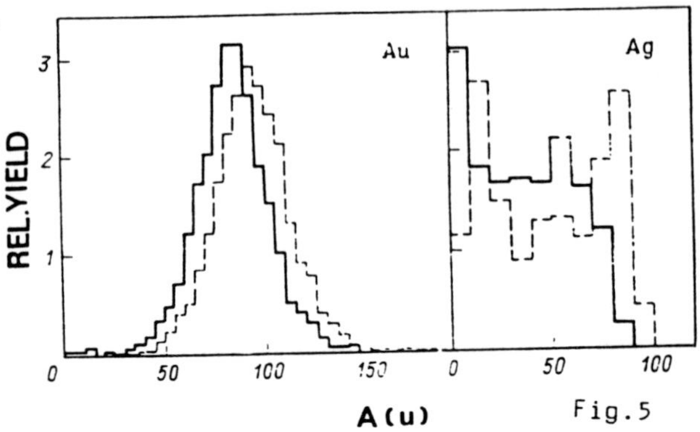

Fig.5

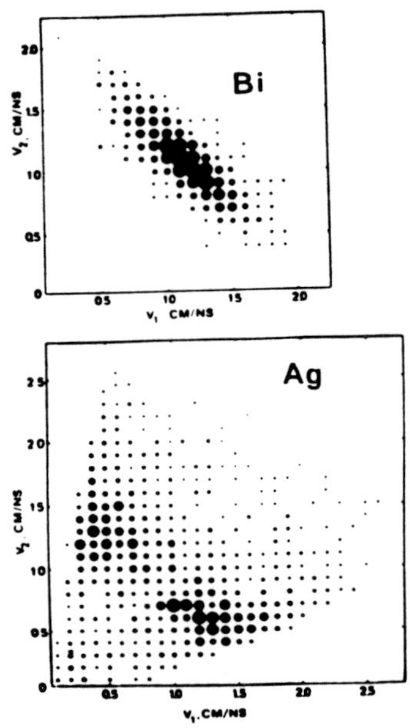

Fig.6

6. KINETIC ENERGIES

Fig.7 shows the average kinetic energy $\langle E \rangle$ of the fragments as function of their masses for various target nuclei. The shape for fragments from the p+Ag interaction appears to be quite similar to the dependences for Bi and W for which the fission process is well identified. To make a comparison we show also the dependence of competing spallation products from interactions of 1...3 GeV protons with Ag (lower solid line). The behaviour displayed in fig.7 agrees qualitatively with the expected one of the Coulomb repulsion energies of deformed two-body systems.

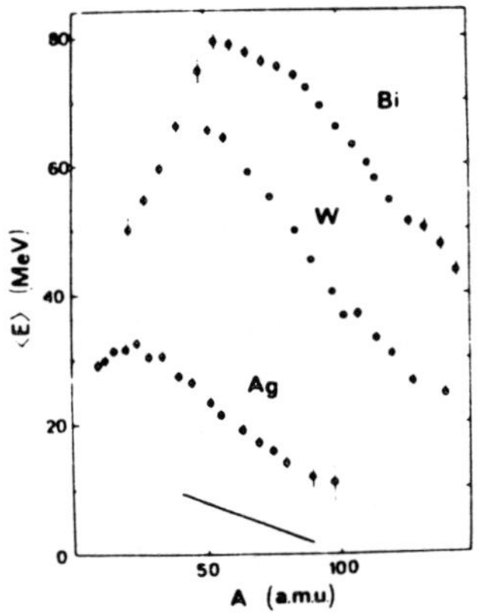

Fig.7

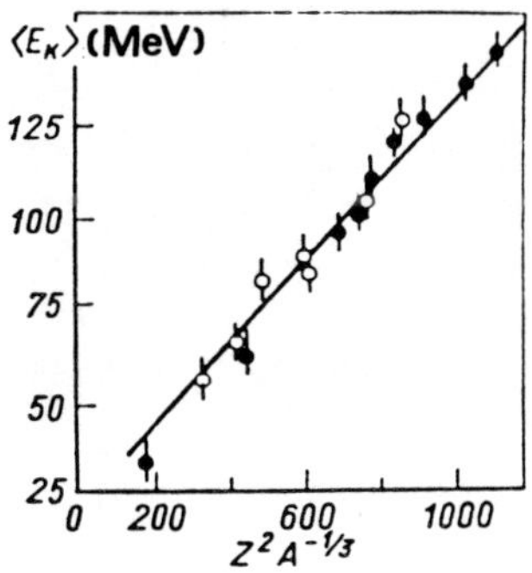

Fig.8

Fig.8 shows the mean total kinetic energies $\langle E_k \rangle = \langle E_1+E_2 \rangle$ as a function of the parameter $Z^2/A^{1/3}$. The TKE are corrected (with exception of the lowest point) by a factor of $(1-\langle n \rangle/\langle A_f \rangle)^{-1}$ where $\langle n \rangle$ is the calculated mean number of emitted nucleons and $\langle A_f \rangle$ the average mass of the nuclei undergoing fission. We used also 0.6 GeV proton data /8/ to complete this systematics. All data points fit the empirical relation $\langle E_k \rangle = 0.1071\ (Z^2/A^{1/3})+22.2$ (MeV) which was deduced from low-energy induced fission /9/. Altogether, the observed kinetic energies confirm that the registered fragments belong to fission events.

7. ANGULAR CORRELATIONS

Folding angle distributions were measured by using the position sensitive PPAC together with the TOF-E arm arranged at the opposite side at θ_1=90 deg. with respect to the beam axis. We observed an increasing deviation of the mean folding angle from collinearity, i.e. $\theta_1+\theta_2$=180 deg., with decreasing target mass. The reason is the increasing longitudinal momentum transferred to the system undergoing fission. In case of symmetric fission the average component of the transferred longitudinal momentum follows from the equation

$$\langle p_{/\!/} \rangle = \langle v_f \rangle \ A_f \ / \ (4\cot^2 \langle \theta_1 + \theta_2 \rangle + 1)^{1/2}$$

where θ_2 is measured by the position sensitive PPAC, A_f stands for the mass of the disassembling nucleus and $\langle v_f \rangle$ is the average velocity of the fission fragments. The mean values $\langle v_f \rangle$ have been evaluated from the systematics of the TKE (fig.8) and the masses A_f were taken from calculations made with a cascade-evaporation model. The influence of asymmetric fission and out-of-plane contributions which obsure the binary characteristics have been estimated to be less than 15% resulting in somewhat underestimated values $\langle p_{/\!/} \rangle$.

On the basis of different models /10/ /11/ it was shown that the mean excitation energy E^* is proportional to $\langle p_{\parallel} \rangle$ via the approximate relation $E^*/E_{max} \simeq 0.8 \langle p_{\parallel} \rangle / p_{max}$ where E_{max} and p_{max} are the energy and the momentum which the system would attain in the complete fusion process. We demonstrated the validity of this relationship in the following way. The folding angle distributions were evaluated for selected bins of the total mass (A_t+1) each of which corresponds to an average mass loss $\langle \Delta A \rangle$ Provided that $\langle \Delta A \rangle$ represents mainly the amount of evaporated nucleons,

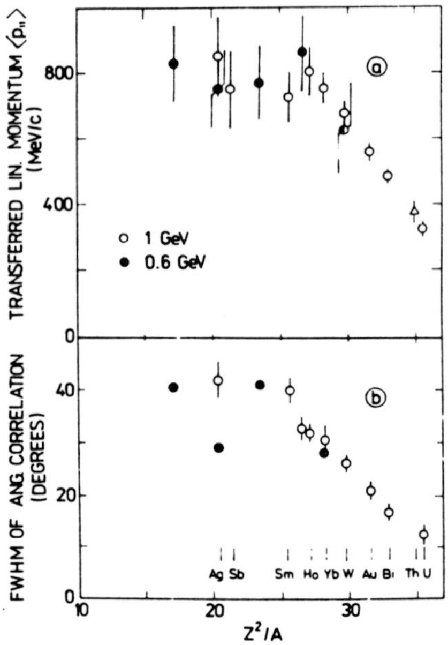

Fig.9

the excitation energy E^* is proportional to $\langle \Delta A \rangle$. In fact, a linear relationship between $\langle p_{\parallel} \rangle_A$ and the mass loss $\langle \Delta A \rangle$ was established for the U+p and Bi+p interactions at 1 GeV incidence energy.

Fig.9 represents the obtained mean transferred momenta $\langle p_{\parallel} \rangle$ as a function of the fissility parameter Z^2/A. This plot is completed by some data for 0.6 GeV (black circles)/8/. The widths of the corresponding folding angle distributions are given in the lower part of this figure. It is seen that $\langle p_{\parallel} \rangle$ exhibits a striking behaviour manifested in a steady increase of the transferred momentum with dropping heavy target mass and a rather abrupt levelling-off at masses around $A \simeq 160$. Fission induced by 1 GeV protons in heavy targets is characteristized by low linear momentum transfer indicating a dominating peripheral interaction. The rather narrow folding angle distribution is consistent with a predominant binary decay. The interaction become more violent with decreasing Z^2/A. The enlarged number of accompanying light particles obscure the binary characteristics as also indicated by the rising widths of the folding angle distributions. Indeed, we found a strong correlation between the widths and the mass loss, i.e. the number of missing particles /12/.

The behaviour of $\langle p_{\parallel} \rangle$ around $Z^2/A \simeq 25$ can be understood qualitatively by the following idea /13/. The momentum transfer for $A > 160$ is mainly controlled by the fission time. The proton momentum loss becomes nearly constant for masses $A < 170$ and the value of the fission time is comparable with that for multifragmentation in this region. For still smaller masses the multifragmentation channel becomes very essential and seems to dominate over the induced fission. This conclusion is also supported by estimates of the deposited excitation energies. The above mentioned relation between $\langle p_{\parallel} \rangle$ and E^* supply excitation energies of $E^* = 320 \pm 30$ MeV, 330 ± 60 MeV and 380 ± 50 MeV for Sm, Sb and Ag targets, respectively. Another check follows from the mean mass losses $\langle \Delta A \rangle = (A_t + 1) - 2\langle A_1 \rangle$ where $\langle A_1 \rangle$ is the average value of the single fragment mass distribution. Provided that the average nucleon binding energy is about 10 MeV we can immediately estimate the corresponding E^*. The resulting excitation energies are in agreement with the above given values within the error limits.

These results for the excitation energy and the adjacent nuclear temperatures 4.8 MeV $\leq T \leq$ 6 MeV are in reasonably

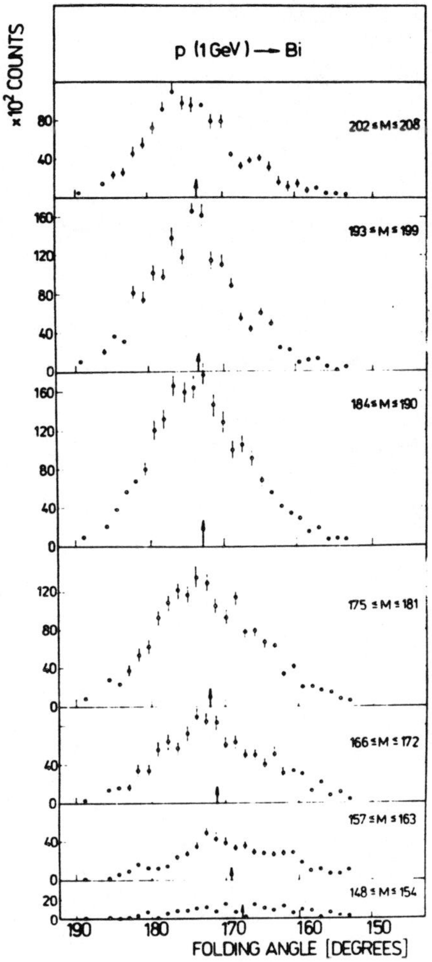

Fig.10

good agreement with recent observations of a limiting temperature in fragmentation reactions /14/. The fragmentation process is to set in rather abruptly at T=5-6 MeV.

Changes in the reaction process are also indicated in the evolution of the angular correlations as function of the involved fragment masses. Fig.10 shows the angular correlations for fragments in 6 ranges of the total mass for a Bi target. The correlations are fairly narrow for the three upper ranges and they are peaked a few degrees forward 180° consistent with formation by binary fission of near-target residues. The angular correlations become much wider for smaller values of the total mass. They clearly extend well beyond the region of collinearity indicating a rising many-body character of the interaction. This behaviour was also found for beams of 5 GeV protons bombarding a Au target /15/.

REFERENCES

/1/ U.L. Businaro, S. Gallone, Nuovo Cimento 5 (1957) 315

/2/ L.A. Vaishnene, L.N. Andronenko et al., Z. Phys. A302 (1981) 143

/3/ A.A. Kotov, G.G. Kovshevny et al., Nucl. Instrum. Meth. 178 (1980) 55

/4/ J.R. Nix, E. Sassi, Nucl. Phys. 81 (1966) 61

/5/ A.S. Iljinov, E.A. Cherepanov, S.E. Chigrinov, Z. Phys. A287 (1978) 37

/6/ S.B. Kaufman, E.P. Steinberg, Phys. Rev. C22 (1980) 167

/7/ W. Neubert, A.A. Kotov et al., Nucl. Instrum. Meth. 204 (1983) 453

/8/ G. Andersson, M. Areskoug et al., Z. Phys. A293 (1979) 241; Phys. Lett. 71B (1977) 279

/9/ V.E. Viola Jr., Nucl. Data 1 (1966) 391

/10/ N.T. Porile, Phys. Rev. 120 (1960) 572

/11/ J.B. Cumming, Phys. Rev. Lett. 44 (1980) 17

/12/ L.N. Andronenko, Thesis, Leningrad 1985

/13/ H.W. Barz, J.P. Bondorf et al., Nucl. Phys. A460 (1986) 714

/14/ S. Song, M.F. Rivet et al., Phys. Rev. Lett. 130B (1983) 14

/15/ S.B. Kaufman, Nucl. Phys. A471 (1987) 163c

Acknowledgements

We are indebted to M.N. Andronenko, Dr. D.M. Seliverstov, G.G. Kovshevny, Dr. H. Schulz, Dr. H.W. Barz, Prof. J.P. Bondorf, Dr. G.E. Solyakin and Dr. B. Schrøder for valuable discussions. We wish to thank Mrs. I. Probst for preparing the typescript.

PI-MESON INDUCED FISSION

S.M. de Barros, I.O. de Souza
Inst. de Fisica da Universidade Federal do Rio de Janeiro, Brasil
R.J. Peterson
Nuclear Physics Laboratory, University of Colorado, Colorado, USA

In the heaviest nuclei, the cross section for fission approaches the reaction cross section,[1] making possible the study of such reaction cross sections by measurements of fission rates. The present work utilizes fission track detector methods for measurements of pion-induced fission near the Δ (3-3) resonance. Like a photon, the pion may be freely absorbed on complex nuclei. With both π^+ and π^- beams, such absorption is very likely[2] and will result in initial highly excited nuclei of charges Z$\pm$1 for a target of atomic number Z. If this charge difference is also present at the stage where fission occurs, then the fissibility parameters Z^2/A will also be different. Since fission probabilities depend quite strongly on this parameter, different fission cross sections will result for π^+ and π^-. A set of heavy targets covering a range of $\tilde{Z}^2$/A, with $\tilde{Z} = Z \pm 1$, has been studied in the present work.

Very little angular momentum is provided by the pion beam, with at most 10 units for the present study. The nuclear excitation energy, however, is very high following pion absorption, being about 270 MeV in the present work. At this same nuclear excitation energy, photon absorption can provide about 12 units of angular momentum. It is therefore of great interest to compare these two reactions and to contrast their results with reactions requiring much higher angular momentum to reach comparable nuclear excitations.

Targets to be studied were placed adjacent to Makrofol E track detector material of 100 μm thickness directly in a momentum-analyzed beam of pions on the EPICS beam line at the Clinton P. Anderson Meson Physics Facility (LAMPF). Three exposures were made, with the same sequence of targets and detectors for each. An ion chamber in the beam was used as the monitor for long exposures, normalized by a ^{11}C radioactivation technique.

Pions entered the foil stack with energies of 131 MeV for π^+, and for two π^- exposures at 131 MeV and 138 MeV, with integrated fluxes of approximately $10^{10}\pi/cm^2$. Targets were ^{209}Bi, ^{231}Pa, ^{233}U, ^{234}U, ^{235}U, ^{237}Np, ^{241}Pu, and ^{242}Pu, with Bi targets $459\mu g/cm^2$ thick and actinides [3] near 20μ g/cm^2 thick. Pion-induced fission yields fission fragments nearly back-to-back in the laboratory frame[4,5] with essentially only binary fission.[5] No double counting of tracks from an event is then expected. Due to the 1mm air gap between target and Makrofol, some spreading of events past the target area wa observed. This was corrected for by a Monte Carlo calculation assuming an isotropic angular distribution and a detector area corresponding to that where counts were measured. Our normalization uncertainties are about $\pm$5%.

Exposures to Makrofol after the pion bombardment were made in order to detect any activation or spontaneous fission events. Very few counts were found, except for ^{242}Pu. Appropriate subtractions were made for that case using the known half-life and spontaneous fission branching ratio[6]; about half the events were due to spontaneous fission for this one example.

The Makrofol material was developed by etching for 30 minutes at 70°C in 6.5 N NaOH solution. Only tracks due to mass greater than about 20 are made visible by this method.[7]

Fission cross sections for the seven actinide targets used for the present study are plotted in the figure against the fissibility variable $\tilde{Z}^2/A$, with the charge and mass of the target as changed by the pion sign. The two π^- beam energies are seen to give very similar yields. No dependence upon odd or even neutron or proton numbers is noted. Data for pion-induced fission of ^{238}U from Refs. 4 and 5 are also shown, although using lower beam energies.

If the pion were absorbed, the initial charge of the system changes to $\tilde{Z}$=Z±1. If all following processes, such as precompound nucleon emission, were the same for both beams, it would be appropriate to compare data to the new fissibility parameter $\tilde{Z}/A$, at least as an organizing principle to compare π^+ and π^-. For the range of targets, the reaction cross section increases as R^2, so we also divide the fission cross sections by $A^{2/3}$. For all but ^{235}U, π^+ cross sections exceed those for π^-. The curve shown is a Gaussian, fit to $\sigma_f/A^{2/3}$=65mb at $\tilde{Z}^2/A$=37.7 and to the Bi cross sections for π^+ and π^- at $\tilde{Z}^2/A$=33.761 and 32.172, respectively. At $\tilde{Z}^2/A$=37.7, 65mb corresponds to a disk of radius $1.438A^{1/3}$fm, for which the charge density of ^{238}U is but 2.5% of the central density. The displacement between π^+ and π^- data follows the trend of this curve.

Other studies comparing reaction and absorption cross sections have been reported for heavy nuclei. For bismuth, Ashery et al. find that about 65% of the reaction cross section is in absorption for 165 MeV π^+,[8] for an absorption cross section of 1585±280mb out of a reaction cross section of 2446mb. At 131 MeV, our π^+ fission cross section is 321mb for Bi.

These results indicate that much of the fission cross section follows absorption of the pion, producing a very high nuclear excitation with low angular momentum. This situation is also indicated in other results by the high probability of ternary fission [5], by the mass distribution of the fragments [4,5,9] and by the linear momentum transfer from the beam to the reaction products.[4] Such pion absorption leads to a hitherto little explored region of very hot nuclei. Few theoretical studies are available for comparison to the present results and to new work in progress.

[1] R. Vandenbosch and J. R. Huizenga, **Nuclear Fission** (Academic Press, New York) 1973

[2] D. Ashery and J. P. Schiffer, Ann. Rev. Nucl. Part. Sci. **36**, 207 (Annual Reviews, Inc., Palo Alto) 1986

[3] Produced by the Chemistry Divison at AERE Harwell, under the sponsorship of the United Nations Development Program.

[4] K. H. Hicks et al., Phys. Rev. C **31**, 1323 (1985)

[5] H. A. Khan, N. A. Khan and R. J. Peterson, Phys. Rev. C **35**, 645 (1987)

[6] E. N. Shurshikov, M. L. Filchenkov, Yu. F. Jaborov and A. I. Khovanovich, Nucl. Data Sheets **45**, 509 (1985)

[7] R. L. Fleischer, B. P. Price and R. M. Walker, **Nuclear Tracks in Solids** (University of California, Berkeley) 1975

[8] D. Ashery et al., Phys. Rev. C **23**, 2173 (1981)

[9] P. David et al., Zeit. Phys. A **328**, 37 (1987).

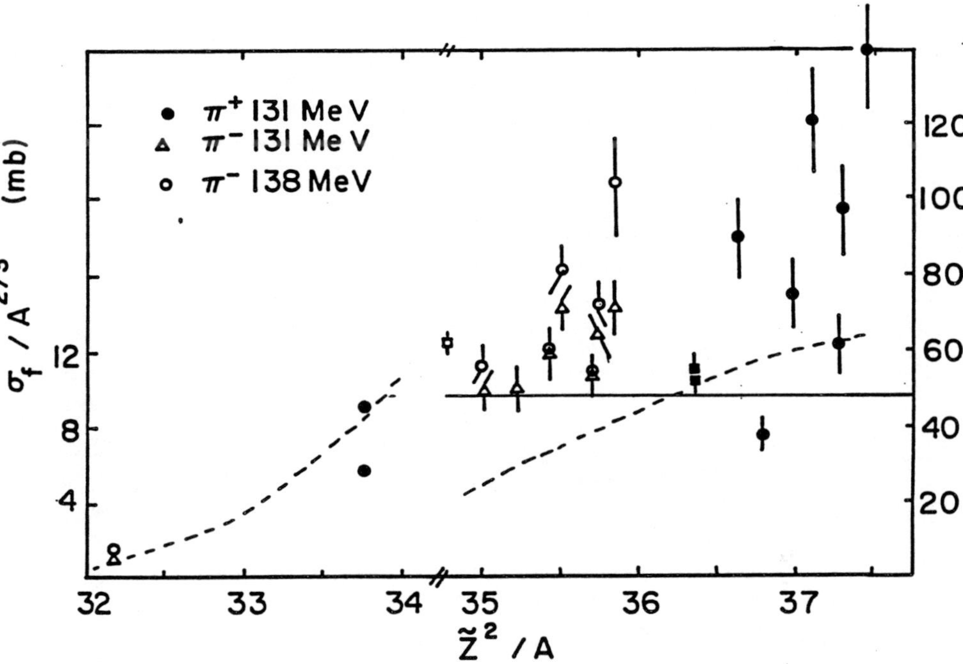

Measured fission cross sections for π^+ and π^- beams of 131 MeV (circles) and 138 MeV (triangles) are shown for a range of heavy elements. Results at 100 MeV for ^{209}Bi and ^{238}U are shown as the circled data points, from Ref. 4. These data have been divided by $A^{2/3}$ and plotted against $(Z\pm1)^2/A$, to reflect the incident pion charge. The curve is from the expression in the text, used simply to guide the eye. A black disk of radius $1.25A^{1/3}$fm yields the horizontal line.

NEW POSSIBILITIES IN THE STUDY OF THE PHOTOFISSION PROCESS USING BACKSCATTERED LASER PHOTONS

G.M. Gurevich, V.G. Nedorezov
Institute for Nuclear Research, Moscow
G.Ya. Kezerashvili
Institute for Nuclear Physics, Novosibirsk

Until now the photofission of nuclei has been studied mainly by means of bremsstrahlung photons having the continuous energy spectrum. At low energies ($E \lesssim 20$MeV) the valuable information was obtained about multipolarity of excitations, quantum numbers of transition states, fission barriers and widths of actinides (see rev.[1]). In the intermediate and high energy region the fission decay becomes possible for any nuclei and the subject of nuclear fission includes the problem of fragmentation, symmetric and asymmetric decay modes probability etc.

To improve the experimental accuracy of photofission experiments many efforts have been made using the quasimonochromatic gamma beams obtained by different techniques. Among them: the annihilation of positrons in flight, coherent radiation from the crystal, energy tagging by means of the coincidence registration of fission fragments and scattered electrons and the backward Compton scattering of laser photons on high energy electrons. The comparison of these methods and the discussion of the intermediate energy fission results one can find in [2].

The backscattered laser photon method has sime advantages: hard photon spectrum, low neutron background, good geometry, complete linear or circular polarization. The high monochromatization can be improved by the tagging system or by the collimation of the beam. Kinematics of the laser photon-electron head-on collision looks as following:

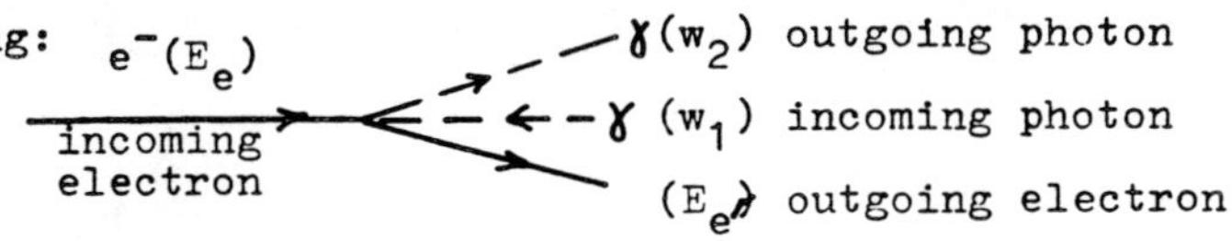

The energy of backscattered photons is definitely connected with the angle of radiation :

$W_c = 4\gamma^2 W_1/(1+n^2+\lambda)$, where W_1 is the energy of laser photons, $\gamma = E_e/m_e$, $n = \theta\gamma$, $\lambda = 2\gamma W_1/m_e$. E_e, m_e - the energy and rest mass of electrons, respectively.

The energy spectrum $d\sigma/dw_2$ and angular distributions $d\sigma/d\theta$ are shown in fig.1.

In 1984 the backscatterd laser photon fasility was realised on the storage ring VEPP-4 (E_e= 5.5 GeV) in Novosibirsk and used for the photofissimn experiments, [3]. Some years earlier fasility of the same kind was created in Frascati on the Adone storage ring (E_e=1.5GeV)[4].

fig.1

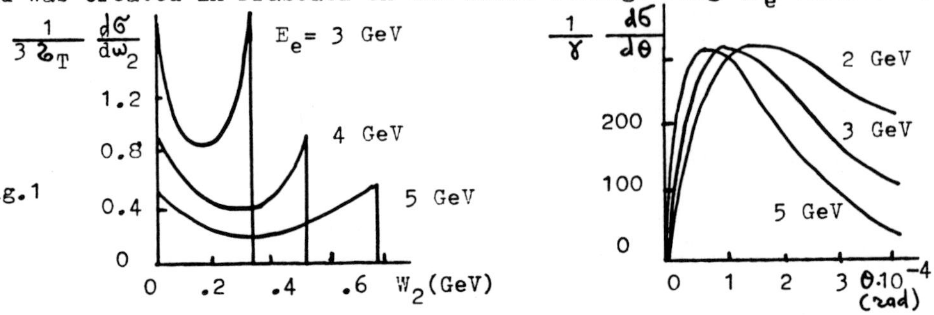

The experiment in Novosibirsk indicates a new mechanism of nuclear excitation followed by fission decay with low energy and momentum transfer, when the nucleous get the small part of the incident photon energy. The probable process which explains the experimental data was suggested as inelastic e^+e^--pair production:

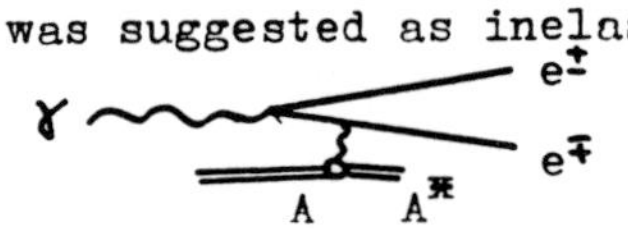

Experimental value of the probability of this process as compared with the total photofission yield for ^{238}U and ^{237}Np nuclei (average) equals 17 ± 4 % for the energy region E_γ = 150 – 710 MeV. The average photofission cross sections of these nuclei σ_f=63±8 mb and 102±11 mb, respectively, for this energy region were obtained [3]. In comparison with the atomic e^+e^--pair production cross section the experimentally measured $\sigma_{e^+e^-A^*}$ cross section is rather small (2.10^{-4})

At present the storage ring VEPP-4 is in reconstruction, so the experiment was interrupted. In 1986-88 the new installation has been created on VEPP-3 (E_e=2 GeV) with new fission detector based on the low pressure multiwire proportional chambers (MWPC) to measure the angular distributions for the particular mass fission fragments [5].

The experimental scheme and parameters of new installation are as following:

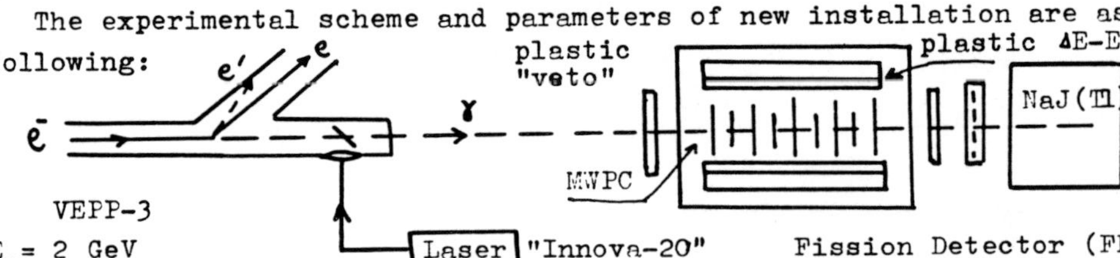

"Innova-20"

E_e= 2 GeV
I_e= 200 mA
f_e= 4 MHz

λ = 514 nm
$\bar{p}$ = 4 Wt

Fission Detector (FD)
10 MWPC
4 ΔE-E plastic telescopes. [5].

γ -beam parameters:
energy range 30-180 MeV, energy resolution 2%, total intensity $10^6 c^{-1}$.

The experimental program for this facility includes at present:

1. Photofission cross section measurements for actinide nuclei in the

energy region 30-180 MeV to get the total photoabsorption cross sections for these nuclei.

2) Registration of charge particle emission (p,d,α etc) in coincidences with fission fragments to study the cascade - evaporation mechanism of nuclear decay.

3) Study of the mentioned above low transfer nuclear excitations in the energy region 30 - 180 MeV.

4) Search for the direct (non statistical) mode of nuclear fission in accordance with theoretical predictions [6].

5) Study of the polarization effects in the photonuclear fission.

Each of these subjects needs a special time for detailed discussion. The first results obtained in VEPP-3 were published already [7].

Now, we would like to discuss the perspectives of the experiments with polarized actinide targets which seems to be very promissible and completely realise the advantages of the backscattered laser Compton beams.

In most experiments with polarized targets performed until now, polarized protons and deutrons were used. The number of works with heavy polarized targets is very scarce. In the field of electromagnetic interactions only two experiments on polarized ^{165}Ho are known (see ref. in [2]). Therefore, phenomena connected with the spin polarization of heavy nuclei are investigated only slightly. The program of polarization experiments can be outlined as following:

1) The study of the optical anisotropy of atomic nuclei, i.e. dependence of the total cross section in the gaint resonance region on the relative orientation of the photon momentum and nuclear spin. According to the hydrodynamic model the frequences of the collective oscillation modes (giant resonances) in deformed nuclei should vary for different orientations, parallel or perpendicular to the symmetry axis of nucleous. Thus, for the photofission reaction whose cross section is near enough to the total nuclear photoabsorption cross section, reaction product yield should depend on the nuclear polarization direction with respect to the wave vector of incident photon.

2) The study of the nuclear fission channels, i.e. the sets of quantum numbers characterizing the stage of fissioning nucleus. Orientation of the target nuclei as well as polarization of the incident photons allows to determine the complete set of quantum numbers (spin, parity, spin projection on the nuclear symmetry axis). Experimental data on the fission of polarized nuclei by photons still absent.

The only polarization experiment performed for ^{232}Th nuclei by linearly polarized bremsstrahlung photons [8]. In this case (unpolarized target) the angular distributuin function looks as following:

$$W(\theta, \varphi) = a + b\sin^2\theta + c\sin^2 2\theta + \lambda P_\gamma \cos 2\varphi\,(d\sin^2\theta - 4c\sin^2 2\theta),$$

where P_γ is a degree of polarization, $\lambda = \pm 1$ for E1 and M1 excitations, θ, φ are the escape angle in the reaction plane and azimuth, respectively.

Up to now several experiments on the fission of the odd A-uranium isotopes by neutrons (see,e.g. [9]) were performed. Obtained data sharpened the contradictions of fairly general character arising when one attempts to discribe by identical structure of fission channels the cross sections and angular distributions of fission fragments for the same nuclei exciting in different reactions. It makes evident the importance of the experiments on the fission of oriented nuclei by polarized photons.

Schematic diagram of the experiment on the photofission of oriented nuclei looks as following:

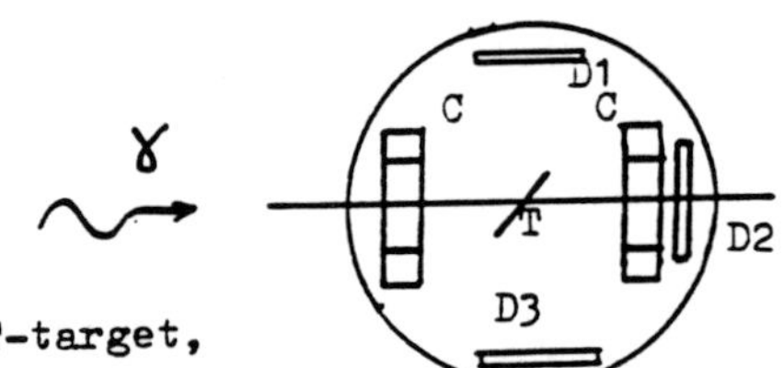

T-target,
D1-D3- detectors,
C - coils

To achieve considerable polarization the temperature T and magnetic field B acting on the nuclei should satisfy the condition: $\mu B/kT \approx 1$, where μ -magnetic moment, k- Boltzmann constant. If $\mu = 1$, $B/T = 10^7$ G/K, e.g. T=10 mK, B= 10^5G. In the case of actinides hyperfine fields created by 5f electrons can achieve 10^6G.

Target materials must have high heat conductivity and short nuclear spin relaxation time, therefore they should be metals. The use of dielectric targets, as in experiments on neutron fission of oriented nuclei, makes the target temperature and orientation uncertain.

The polarization of ^{241}Am nuclei in ferromagnetic $ZrFe_2$ was realised for the first time by purely magnetic hyperfine interaction using the NORD fasility [10] (INR, Moscow), which comprise ^{3}He-^{4}He dilution refrigerator and two Helmholts paires of superconductor coils for target magnetization. The nuclear polarization was detected by the angular anisotropy of ^{241}Am α-radiation. The results obtained show preferential $(\tau(\varphi=0)/\tau(\varphi=\pi/2) = 1.15 \pm 0.02)$ α-emission along nuclear symmetry axis.

1. V.V.Varlamov e.a. Photonuclear Data. Moscow University, 1983,
2. V.G.Nedorezov, Yu.N.Ranyuk. Photofission of nuclei above the giant resonance. Naukova Dumka, USSR, 1988, 209 p.
3. V.A.Zapevalov, D.I.Ivanov, e.a. Proc. Int.school on Int.Energy Nucl. Phys. Venice, Italy, 1988, 8p. A.A.Kazakov e.a. JETP Lett.40,1984, 145.

Referenses 3,4,5,6,8,9,10 see in ref [2].

VIII. HEAVY-ION REACTIONS AND FISSION

PROXIMITY FRICTION BY INELASTIC EXCITATION IN PERIPHERAL HEAVY-ION COLLISIONS

H.J. Krappe
Hahn-Meitner-Institut Berlin GmbH, D-1000 Berlin 39, Germany

Abstract: The linear response expression for the energy dissipation rate in heavy-ion collisions through the inelastic excitation mechanism is evaluated in Thomas-Fermi approximation. We discuss the relative strength of this mechanism compared to the transfer-related energy dissipation, the temperature dependence of friction coefficiednts, and the condition for time-local equations of motion.

It might appear that the energy dissipating mechanism in heavy-ion collisions has long been explained in terms of the recoil effect of exchanged nucleons and described satisfactorily by Randrup's proximity friction formula[1]). Recent time-dependent shell-model calculations[2]) showed however that the inelastic excitation mechanism can not be neglected compared to particle transfer as a source of energy dissipation. We therefore re-investigate the linear response expression for the energy-dissipation rate[3])

$$\dot{E}(t) = -\int_{-\infty}^{0} d\iota \sum_{ph}{}' P_h(1-P_p) <\psi_p \left| \dot{V}_2(t) \right| \psi_h> <\psi_h \left| \dot{V}_2(t+\iota) \right| \psi_p> \frac{1}{\omega} \cdot exp\left(\frac{i}{\hbar}\omega\iota\right) + c.c. + 1 \leftrightarrow 2 \,, \tag{1}$$

through the inelastic excitation mechanism. Here P_h and P_p are the temperature-dependent occupation probabilities for hole and particle states, V_2 is the single-particle potential of nucleus 2 moving (without deformation) in the restframe of nucleus 1, whose particle and hole states are ψ_p and ψ_h, and ω is the excitation energy. The symbol $1 \leftrightarrow 2$ is to indicate that a similar integral should be added where the roles of nuclei 1 and 2 are exchanged.

Although expression (1) could be evaluated numerically in a two-center shell model, it is easier to exhibit its salient physical features by a largely analytical evaluation. This can be achieved [4]) by the following approximations:

1. Substitution of sums over particle and hole states by integrals over energies after introduction of an appropriate Thomas-Fermi level density.
2. Description of the particle-hole density in Helm's model [5]). There, for particle-hole states with angular momentum quantum numbers L,M,

$$|\psi_p^*(\mathbf{r})\,\psi_h(\mathbf{r})|_{LM} \propto \partial_{R_p}\,\theta(R_p - r)\,Y_{LM}(\hat{\mathbf{r}}) * S(a) \,, \tag{2}$$

where θ is the step function , S(a) is a Yukawa function of range a, and the asterisk indicates a folding product. Similarly the single-particle potential is represented by the folding product

$$V(\mathbf{r}) = V_0 \theta(R_v - r')^* S(a) \qquad (3)$$

with $\mathbf{r}' = \mathbf{r} - \mathbf{d}(t)$, $\mathbf{d}$ being the distance between the centers of the two nuclei.

I omit any details of the calculation, which will be published elsewhere[4]), except for the evaluation of the τ-integral in (1). Writing only the τ-dependent factors of the integrand, we consider

$$I = \int_{-\infty}^{0} d\tau \left(e^{i\omega\tau/\hbar} + c.c. \right) \langle \psi_p \left| \dot{V}_2(t+\tau) \right| \psi_h \rangle_{LM} .$$

where V_2 depends on the time only via $\mathbf{d}(t)$. Using the folding products (2) and (3) the d-dependent part of the matrix element can be written

$$\langle \psi_p \left| \dot{V}_2(t+\tau) \right| \psi_h \rangle_{LM} \propto P_L\left(\frac{a}{d(t+\tau)} \right) e^{-d(t+\tau)/a} ,$$

where P_L is a polynomial of order 2L. Introducing the new dimensionless integration variable $x = -|\mathbf{d}(t+\tau) - \mathbf{d}(t)|/a$, the τ-integral takes the form

$$I = \int_0^{\infty} \left(e^{-(1-i\frac{\omega}{\Gamma})x} \cdot e^{i\phi(x)} + c.c. \right) P_L\left(\frac{a}{d(t+\tau(x))} \right) \frac{d\tau}{dx}\, dx$$

with the width $\Gamma = v(t)\hbar/a$ and $\phi(x) = -1/2(\omega/\Gamma)(a|\ddot{\mathbf{d}}(t)|/v^2)\cos[\dot{\mathbf{d}}(t),\ddot{\mathbf{d}}(t)]\, x^2 + O(x^3)$. Since the first factor of the integrand is more rapidly varying with x than the last two factors, a moment expansion of the integral is possible and yields

$$I = P_L\left(\frac{a}{d(t)} \right) \frac{d\tau}{dx} \Big|_{x=0} \int_0^{\infty} e^{-(1-i\frac{\omega}{\Gamma})x + i\phi(x)} dx + c.c. + \textit{higher order terms}$$

$$\approx P_L\left(\frac{a}{d(t)} \right) \frac{d\tau}{dx} \Big|_{x=0} \frac{2\Gamma^2}{\Gamma^2 + \omega^2} + \textit{higher order terms} ,$$

if $\phi(x)$ is negligible compared to $\omega x/\Gamma$. This is true for moderate accelerations: $a|\ddot{\mathbf{d}}|/v^2 < 1$. It yields an estimate for friction coefficients not to depend on the acceleration so that a time-local description of the orbital motion is possible. The condition is usually violated near the distance of closest approach if the impact parameter is sufficiently small. It is always violated for Coulomb excitation because of the infinite range of the Coulomb field ($a \rightarrow \infty$).

The appearance of the Lorentzian with width Γ resulting from the τ-integration is very instructive. Had we neglected the τ-dependence of the matrix element, as it is often done, one would have obtained $\pi\delta(\omega)$ instead. This causes an extreme sensitivity of friction to the level density at zero excitation energy. Since in the shell model the level density vanishes at zero energy and zero temperature, it was argued that finite temperature and residual interactions are mainly responsible for the observed energy dissipation. In view of the fact that $\Gamma = 12$ MeV for a velocity corresponding to 1 MeV/A, the width of the Lorentzian is seen to be usually at least as large as typical shell energies. Therefore there is no sensitivity of friction to the details of the local density nor does the temperature dependence amount to more than a slight increase of friction in the temperature range of interest.

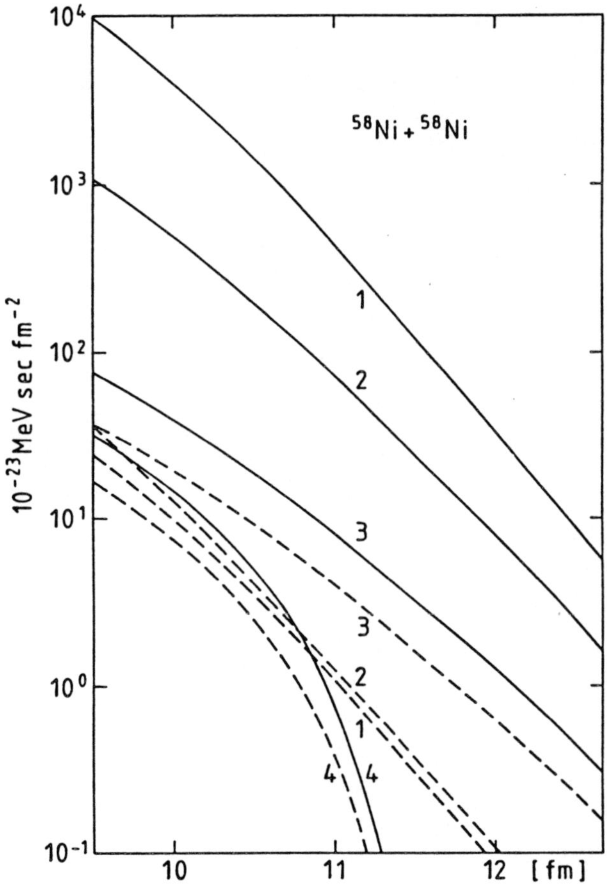

FIGURE 1

Radial (full lines) and tangential (dashed lines) friction form factors for the system ^{58}Ni + ^{58}Ni as function of the center of mass distance d, starting with s = O. (1) Gross-Kalinowski model, (2) present model (3) proximity model including tunnelling effects, (4) Randrup's original proximity model.

Evaluation of (1) with the Thomas-Fermi approximations mentioned above leads to radial friction coefficients which are smaller than the phenomenologically fitted coefficients of the Groß-Kalinowski[6]) model, but orders of magnitude larger than Randrup's proximity friction coefficients[4]). Tangential friction coefficients are still larger than in Randrup's model, but smaller than exchange-related tangential friction coefficients if tunneling effects are considered in evaluating the one-sided flux of Randrup's model. The situation is illustrated in Fig. 1 for the system ^{58}Ni + ^{58}Ni.

REFERENCES

1. J. Randrup, Ann. Phys. 112 (1978) 356
2. S. Pal and D.H.E. Gross, Z. f. Phys. A329 (1988) 349
   M. Baldo, A. Rapisarda, R.A. Broglia, and A. Winther, Nucl. Phys. A472 (1987) 333, and report NBI-88-05
3. D.H.E. Gross, Phys. Lett. 68B (1977) 412
4. H.J. Krappe, to be published
5. R.H. Helm, Phys. Rev. 104 (1956) 466
6. D.H.E. Gross and H. Kalinowski, Phys. Lett. 48B (1974) 302; 45C (1978) 175

RECENT EXPERIMENTS ON THE DYNAMICS OF HEAVY- AND LIGHT-ION INDUCED NUCLEAR FISSION

K.D. Schilling
Zentralinstitut für Kernforschung Rossendorf, 8051 Dresden, Germany

ABSTRACT: We report on our recent experimental activities in Dubna and Rossendorf in the field of heavy- and light-ion induced nuclear fission. The results obtained with multiparameter detection equipments are only briefly summarized here, because most of them have been published previously. The topics concerned are heavy-ion induced quasi-fission, fission of superdeformed nuclei and charged-particle induced ternary fission.

INTRODUCTION

The unifying idea of the present investigations - i.e. the common physical basis - is the exciting phenomenon of the nuclear fission process regarded under different aspects. A common feature of the systems under consideration is the importance of the influence of nuclear shell effects on their dynamical evolution.

Why is fission so interesting ?

(i) Fission is one of the fundamental phenomena in nuclear physics and, beyond this, in other spheres of our material world.

(ii) Although we celebrate in 1988 already the 50th anniversary of the discovery of the nuclear fission process by O. Hahn, L. Meitner and F. Straßmann, a great many problems in its detailed understanding remained.

(iii) Fission, especially cold fragmentation, can be regarded as the reverse process to heavy-ion fusion. The understanding of nuclear fission may also help us to better understand heavy-ion reaction mechanisms.

This contribution is divided into three topics :

1. Quasi-fission in heavy-ion induced reactions.
2. Fission of nuclei residing in superdeformed states ?
3. Charged-particle induced ternary fission.

& Invited talk

1. QUASI-FISSION IN HEAVY-ION INDUCED REACTIONS

The formation of two heavy fragments in reactions between complex nuclei at incidence energies below about 10 MeV/amu is the result of the evolution of the mass asymmetry degree of freedom of the collision system. Depending on the entrance channel parameters, different degrees of kinetic-energy dissipation and mass relaxation can be observed, which are usually related to different reaction mechanisms. Especially important for the present collision systems are the so-called fusion-like processes /1/ located in between the damped (dissipative, deep-inelastic) processes and compound nucleus (CN) formation. We looked, in particular, for an influence of nuclear structure effects on the dynamics of the fragmentation process. Our emphasis is placed on total kinetic energy (TKE) vs. fragment mass (M) distributions,which provide an important tool in such investigations.

The experiments were performed at the heavy-ion cyclotron U-300 of the JINR Laboratory of Nuclear Reactions in Dubna. Based on the kinematic coincidence method, the binary reaction products were detected and identified using the double-arm time-of-flight spectrometer DEMAS, which has extensively been described elsewhere /2/. By measuring the velocity vectors of both correlated final fragments, the pre-evaporative mass and total kinetic energy distributions for two-body events were deduced /2/. Thereby, the two-body assumption was verified for every event by demanding collinearity of the two fragment velocity vectors in the centre-of-mass frame.

The motivation for the present investigations was twofold : First, to study the recently discovered reaction mechanism termed either fusion-like processes /1/ or quasi-fission /3,4/. Second, to look for a possible influence of nuclear shell effects on the evolution of the mass asymmetry degree of freedom.

A few typical signatures of the quasi-fission process are shortly compiled here (cf. ref. /4/) :

(a) Large mass rearrangements between the reaction partners occur on a time scale ($2...10\times10^{-21}$ s) comparable with the rotational period of the intermediate complex.

(b) The mass flow is directed towards symmetry. Varying degrees of mass relaxation are observed. The mass drift is clearly incomplete for heavier collision systems.

(c) The various degrees of mass drift are combined with the maximum possible damping of the kinetic energy degree of freedom in the exit channel.

(d) The widths of the experimental mass distributions deviate progressively from the statistical equilibrium estimate with decreasing mass asymmetry of the heavy collision systems.

It was the special aim of the present investigations to search for nuclear shell effects in the dynamics of heavy-ion collisions. We obtained additional support in this effort by fission-barrier calculations /5/ for excited heavy nuclei. These calculations predict that shell effects remain noticeable in the fission-barrier structure of the nucleus $^{272}108$ up to excitation energies $E^{*} \approx 50$ MeV. Moreover, nuclear structure effects may influence the heavy-ion fusion process at excitation energies up to about 25 MeV /5/.

We investigated experimentally four collision systems with different mass asymmetries in the entrance channels and the same total nuclear charge number $Z_p + Z_t = 108$. The results have been published and extensively discussed in /6/. Additionally, three projectile-target combinations with $Z_p + Z_t$ between 100 and 110 have been studied /7/. The experimental TKE vs. M distributions as well as their projections onto the mass axis have been analysed in terms of the initial mass asymmetry and of the excitation energy of the composite systems. As has been pointed out in /6/, the intrinsic excitation energy increases during the evolution of the mass asymmetry of the system towards the CN configuration and has been evaluated in /6/ for two extreme cases of the mass asymmetry (initial and CN configuration). The main results and conclusions drawn in /6,7/ can be summarized here as follows :

For the collision systems ^{32}S(192 MeV) + ^{238}U and ^{40}Ar(220 and 206 MeV) + ^{232}Th, the maximum possible values of the excitation energy were estimated to be below or near the theoretical limits /5/ deduced for a possible influence of shell effects on the fission barrier of a compound nucleus with $Z = 108$, as well as on the preceding fusion process. The experimental results indicate - in particular, for the two above systems - that the fully equilibrated CN configuration will, probably, not be reached during the evolution of the mass asymmetry, as it is characteristic of a quasi-fission process. Thus, the actual excitation energies of the composite systems during their evolution will develop between the two estimated extreme values. That means the conditions to observe an influence of the nuclear shell structure on the dynamics of the considered reactions should be even more favourable compared to the estimates /6/. Therefore, the relative maxima observed in the integrated M-distributions around $M = 200$ (Pb shell) of the above reactions have been interpreted /6,7/ as being due to the influence of nuclear shell effects in correspondence with the predictions of /5/. It should be emphasized that these effects may already occur during the evolution of the mass asymmetry degree of freedom towards symmetry and may thus influence the dynamics of the whole collision process. This interpretation is supported by calculations of the corresponding driving potentials performed by A. Pop (cf. /7/).

Furthermore, the energy-integrated M-spectra of the collision systems ^{32}S(192 MeV) + ^{238}U and ^{40}Ar(220 MeV) + ^{232}Th are very different in shape, although the initial mass asymmetries as well as the estimated excitation energies of both systems are very similar. We draw in /6/ the conclusion that the dynamical evolution of both collision systems towards symmetric fragmentation differs considerably from each other.

2. FISSION OF NUCLEI RESIDING IN SUPERDEFORMED STATES ?

The recent discovery of superdeformed (SD) structures at high angular momenta in the nucleus ^{152}Dy by Twin et al. /8/ is of outstanding importance for the nuclear structure physics. The observation of discrete rotational states up to a spin of 60 $\hbar$ in the reaction $^{108}Pd(^{48}Ca, 4n)^{152}Dy$ is connected with a large prolate nuclear deformation of $\varepsilon \approx 0.6$, i.e. a major-to-minor axis ratio of 2 : 1. Such large deformations have only been known up to now in the actinides (fission isomers). Unlike there, the existence of SD states in the rare earth nuclei is strongly dependent on the angular momentum. It opens up the possibility of studying the properties of strongly deformed nuclear matter and raises many exciting questions /9/. For instance, how is the fission limit of the nucleus affected ?

Very few cases of superdeformed structures have been found so far in some neighbouring nuclei in the A=150 region. In contrast, several theoretical studies predict many nuclei in this region to be superdeformed at high spin.

The decisive impulse for our experimental activities was given by the author of ref. /9/ during the Dubna Heavy-Ion Conference in 1986. The following questions arose : Is there any possibility for the SD nucleus to undergo fission out of those highly excited states ? Is there any experimental chance to observe the corresponding fission fragments ? We decided to look for them - even if the chances were very small - in a first approach with our experimental set-up /2/ described above. Additional theoretical support was obtained by a recent work /10/, which demonstrates qualitatively that the fission barrier for ^{152}Dy becomes double-humped and decreases with increasing angular momentum, whereas the shell energy at the 2:1 deformation is not much affected by the rotation. The same conclusions are drawn in /11/. Moreover, it is pointed out in /11/ that the nucleus ^{152}Dy becomes unstable against scission at angular momenta beyond about 60 $\hbar$.

Starting from these arguments, the basic idea of the present experiment consisted in the following : We tried to separate the fission fragments, which originate from the fission out of SD states, from those resulting from usual fast-fission events (cf. /6/ for terminology). If we suppose that the effective lifetime of the SD nucleus Dy (which constitutes an evaporation residue of a heavy-ion fusion reaction) is comparable to or larger than the time needed for the slowing-down process in the target, then the Dy recoil nucleus would undergo fission at rest ! The correlation angle between the corresponding fission fragments should then be about 180° in the laboratory frame. In contrast, the correlation angle of fast-fission events resulting from recoiling residual nuclei is expected to be about 110°. An angular separation of both kinds of fission fragments should then be possible.

We irradiated a ^{116}Cd target with ^{40}Ar projectiles at different laboratory energies (E_{lab} = 185 - 225 MeV). The target consisted of 300 $\mu g/cm^2$ enriched (90 %) ^{116}Cd with a 3 μm Al backing. It had been positioned under an angle of 10° between target plane and beam direction. Thus, the target was, on the one hand, sufficiently thick in beam direction in order to stop the recoil nuclei and, on the other hand, thin enough to enable the fission fragments to escape sidewards. The critical angular momentum for fusion amounts in the present case to $\approx 70\ \hbar$. The characteristic slowing-down time in the target layer was estimated on the basis of realistic stopping power values to be $\tau \approx$ 2 - 4 ps. Finally, we obtained the following result :

We did not observe any fission event at 180° in the lab frame under the given experimental conditions , but a lot of fast-fission events at forward direction. This result can be expressed in the following upper limit for finding fission events from SD states in the present reaction and experimental geometry :

$$\frac{d\sigma}{d\Omega}(\theta_{lab} = 90^{\circ} \pm 7.5^{\circ}) < 2\ \mu b/sr.$$

The result is understandable if we assume that the effective lifetime for fission out of the SD states is less than the stopping time. This assumption has been confirmed - after our experiments had been finished - by a recent work of Bentley et al. /12/. They measured the lifetimes of the SD states in ^{152}Dy by the Doppler-shift-attenuation method. The lifetimes of states above 50 $\hbar$ are, on the average, less than 10 fs (former estimates /8/ resulted in values less than 100 fs). Consequently, if fission fragments from SD states exist, they would be superimposed by the usual fast-fission events at forward angles. The problem arises, how to extract them from the background of undesired events ? Other open questions remain, as : How is the fission probability of the SD states and how are they populated in dependence on the angular momentum and on the mass asymmetry in the entrance channel of the reaction ? A further essential problem is to select unambiguously the desired reaction channel.

3. CHARGED-PARTICLE INDUCED TERNARY FISSION

Nuclear fission accompanied by long-range alpha particles (lra) has been investigated. The results to be presented here have already been published in /13,14/. The equatorial emission constitutes the main component of lra and occurs in a plane perpendicular to the fission axis in a narrow angular interval indicating that the emission takes place very near the scission point. The study of lra accompanied fission should, therefore, provide information on the fission dynamics.

The in-beam measurements were performed at the Rossendorf cyclotron U-120 by irradiating a natural uranium target with 13.5 MeV deuterons. The fission fragments (ff) were recorded with a large position-sensitive double-grid avalanche counter (DGAC) /15/ and a small parallel-plate avalanche counter (PPAC) with both detectors arranged perpendicular to the beam axis in 180° geometry. The coincident light charged particles were identified by a time-of-flight (TOF) and energy measurement performed by a small transmission PPAC and a large surface barrier detector as start and stop/E detectors, respectively. The orientation of the ff axis with respect to the lra direction was measured with the DGAC. The ff mass was deduced from the TOF difference of both ff making use of the Viola systematics /16/. For more details we refer to /13,14/.

The $4 \cdot 10^{4}$ recorded six-parameter events were analysed by generating different two-dimensional distributions . Besides the huge background of random coincidences, about 10^{3} true lra events with $E_{\alpha} \geqq 11$ MeV could be identified. For the lra angle with respect to the light fragments' (lf) direction, a most probable value of $\bar{\vartheta}_{lf-\alpha} = 82.1^{\circ} \pm 0.9^{\circ}$ and a dispersion of $\Delta\vartheta = 18.4^{\circ} \pm 1.2^{\circ}$ (FWHM) were determined. The corresponding parameters of the energy distribution are $\bar{E}_{\alpha} = 14.8 \pm 1.0$ MeV and $\Delta E = 9.1 \pm 1.2$ MeV (FWHM). These values are in surprisingly good agreement with thermal-neutron induced or spontaneous fission results (cf. /14/). The relative stability of the mean values and widths over a wide range of fissioning systems indicates that lra emission is a unique process in the fission of the actinides. The systematic increase of the mean lra emission angle when moving from uranium to californium is well understood in terms of small variations of the focusing Coulomb field as a result of the shell stabilization of the heavy fragments.

Fission fragment mass distributions for different α-energies have been derived in /13,14/. Despite of a limited mass resolution ($\delta M \approx$ 15 - 20 amu), a remarkable increase of the peak-to-valley ratio (PVR) with increasing α-energy is observed. From binary fission investigations, it is well known that the symmetric component of the double-humped mass distribution increases with increasing excitation energy. This is due to the decrease of the role of shell corrections, which lead to asymmetric fission. The PVR increase with increasing α-energy indicates that the α-particle arising from ternary fission cools down the fissioning system leaving it with lower free energy (deformation + internal heating) than in binary fission. This observation is confirmed by direct measurements of the TKE release in ternary fission /17/.

The drastic increase of the dispersion of the lra angular distribution for near-symmetric fragmentation observed in /13,14/ supports the multichannel fission model of Brosa et al. /18/ predicting three different paths in the fission of ^{236}U what leads to a superposition of several lra angular distributions in the overlap region of the individual fission channels near mass-symmetry.

The present observations suggest the conclusion that lra emission is an outstanding phenomenon characterizing the fission process in the entire actinide region, the relevant features of which can be observed independently of the entrance channel configuration. The PVR dependence on the α-energy indicates that the α-particle observed in ternary fission turns out to be not only a spectator, but an active participant in the dynamics of the fission process.

I am very much indebted to all of my colleagues in Rossendorf, Dresden and Dubna for a fruitful collaboration, the results of which have been presented here.

REFERENCES

/1/ W.U. Schröder and J.R. Huizenga, in: Treatise on heavy-ion science, vol. 2, ed. D.A. Bromley (New York, London, 1984) p. 115

/2/ K.D. Schilling, P. Gippner, W. Seidel, F. Stary, E. Will, K. Heidel, S.M. Lukyanov, Yu.E. Penionzhkevich, V.S. Salamatin, H. Sodan and G.G. Chubarian, Nucl. Instr. and Meth. in Phys. Res. A 257 (1987) 197

/3/ J. Tōke, R. Bock, G.X. Dai, A. Gobbi, S. Gralla, K.D. Hildenbrand, J. Kuzminski, W.F.J. Müller, A. Olmi, H. Stelzer, B.B. Back and S. Bjørnholm, Nucl. Phys. A 440 (1985) 327

/4/ W.Q. Shen, J. Albinski, A. Gobbi, S. Gralla, K.D. Hildenbrand, N. Herrmann, J. Kuzminski, W.F.J. Müller, H. Stelzer, J. Tōke, B.B. Back, S. Bjørnholm and S.P. Sørensen, Phys. Rev. C 36 (1987) 115

/5/ Z. Lojewski, V.V. Pashkevich and S. Cwiok, Nucl. Phys. A 436 (1985) 499

/6/ P. Gippner, K.D. Schilling, W. Seidel, F. Stary, E. Will, H. Sodan, S.M. Lukyanov, V.S. Salamatin, Yu.E. Penionzhkevich, G.G. Chubarian and R. Schmidt, Z. Phys. A 325 (1986) 335

/7/ P. Gippner, K.D. Schilling, W. Seidel, F. Stary, E. Will, H. Sodan, S.M. Lukyanov, G.G. Chubarian, Yu.E. Penionzhkevich, V.S. Salamatin, R. Schmidt and A. Pop, Proc. XVIth Int. Symp. on Nuclear Physics, Gaußig (Nov. 1986) ZfK-610, p. 42

/8/ P.J. Twin, A.H. Nelson, B.M. Nyako, D. Howe, H.W. Cranmer-Gordon, D. Elenkov, P.D. Forsyth, J.K. Jabber, J.F. Sharpey-Schafer, J. Simpson and G. Sletten, Phys. Rev. Lett. 55 (1985)1380; P.J. Twin, B.M. Nyako, A.H. Nelson, J. Simpson, M.A. Bentley, H.W. Cranmer-Gordon, P.D. Forsyth, D. Howe, A.R. Mokhtar, J.D. Morrison, J.F. Sharpey-Schafer and G. Sletten, Phys. Rev. Lett. 57 (1986) 811

/9/ J.F. Sharpey-Schafer, invited talk at the Int. Symp. on Weak and Electromagnetic Interactions in Nuclei, Heidelberg (1986)

/10/ S. Åberg, Lund preprint MPh-87/06

/11/ V.M. Strutinsky, Z. Phys. A 326 (1987) 261

/12/ M.A. Bentley, G.C. Ball, H.W. Cranmer-Gordon, P.D. Forsyth, D. Howe, A.R. Mokhtar, J.D. Morrison, J.F. Sharney-Schafer, P.J.

Twin, B. Fant, C.A. Kalfas, A.H. Nelson, J. Simpson and G. Sletten, Phys. Rev. Lett. 59 (1987) 2141

/13/ R. Kotte, E. Hentschel, H.G. Ortlepp, W. Seidel and F. Stary, Z. Phys. A 328 (1987) 495

/14/ R. Kotte, W.D. Fromm, E. Hentschel, H.G. Ortlepp, K.D. Schilling, W. Seidel, F. Stary and G. Zwicker, Z. Phys. A 330 (1988) 189

/15/ W. Seidel, H.G. Ortlepp, F. Stary and H. Sodan, Nucl. Instr. and Meth. in Phys. Res. A 273 (1988) 536

/16/ V.E. Viola, Nucl. Data A1 (1966) 391

/17/ J.P. Theobald, TH Darmstadt preprint IKDA 87/35 (1987)

/18/ U. Brosa, S. Grossmann and A. Müller, Z. Naturforsch. 41a (1986) 1341 and GSI Darmstadt Nachrichten, GSI 08-87 (1987) 13

CURRENT AND DENSITY ALGEBRA APPROACH TO LOW-ENERGY HEAVY ION COLLISIONS

S.P. Ivanova and R.V. Jolos
Joint Institute for Nuclear Research, Dubna, USSR

If the projectile velocity is smaller than the Fermi velocity, the main mechanism of nuclear excitation during collision will be the single particle one which is connected with the fast rearrangement of the nuclear mean field. For this reason the consideration of the nuclear collisions in the framework of the TDHF method will be more adequate to the considered problem. However, application of this method requires a significant amount of calculations allowing for the self-consistency of the density and average potential. On the other hand, whether the self-consistency requirement is important for the predictions of the collision dynamics is unclear yet. There are well-known examples when disregarding the self-consistency one can simplify the problem thus getting a deeper insight into the process. This can be exemplified by nuclear shell model.

It means that the task should be formulated in terms of quantities which have macroscopic analogs. In our case these are nucleon density and current operators.

How to realized this approach based on the following Hamiltonian of the many-fermion system?

$$\hat{H} = \frac{\hbar^2}{2m} \int d^3x\, \nabla \hat{\psi}^+(x) \cdot \nabla \hat{\psi}(x) + \int d^3x\, d^3y\, \hat{\psi}^+(x)\, \hat{\psi}(x)\, F(x,y)\, \hat{\psi}^+(y)\, \hat{\psi}(y)\,. \tag{1}$$

It is known [1] that the density $\hat{\rho}(x)$ and the current $\bar{j}(x)$ operators form a closed algebra

$$[\rho(x), \hat{j}_\alpha(y)] = -\frac{i\hbar}{m} \frac{\partial}{\partial x_\alpha}(\rho(x)\,\delta(\bar{x}-\bar{y}))\,, \tag{2}$$

$$[\hat{j}_\alpha(x), \hat{j}_\beta(y)] = -\frac{i\hbar}{m}\left(\frac{\partial}{\partial x_\beta}(j_\alpha(x)\,\delta(\bar{x}-\bar{y})) - \frac{\partial}{\partial y_\alpha}(j_\beta(y)\,\delta(\bar{x}-\bar{y}))\right). \tag{3}$$

For the Fermi systems this algebra must be extended by including the tensor of the kinetic energy density $\hat{\mathfrak{T}}_{\alpha\beta}(x) \equiv \frac{\partial\hat{\psi}^+}{\partial x_\alpha} \cdot \frac{\partial\hat{\psi}}{\partial x_\beta} + \frac{\partial\hat{\psi}^+}{\partial x_\beta} \cdot \frac{\partial\hat{\psi}}{\partial x_\alpha}$:

$$[\hat{\mathfrak{T}}_{\alpha\beta}(x), \hat{\rho}(y)] = \frac{2mi}{\hbar}\left(\frac{\partial}{\partial y_\beta}(j_\alpha(y)\,\delta(\bar{x}-\bar{y})) + \frac{\partial}{\partial y_\alpha}(j_\beta(y)\,\delta(\bar{x}-\bar{y}))\right), \tag{4}$$

$$[\hat{\mathfrak{T}}_{\alpha\beta}(x), \hat{j}_\gamma(y)] = \frac{i\hbar}{m}\left(\frac{\partial}{\partial y_\alpha}(\hat{\mathfrak{T}}_{\beta\gamma}(y)\,\delta(\bar{x}-\bar{y})) + \frac{1}{2}\hat{\mathfrak{T}}_{\alpha\beta}(y)\frac{\partial}{\partial y_\gamma}\delta(\bar{x}-\bar{y}) - \frac{1}{2}\frac{\partial^2}{\partial y_\alpha\,\partial y_\gamma}\left(\frac{\partial\hat{\rho}(y)}{\partial y_\beta}\delta(\bar{x}-\bar{y})\right) + (\alpha \rightleftarrows \beta)\right). \tag{5}$$

The analysis of the commutation relations (2-5) shows that $\hat{\mathfrak{T}}_{\alpha\beta}(x)$ may be represented as

$$\hat{\mathfrak{T}}_{\alpha\beta}(x) = \frac{m^2}{\hbar^2}\left(\hat{j}_\alpha(x)\frac{1}{\hat{\rho}(x)}\hat{j}_\beta(x) + \hat{j}_\beta(x)\frac{1}{\hat{\rho}(x)}\hat{j}_\alpha(x)\right) + \hat{t}_{\alpha\beta}(x)\,, \tag{6}$$

when the operator $\hat{t}_{\alpha\beta}(x)$ commutes with $\hat{\rho}(x)$ and can be found by (5) step by step. We will use the following approximation for $\hat{t}_{\alpha\beta}(x)$ [2]:

$$\hat{t}_{\alpha\beta} = \frac{1}{2\hat{\rho}(x)}\frac{\partial\hat{\rho}}{\partial x_\alpha}\cdot\frac{\partial\hat{\rho}}{\partial x_\beta}\,. \tag{7}$$

This approximation is insufficient to calculate the phonon energies in the RPA but we shall use it only to illustrate the main ideas of our approach.

Using (1-5) we get the following equations of motion for $\hat{\rho}$ and $\hat{j}_\alpha$:

$$[\hat{H}, \hat{\rho}(x)] = i\hbar \sum_\alpha \frac{\partial}{\partial x_\alpha} \hat{j}_\alpha(x), \tag{8}$$

$$[\hat{H}, \hat{j}_\alpha(x)] = i\hbar \Big[\frac{\hbar^2}{2m^2} \sum_\beta \frac{\partial}{\partial x_\beta} \hat{\mathfrak{J}}_{\alpha\beta}(x) - \frac{\hbar^2}{4m^2} \frac{\partial}{\partial x_\alpha} \Delta \hat{\rho}(x) + \\ + \frac{2}{m} \hat{\rho}(x) \frac{\partial}{\partial x_\alpha} \int d^3y\, F(x,y)\, \hat{\rho}(y) \Big]. \tag{9}$$

The Hamiltonian (1) can be written as

$$\hat{H} = \frac{\hbar^2}{4m} \sum_\alpha \int d^3x\, \hat{\mathfrak{J}}_{\alpha\alpha}(x) + \int d^3x\, d^3y\, \hat{\rho}(x)\, F(x,y)\, \hat{\rho}(y). \tag{10}$$

Let us describe the dynamics of the nuclear collisions. For this purpose, it is necessary to select from the full set of the nuclear degrees of freedom the relative distance $\bar{R}$, conjugate momentum $-i\hbar\frac{\partial}{\partial \bar{R}}$ and the phonon operators b_s^+, b_s describing the excited nuclear states. Hence, we represent the operators $\hat{\rho}(x)$ and $\hat{j}(x)$ as expansions in degrees of $-i\hbar\frac{\partial}{\partial \bar{R}}$, b_s^+ and b_s

$$\rho(x) \equiv \rho_0 + \rho' + \dots = \rho_0(x) + \sum_s f_s(x,R)(b_s^+ + \sigma_s b_{\bar{s}}) + \dots, \tag{11}$$

$$j_\alpha(x) = \frac{\hbar}{2mi}\Big[\sum_\beta \Big(\frac{\partial}{\partial R_\beta} A_{\beta\alpha}(x,R) + A_{\beta\alpha}(x,R) \frac{\partial}{\partial R_\beta} \Big) + \sum_s B_\alpha^s(x,R)(b_s^+ - \sigma_s b_{\bar{s}}) \Big] + \dots, \tag{12}$$

where σ_s is a phase factor.

Substituting (11), (12) and (6) into (10) we get the Hamiltonian

$$\hat{\mathcal{H}} = -\frac{\hbar^2}{2} \sum_{\alpha,\beta} \frac{\partial}{\partial R_\alpha} \mu_{\alpha\beta}^{-1}(R) \frac{\partial}{\partial R_\beta} + U(R) + \sum_s \omega_s b_s^+ b_s + \\ + \sum_s V_s(R)(b_s^+ + \sigma_s b_{\bar{s}}) + \sum_{s,\alpha} \Big(G_\alpha^s(R) \frac{\partial}{\partial R_\alpha} + \frac{\partial}{\partial R_\alpha} G_\alpha^s(R) \Big)(b_s^+ - \sigma_s b_{\bar{s}}) + \dots. \tag{13}$$

The functions $\mu_{\alpha\beta}^{-1}$, U, V_s, G_α^s and ω_s are expressed in terms of ρ_0, f_s, $A_{\alpha\beta}$ and B_α^s. These functions can be found from the equations of motion (8,9). For this we substitute (11-13) into (8,9) and compare the coefficients at the same operators in right- and left-hand sides of the equations.

It is useful to find a solution of the equation for ρ_0, $A_{\alpha\beta}$, $\mu_{\alpha\beta}$, G_s and V_s in the asymptotic region $R \gg R_1 + R_2$. In this case,

$$\rho_0(x,R) = \rho_1\Big(\bar{x} - \frac{A_2}{A}\bar{R}\Big) + \rho_2\Big(\bar{x} + \frac{A_1}{A}R\Big), \\ A_{\alpha\beta}(x,R) = \delta_{\alpha\beta}\Big(\frac{\rho_1}{A_1} - \frac{\rho_2}{A_2}\Big), \quad \mu_{\alpha\beta} \equiv \delta_{\alpha\beta}\,\mu, \quad \mu = m\frac{A_1 A_2}{A_1 + A_2}, \tag{14}$$

$$G_\alpha^s(R) = -\frac{\hbar^2}{2\mu} \frac{1}{\omega_s} \frac{\partial V_s(R)}{\partial R_\alpha}, \tag{15}$$

$$V_s(R) = \int d^3x\, f_s(x,R)\Big(2\int d^3y\, F(x,y)\,\rho_0(y) - \frac{\hbar^2}{4m}\frac{\Delta\rho}{\rho_0} + \frac{\hbar^2}{8m}\frac{(\nabla\rho_0)^2}{\rho_0^2}\Big). \tag{16}$$

The last relation is correct for arbitrary R.

For $R \gg R_1 + R_2$, when we can use (14), the expression in the parenthesis under the integral in (16) equals zero because it coincides in this limit with the self-consistency condition for the densities of noninteracting nuclei

$$2 \int d^3y F(x,y)\rho_{1,2}(y) - \frac{\hbar^2}{4m}\frac{\Delta\rho_{1,2}}{\rho_{1,2}} + \frac{\hbar^2}{8m}\frac{(\nabla\rho_{1,2})^2}{\rho_{1,2}^2} = 0. \qquad (17)$$

Equation (17) is correct if the assumption (7) is correct too.

The system of equations for f_s and B^s_α describes among other the vibrational excitations in the dinuclear system or in every nucleus (for $R \gg R_1 + R_2$). If we approximate $F(x,y)$ by

$$F(x,y) = \frac{mc_v^2}{\rho_0}\delta(x-y)$$

and neglect the diffuseness of ρ_0 we get the wave equation for f_s:

$$\omega_s^2 f_s + \hbar^2 c_v^2 \Delta f_s = 0,$$

the eigenvalues of which are the vibrational frequences. The form of ρ_0 determines the boundary conditions for the solutions of the wave equation.

In the general case the calculation of the equations, obtained above, is very complicated. The problem can be simplified essentially if we disregard the self-consistency requirement in determining ρ_0 and fix ρ_0 on the basis of the macroscopic picture of the process or calculate ρ_0 based on the solutions of the single particle Schrödinger equation with the two-center potential. In the last case

$$\rho_0 = \sum_h |\psi_h(x,R)|^2,$$

where h means the occupied single-particle states. This single particle basis is useful for the solution of the equations for f_s and B^s_α. The functions f_s and B^s_α can be represented in the form

$$f_s(x,R) = \sum_{p,h} \psi_p^*(x,R)\,\psi_h(x,R)\,(\psi^s_{ph} - \phi^s_{ph}),$$

$$B^s_\alpha(x,R) = \sum_{p,h} \left(\psi_p^*(x,R)\frac{\partial}{\partial x_\alpha}\psi_h(x,R) - \frac{\partial}{\partial x_\alpha}\psi_p^*(x,R)\cdot\psi_h(x,R)\right)\cdot(\psi^s_{ph} + \phi^s_{ph}).$$

Here indices p mean unoccupied states (we do not take into account the pair correlations). The equations analogous to the RPA equations are obtained for the amplitudes ψ^s_{ph} and ϕ^s_{ph}. The first of them follows from the operator equation of continuity (8):

$$(E_p - E_h)(\psi^s_{ph} + \phi^s_{ph}) + 2\sum_\alpha G^s_\alpha(R) \langle h|\frac{\partial}{\partial R_\alpha}|p\rangle = \omega_s(\psi^s_{ph} - \phi^s_{ph}).$$

It we consider the excitations of noninteracting nuclei, the second term in the left-hand side equals zero.

For the derivation of the second RPA equation the approximation (7) for $t_{\alpha\beta}$ is unsatisfactory and must be improved.

The Hamiltonian (13) can be used to study the particle emission in heavy ion collisions. The last two terms in (13) describing the coupling of the intrinsic and relative motion degrees of freedom are responsible for the transition of a part of nucleons from bound states into a continuum. In the coordinate representation

$$H_1 = \int d^3x\, U_T(x)\,\rho'_\rho(x), \quad H_2 = m\int d^3x\, \bar{j}_{col.}(x)\frac{1}{\rho_0(x)}\bar{j}_{in}(x). \qquad (18)$$

Here U_T is the potential generated by the target-nucleus. The collective current $\bar{j}_{col.}$ has the form

$$\bar{j}_{col.} = f(x)\,\bar{v}(t)\ ,$$

where $\bar{v}(t)$ is the relative motion velocity, $f(x) = \rho_p(x)$ in the laboratory system, ρ_0 is the density of the dinuclear system produced, ρ'_0 is the fluctuating part of the nucleon density in the projectile.

Theoretical calculations of the nucleon yield include those of emission of preequilibrium particles at the initial stage of the reaction and statistical particles. The emission of statistical nucleons is calculated in the framework of the standard cascade-evaporative model. The details of calculation of the preequilibrium particles yield are discussed in [3].

The results of calculation of the double differential cross-section $\frac{d^2\sigma}{dE\,d\Omega}$ for neutron emission in the reaction ^{20}Ne (180 MeV) + ^{181}Ta at = 0^0 in the laboratory system are shown in fig. It is seen that the model describes the absolute value and the energy dependence of the cross-section. However, with increasing emission angle the theoretical cross section descreases more steeply than experimental ones.

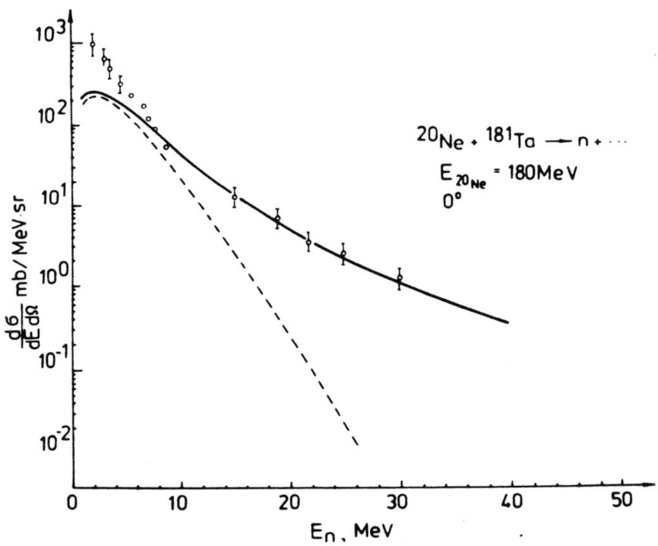

REFERENCES

1. Dashen R.F. and Sharp D.H. Phys.Rev., 1968, 165, p. 1957.
2. Jolos R.V., Kartavenko V.G., Permjakov V.P. Jad. Fiz., 1981, 34, p. 1857.
3. Jolos R.V., Ivanova S.P. In: Proc. Intern. School of Nuclear Structure, D4-85-851, Dubna, JINR, 1985, p. 304;
   Jolos R.V., Ivanova S.P. Jad.Fiz., 1986, 43, p.1463;
   Ivanova S.P., Tomas H. Preprint JINR, E4-88-780, Dubna, 1988.

FISSION DYNAMICS OF HOT NUCLEI

D. Hilscher, D.J. Hinde, and H. Rossner
Hahn-Meitner-Institut Berlin GmbH, D-1000 Berlin 39, Germany

ABSTRACT

Pre- and post-scission neutron multiplicities in fusion-fission reactions are used to determine prescission lifetimes which are found to be in the order of a few 10^{-20} sec. Most of the excitation energy in highly excited composite systems is removed by evaporation of light particles <u>prior</u> to scission. Nuclei are rather cold at the scission point.

1. INTRODUCTION

The dynamics of fission is one of the most delicate problems in nuclear physics. To solve it theoretically the essential input has to be the nuclear dissipation or viscosity - the coupling between the collective motion and the nuclear heat bath - as a function of temperature and the potential energy surfaces for different angular momenta driving the system apart. Originally this problem has been partly attacked by Davies et al.[1)] and more recently by Grangé and Weidenmüller[2)] and Nix et al.[3)]. Independent of theoretical modelling of the fission dynamics, however, it is necessary to have systematic experimental data of the timescales involved in nuclear fission as a function of angular momentum, temperature, mass-split and fissility. In order to measure or deduce the "fission lifetime" the following method has been applied by several groups[4-12)]. The basic idea of this method is to measure the number of neutrons evaporated prior to (M_n^{presc}) and post (M_n^{post}) scission. By using the relation

$$\tau_{presc} = \sum_{i=1}^{M_n^{presc}} \frac{\hbar}{\Gamma_{n,i}} \quad : \quad \Gamma_n \propto (2S_n+1) \sum_{l=1}^{\infty} \sum_{I-l}^{I+l} (2I+1) \int_o^{E-E_{rot}(I)-B_n} \rho(E - E_{rot}(I) - B_n - \epsilon) T_n^l(\epsilon) d\epsilon$$

the prescission lifetime τ_{presc} is traced back to the level densities and thus any given absolute time depends heavily on our knowledge of the nuclear level densities which are not well known at high excitation energies[13,14)]. However, a relative timescale still can be established by employing this method.

The experimental prescission time τ_{presc} as determined by the above description is essentially the time difference from the moment when the system has obtained enough excitation energy to start to evaporate neutrons until shortly after the moment of scission, when the fragments have gained enough kinetic energy so that the kinematic focussing into the direction of the final velocity vectors of the fragments is large enough in order to separate experimentally the prescission from postscission neutrons.

2. PRESCISSION LIFETIME

2.1 Prescission Neutrons at Intermediate Temperatures ($\approx 2.7 MeV$)

We will discuss in the following a special case[7)] in which approximately the same composite nucleus with the same excitation energy or temperature of about 2.7 MeV was produced by different mass asymmetries in the entrance channel: ^{175}Lu + 192 MeV ^{12}C, ^{165}Ho + 220 MeV ^{20}Ne, and ^{141}Pr + 316 MeV ^{40}Ar. The neutrons were measured in coincidence with one fission fragment which was identified by its velocity and energy. The number of evaporated neutrons was obtained by fitting the measured coincident neutron energy-spectra in the laboratory system with evaporation-like spectra from three different sources: the hot composite nucleus resulting in the prescission neutron multiplicity M_n^{presc} and the two fission fragments resulting in the postscission multiplicity M_n^{post} from the much colder fission fragments. In figure 1 the experimental angular correlation of the neutron yield integrated between 2 and 10 MeV is shown and compared to the fit. It is obvious that without a large contribution from the composite system it is not possible to describe the measured angular correlation. The resulting pre (post) scission neutron multiplicity are 3.6 (4.4), 5.6 (3.7), and 6.3 (3.6) neutrons for the reactions on Pr, Ho, and Lu, respectively. The prescission neutron multiplicities result in prescission lifetimes of about 1 to 6×10^{-20} sec. For more fissile systems Hinde et al[8,11] find that the extra delay time is between 3 and 10×10^{-20} sec. Thus we can summarize that independent of the system and to some extent also independent of the assumptions made in the analysis the prescission lifetime deduced from the prescission neutron multiplicity is a few times 10^{-20} sec.

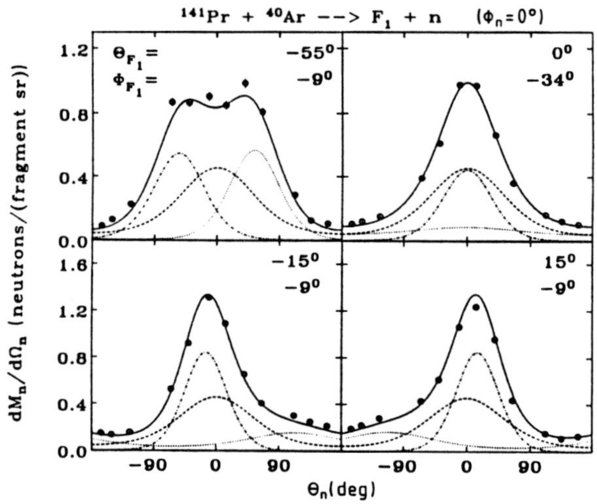

Figure 1: Measured (points) and fitted (curves) angular correlations for neutrons in coincidence with fission fragments detected at different fixed trigger angles Θ_{F_1}, Φ_{F_1} in 314 MeV argon induced fission on Pr. The given neutron yield was integrated between 2 and 10 MeV. The dotted and dotted-dashed curves are the contributions from the fully accelerated fission fragments, the dashed curve represents the prescission neutrons. The solid line is the sum of all three contributions (ref.[7]).

2.2 Prescission Neutrons at High Temperatures ($\approx 5MeV$)

In order to further investigate the prescission lifetime of highly excited composite systems at temperatures of about 5 MeV the neutron emission was measured[15] in the systems $^{144,154}Sm$ + ^{32}S at 838 MeV bombarding energy. The velocity vectors of the two heavy fragments were measured in coincidence by two multi-wire chambers. Neutrons were measured in coincidence with both fragments. The data[15] show that for asymmetric mass-splits considerably less (more) neutrons are emitted prior to (post) scission. The absolute number of neutrons is larger for the neutron rich Sm isotope. But in particular the data show that prior to scission for symmetric mass-splits 12-14 neutrons are emitted in the reaction $^{154}Sm + ^{32}S$ at these high temperatures. The corresponding prescission lifetime is about 10^{-20} s. The postscission neutron multiplicity, independent of excitation energy, is about 5 neutrons. This is very close to the postscission neutron multiplicities found for the lower excitation energies in section 2.1.

3. Systematics of Postscission Neutron Multiplicities.

Since the mean evaporation time increases exponentially with decreasing temperature the prescission lifetime as deduced with the above described method is essentially determined by the longest evaporation time or the last neutrons in the prescission evaporation cascade. This then leads us to the conclusion that the excitation energy or temperature at or close to scission is the essential quantity. The same quantity, in turn, determines the postscission neutron multiplicity. Consequently if there were no neutron binding energy effects for different systems a constant prescission lifetime would result in a constant postscission neutron multiplicity. Thus the emission of the last neutron prior to scission will always occur at approximately the same temperature. The postscission neutron multiplicity will depend, however, on the neutron binding energies and fission Q-values for different fissioning systems.

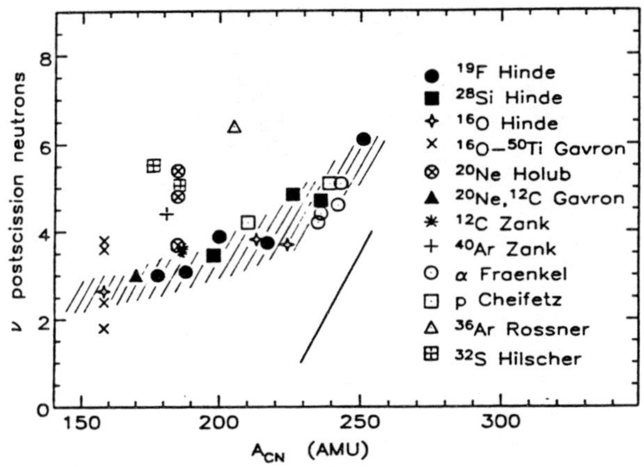

Figure 2: Compilation of postscission neutrons ν. The data was taken from Gavron et al.[5,10], Hinde et al.[8,11], Holub et al.[6], Zank et al.[7], Rossner et al.[16], Hilscher et al.[15], Cheifetz et al.[17], and Fraenkel et al.[18]. The solid line represents the average number of neutrons emitted in spontaneous fission or thermal induced fission corrected for zero excitation energy[19].

A compilation of a large body of published and unpublished postscission neutron multiplicity data (total number of neutrons ν emitted from both fission fragments) is shown in figure 2 as a function of the mass $A_{CN} = A_P + A_T$ of the composite system. The excitation energies of the initially produced composite system range between 35 and ≈550 MeV. Our above discussed expectation of a constant postscission neutron multiplicity is almost fulfilled, the data cluster (hatched area in fig. 2) at almost constant multiplicity with a few exceptions having higher postscission neutron multiplicity. Furthermore we observe a multiplicity slowly increasing with increasing A_{CN}. This is mainly due to the increase of available energy released in fission reduced by the total kinetic energy of the fission fragments (Q_{fiss} - TKE) and to a smaller extent also due to the increase of neutron richness of the fission fragments. This is indicated by the empirical solid line which represents an approximation to the average number of neutrons $\bar{\nu}$ per spontaneous fission for thermal neutron induced fission corrected to zero excitation energy[19]. Thus the solid line represents the number of postscission neutrons of fissioning nuclides with zero excitation energy. The additional number of postscission neutrons in induced fission represents the excitation energy of these nuclei at scission. For lower A_{CN} the value of (Q_{fiss} - TKE) becomes negative and the neutron richness of the fission fragments is also decreasing so that the postscission neutron multiplicity should and does further decrease to smaller A_{CN}. For $A_{CN} = 220$ where $\bar{\nu}(E^* = 0) \approx 0$, $\nu_{post} \approx 4$ from which we deduce a mean excitation energy at scission of about 40 to 50 MeV assuming that each neutron carries away about 10 to 12 MeV. The systematics shown in figure 2 points to a rather constant prescission lifetime for all true fusion-fission events.

4. CONCLUSIONS

All the pre- and post-scission neutron data in heavy-ion induced fission discussed in this contribution are consistent with a prescission lifetime of approximately a few times 10^{-20} sec. However, for asymmetric mass-splits in the ^{32}S induced fission on ^{154}Sm the prescission neutron multiplicity is smaller than for symmetric mass-splits. This finding clearly indicates that asymmetric mass-splits in fusion-fission like reactions, probably correlated with higher angular momentum, have shorter lifetimes than fission giving a symmetric mass-split.

We have compiled all postscission neutron multiplicity data and found that essentially all fusion-fission reactions have a rather constant postscission neutron multiplicity which is modified only by the fission Q-values. These data show clearly that the composite systems are essentially cold at the scission point with excitation energies of only $\approx$ 40 to 60 MeV. The physical reason for this finding is that for higher excitation energies the evaporation times are becoming so short that the additional time needed for evaporation or cooling is negligible compared to the time needed for any collective motion.

REFERENCES

1) K.T.R Davies, A.J. Sierk, and J.R. Nix, Phys.Rev. C13, 2385 (1976).
2) P. Grangé, Li Jun-Qing, and H.A.Weidenmüller, Phys. Rev. C27, 2063 (1983)
3) J.R. Nix, A.J. Sierk, Proceedings of the 6th Adriatic International Conference on Nuclear Physics: Frontiers of Heavy-Ion Physics, Dubrovnik, Yugoslavia, 1987.
4) D. Hilscher et al, Proceedings of the 3rd Adriatic Europhysics Conference on Dynamics of Heavy-Ion Collisions, Hvar, Yugoslavia, 1981, p. 225.
5) A. Gavron et al. Phys. Rev. Lett. 47, 1255 (1981), erratum: 48, 835 (1982).
6) E. Holub et al. Phys. Rev. C28, 252 (1983).
7) W.P. Zank et al. Phys. Rev. C33, 519 (1986).
8) D.J. Hinde et al. Nucl. Phys. A452, 550 (1986).
9) A. Gavron Phys. Lett. 176B, 312 (1986).
10) A. Gavron et al. Phys. Rev. C35, 579 (1987).
11) D.J. Hinde et al. to be published.
12) D.J. Hinde unpublished data.
13) G. Nebbia et al, Physics Lett. B176, 20 (1986).
14) D. Hilscher Nucl. Phys. A471, 77c (1987).
15) D. Hilscher et al. to be published.
16) H. Rossner et al, to be published.
17) E. Cheifetz et al. Phys. Rev. C2, 256 (1970).
18) Z. Fraenkel et al, Phys. Rev. C12, 1809 (1975).
19) R. Vandenbosch, R., and J.R. Huizenga Nuclear Fission. Academic Press, 1973.

IX. FISSION STUDIES FOR APPLICATIONS

LARGE SCALE FISSION PRODUCT SEPARATION FOR NUCLEAR MEDICINE

R. Münze, O. Hladik, G. Bernhard, and W. Boessert
Dresden, Germany

ABSTRACT

A procedure is described for the production of fission Mo-99. Starting from short period irradiated fuel elements of a research reactor the following steps are included: dissolution of the aluminium cladded fuel element in 6 N HNO_3, separation of Mo-99 from uranium and the bulk of other fission products by adsorption/desorption on alumina, and purification of the molybdenum fraction by means of high temperature volatilization.

1. INTRODUCTION

The discovery of Otto Hahn and Fritz Strassmann has not only opened the door for electricity production in Nuclear Power Stations and to a nearly unlimited access to lots of useful isotopes for application in several fields. But, even some of the fission products itself have found many applications, in particular in nuclear medicine.

Medically useful radioisotopes of the fission products are Sr-89 and Y-90 for therapy and I-131, Xe-133 and Mo-99/Tc-99m (Tucker 1962). Particulary, the Mo-99/Tc-99m system achieved an important role in in vivo diagnostics. More than 90 % of currently diagnostic nuclear medical in vivo examinations have been accomplished with Tc-99m as the radioactive indicator.

This outstanding role is due to the physical and chemical properties of this short lived nuclear isomer which are exceptionally favourable for this purpose. These are:

- the very low radiation dose to the patient because of the absence of highly ionizing particle radiation
- the optimal, nearly monochromatic photon energy of 140 keV
- the unlimited availability of this short lived isotope by using the longer living parent nuclide Mo-99 (67 h) in form of an easily applicable generator system for separating Tc-99m at the site of application.

2. LARGE SCALE PRODUCTION OF Mo-99

Since about 15 years the main source for Mo-99 is nuclear fission, because the specific activity of Mo-99 produced by neutron activation of Mo-98 is to low for loading high activities on very small generator columns. The high fission yield of about 6 % allows the production of kilocuries of Mo-99 by irradiating gram amounts of U-235 for some hours in a nuclear reactor. The main problem is the separation of this nuclide from the extraordinarily high activity of the other fission products, and, in particular, the management of the high volatile activities, like iodine and xenon isotopes. These appear in the range of several kilocuries in any production run. At the present, this technology is managed by producers in Canada, the United States, Belgium and the German Democratic Republic.

2.1 TECHNNOLOGY

The process and technology which are in use in the Central Institute of Nuclear Research Rossendorf (Münze 1984) should be outlined in brief.

Fig 1 shows the technological flow sheet of the process. Some more generally interesting details shall be discussed. The target consists of an usual fuel element of the research reactor in which core the irradiation is performed. The fissile material is 100 gm of 36 % enriched uranium in form of an uranium-aluminium-alloy. The chemical separation process is, unlike to the usual fuel reprocessing cycle, not directed to gain the remaining fissile material but to separate specifically fission molybdenum.

Therefore, immediately after dissolution of the irradiated fuel element in boiling nitric acid, the Mo-99 is separated from most of the other fission products by adsorption on alumina. This material specifically negatively charged ions from aqueous solution, such as molybdate, ruthenate, tellurate and iodide, whereas the uranium and most of the other fission products as cationic molecules pour through this material. This part of the material is discarded as radioactive waste and stored for a later reprocessing.

The following steps to process the raw Mo-99 fraction are dictated by the extremely high radionuclidic purity which is required for the application of this isotope in Nuclear Medicine (table 1).

Table 1: PRODUCT SPECIFICATION OF FISSION MOLYBDENUM

specific activity 100 TBq/g Mo

Radiochem. purity Mo-99 as molybdate 99 %

Radionuclidic purity (with reference to Mo-99 activity)

I-131	5.10^{-4}	Zr-95	4.10^{-4}
Ru-103	6.10^{-4}	Sb-125	4.10^{-4}
Rh-106	3.10^{-5}	total alpha	1.10^{-5}
Ce-141	9.10^{-7}	beta	4.10^{-6}

This purification process includes another anion exchange process on alumina followed by a high temperatur treatment (sublimation) which reduced the content of the product of volatile impurities, in particular iodine and ruthenium compounds. Finally, the Mo-99 remains as some milligrams of molybdenum trioxide with an activity of about 30 TBq (0.8 kCi). This material is dissolved and adsorbed on alumina which is filled in small glass columns of the so called Mo-99/Tc-99m generators.

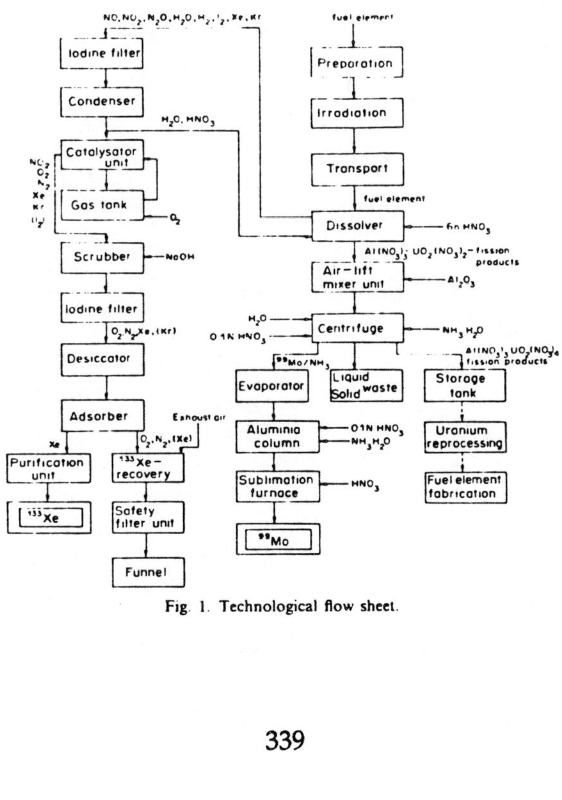

Fig. 1. Technological flow sheet.

Technetium-99m which is continously produced in these columns by the decay of Mo-99 is not adsorbed and thus can be washed out with sodium chloride solution. In this form the material is ready for use either directly for imaging the thyroid or for preparing technetium compounds which are useful for a lot of different nuclear medical examinations.

A view on the separation scheme shows that a considerable part of the technology is devoted to the processing of gaseous effluents. The reason for this big effort is the high radioactivity of iodine and xenon isotopes which are deliberated during the dissolution of the fuel element. This activity is much higher than in case of the usual reprocessing. Especially, the high iodine activity of about 100 TBq (2.6 kCi) must be reduced by a factor of 10^5 before the off gas can leave the stack of the plant. Several chemical processing and adsorption steps on highly efficient sorption materials (silica gel/silver nitrate) are needed.

The required reliability and the measures for nuclear and radiation safety of such a plant are considerable. The whole process is included in a system of hot cells and is operated by remote controlled tools. Such a plant is operating in the Central Institute of Nuclear Research since 1981. Its Mo-99 output covers the whole demand for nuclear medical application in the German Democratic Republic as for exportation in several european countries.

REFERENCES

Münze, R., O. Hladik, G. Bernhard, W. Boessert and R. Schwarzbach
Int.J.Appl.Radiat.Isot. 35 749 (1984)

Tucker, W.D., M.W. Greene and A.P. Murrenhoff
Atompraxis 8, 163 (1962)

FISSION PRODUCT NUCLEAR DECAY SCHEME DATA MEASUREMENTS AND EVALUATION

M.U. Rajput and T.D. MacMahon
Imperial College Reactor Centre, Ascot, England

1. Introduction

Fission product decay schemes have been studied at Imperial College Reactor Centre over a period of several years. Earlier publications have reported the results of measurements of the half-life of ^{140}La (1) and of gamma ray emission probabilities in the decay of ^{140}Ba and ^{140}La (2) and in ^{144}Ce and ^{144}Pr (3).

2. Measurements

2.1 The half-life of ^{154}Eu

Two sources of ^{154}Eu were produced in 1980; one source from the ^{154}Sm(p,n)^{154}Eu reaction, and the other from the ^{153}Eu(n,γ)^{154}Eu reaction. Source strengths were of the order of 10^4 Bq. Each source was mixed with a ^{60}Co solution of similar activity and sealed in polythene containers. The half-life of ^{154}Eu is then determined by comparing the intensity of the 1.27 MeV ^{154}Eu gamma ray peak to those of the 1.17 MeV and 1.33 MeV ^{60}Co gamma ray peaks as a function of time. Gamma ray spectra have been recorded at intervals over a period of eight years. Assuming a ^{60}Co half-life of 5.271y leads to a ^{154}Eu half-life of 8.60 ± 0.16y (3143 ± 59d).

2.2 Gamma ray emission probabilities in the decay of ^{125}Sb

Gamma ray emission probabilities are determined at the Reactor Centre by a combination of high resolution gamma ray spectroscopy (yielding gamma ray emission rates, N_γ) and $4\pi\beta$-γ coincidence measurements (yielding absolute disintegration rates of sources, N_o). The ratio N_γ/N_o is the gamma ray emission probability P_γ. Full details of the technique can be found in references 2 and 3.

Measurements of P_γ values in ^{125}Sb decay are currently in progress. The results shown in Table 1 must be regarded therefore as preliminary results of the present study.

3. Evaluation

3.1 Half-lives of ^{90}Sr and ^{137}Cs

Significant discrepancies exist in published half-life data for both ^{90}Sr and ^{137}Cs, see Tables 2 and 3. These discrepancies make the estimation of evaluated half-lives for these isotopes difficult and contentious.

The weighted mean of the nine values of the ^{90}Sr half-life reported since 1955 and listed in Table 2 is 28.48 ± 0.03y with a total chi-squared of 238 for eight degrees of freedom indicating a high degree of inconsistency. Two values (refs. 8 and 11) contribute together 91% of the total chi-squared. One can only conclude that one or both of these measurements have associated uncertainties that have been seriously underestimated. In order to arrive at an evaluated half-life that is not dominated by these two discrepant values their associated uncertainties have been each increased to 0.2 years. The new weighted mean value then becomes 28.67y with a chi-squared of 26 of which now only 32% is contributed by the two measurements (8 and

11). The data set still remains inconsistent, and, to reduce chi-squared to its expected value, 8, it is necessary to inflate all uncertainties by a factor of 1.82 to obtain the final recommended value 28.67 ± 0.18y (10473 ± 66d).

Table 3 shows that the situation for the half-life of ^{137}Cs is similarly unsatisfactory. The weighted mean of the 18 values shown is 30.24 ± 0.01y with a chi-squared of 783. It is clear that this data is also highly inconsistent and that the major contributors to chi-squared are those values with quoted uncertainties less than 0.1y. In order to arrive at a recommended value, these uncertainties have been increased to 0.1y. This leads to a new mean of 30.11y and reduces chi-squared to 114. To remove the remaining inconsistency it is then necessary to inflate all uncertainties by a factor of 2.6 leading to a final recommended value of 30.11 ± 0.08y (10,999 ± 31d).

3.2 Gamma ray emission probabilities in the decay of ^{125}Sb

Table 1 shows relative gamma ray emission probabilities which have been evaluated from 7 sets of published experimental data (28-34). The data for each gamma ray were treated in a way similar to that indicated above for the ^{90}Sr and ^{137}Cs half-life data. Also shown in Table 1 is the result of converting the evaluated relative intensities of column 1 into absolute intensities by forcing the sum of all transitions feeding the lowest 3 levels of ^{125}Te to be 100%. Finally, column 4 of Table 1 shows the measured absolute intensities published by Walters and Meyer 28).

References

1. Olomo, J.B. and Mac Mahon, T.D., J.Phys.G:Nucl.Phys. 6(1980)367-368.
2. Olomo, J.B. and Mac Mahon, T.D., J.Phys.E:Sci.Instrum. 17(1984)124-125.
3. Olomo, J.B. and Mac Mahon, T.D., Nucl. Energy, 20(1981)237-240.
4. Wiles, D.M. and omlinson, R.H., Can.J.Phys. 33(1955)133-137.
5. Anikina, M.P. et al. Atomn.Energiya 4(1958)198.
6. Flynn, K.F. et al. J.Inorg.Nucl.Chem. 27(1965)21-23.
7. Hoppes, D. NBS private comm. (1977).
8. Lagoutine, F. et al. Int.J.App.Rad. & Isot. 29(1978)269-272.
9. Ramthun, H., Nucl.Instr. & Meths. 207(1983)445-448.
10. Schrader, H., PTB private comm. (1987).
11. Martin, R.H., AECL private comm. (1987).
12. Brown, F. et al. J.Inorg.Nucl.Chem. 1(1955)241-247.
13. Farrar, A.K. et al. Can.J.Chem. 39(1961)681.
14. Cook, H.D. Gatlinburg Conf. on Anal. Chem. in Nucl. Reactor Tech. 1962.
15. Fleishman et al. At.Energy 10(1962)622.
16. Gorbics, S.G. et al. Nucleonics 21(1963)63-67.
17. Rider, B.F. et al. Nucl.Sci.Eng. 15(1963)284-287.
18. Lewis, R.E. et al. Trans.Am.Nucl.Soc. 8(1965)79-80.
19. Harbottle, G. Radiochem. Acta, 13(1970)132-134.
20. Walz, K.F. and Weiss, H.M. Z. Naturforsch. 25a(1970)921-927.

21. Emery, J.F. et al. Nucl.Sci.Eng. 45(1972)319-323.

22. Dietz, L.A. and Pachuki, C.F. J.Inorg.Nucl.Chem. 35(1973)1769-1776.

23. Corbett, J.A. Nucl.Eng.Int. 18(1973)715-716.

24. Gries, W.H. and Steyn, J. Nucl.Instr.Meth. 152(1978) 59-462.

25. Houtermans, H. et al. Int.J.App.Rad.Isot. 31(1980)153-154.

26. Hoppes, D.D. and Schima, F.J. NBS-SP 626(1982).

27. Martin, R.H. AECL-9503 (1987).

28. Walters, W.B. and Meyer, R.A. Phys.Rev. C14 (1976)1925-1934.

29. Iwata, Y. et al. Nucl.Instr.Meth. 219(1984)123-133.

30. Singh, K. and Sahota, H.S. Ind.J.Pure & App.Phys. 21(1983)19-24.

31. Gehrke, R.J. et al. Nucl.Instr.Meth. 147(1977)405-423.

32. Prasad, R., Czech.J.Phys. B29(1979)737.

33. Roney, W.M. and Seale, W.A. Nucl.Instr.Meth. 171(1980)389.

34. Nagpal, T.S. and Gaucher, R.E. Can.J.Phys. 48(1970)2978.

TABLE 1 : Gamma Ray Probabilities in the Decay of ^{125}Sb

Energy (keV)	(1) Relative Intensities Evaluated	(2) Relative Intensities Present Measurement	(3) Absolute Intensities Evaluated	(4) Absolute Intensities Measured (Ref 28)
19.9	0.068±0.002		0.020±0.004	0.02 ±0.01
35.50	14.53 ±0.35		4.22 ±0.35	
58.29	0.091±0.004		0.027±0.005	
109.27	0.233±0.005		0.068±0.006	
117.0	1.066±0.035	0.85±0.008	0.313±0.035	0.255±0.004
172.62	0.75 ±0.05		0.22 ±0.05	0.182±0.003
176.33	23.06 ±0.14		6.78 ±0.14	6.79 ±0.02
178.68	0.110±0.009		0.032±0.009	0.027±0.004
198.6	0.054±0.011		0.016±0.011	0.013±0.003
204.12	1.105±0.011	1.08±0.06	0.32 ±0.011	0.323±0.004
208.08	0.808±0.009		0.238±0.009	0.236±0.004
227.91	0.437±0.008	0.42±0.06	0.127±0.009	0.132±0.004
321.03	1.40 ±0.02	1.48±0.07	0.41 ±0.02	0.410±0.004
380.43	5.13 ±0.04	5.08±0.14	1.51 ±0.04	1.52 ±0.01
408.1	0.611±0.012		0.180±0.012	0.183±0.006
427.88	100	100	29.4 ±0.3	29.44 ±0.09
443.49	1.03 ±0.02	0.95±0.06	0.30 ±0.02	0.303±0.007
463.38	35.47 ±0.05	35.69±0.8	10.43 ±0.05	10.45 ±0.02
600.55	60.36 ±0.11	59.82±1.20	17.75 ±0.11	17.78 ±0.03
606.64	17.03 ±0.03		5.01 ±0.03	5.02 ±0.01
635.89	38.36 ±0.15	38.30±0.79	11.28 ±0.15	11.32 ±0.02
642.14	0.160±0.009		0.047±0.009	
671.40	6.06 ±0.02	6.30±0.14	1.78 ±0.02	1.80 ±0.04

TABLE 2 : ^{90}Sr Half-Life Data

Reference	Half-Life (years)	Chi-Squared contribution	Adjusted uncertainty	Adjusted Chi-Squared
4	27.7 ±0.4	3.8		5.9
5	29.3 ±1.6	0.3		0.2
6	28.0 ±0.4	1.5		2.8
6	28.5 ±0.9	0.0		0.0
7	29.12 ±0.24	7.0		3.4
8	28.15 ±0.034	96.7	0.2	6.9
9	28.99 ±0.25	4.1		1.6
10	29.58 ±0.45	5.9		4.0
11	28.92 ±0.04	118.6	0.2	1.5
		237.9		26.4

TABLE 3 : ^{137}Cs Half-Life Data

Reference	Half-Life (years)	Chi-Squared contribution	Adjusted uncertainty	Adjusted Chi-Squared
12	30.0 ±0.4	0.4		0.1
13	30.4 ±0.4	0.2		0.5
14	29.40 ±0.18	21.7		15.6
15	30.1 ±0.7	0.0		0.0
16	29.68 ±0.05	124.9	0.1	18.6
17	29.2 ±0.3	12.0		9.2
18	30.72 ±0.13	13.7		21.9
6	29.9 ±0.5	0.5		0.2
6	30.9 ±0.7	0.9		1.3
19	30.64 ±0.43	0.9		1.5
20	29.901±0.045	56.4	0.1	4.4
21	30.18 ±0.10	0.3		0.5
22	30.174±0.011	34.8	0.1	0.4
23	30.21 ±0.08	0.1	0.1	1.0
24	29.86 ±0.09	17.7	0.1	6.3
25	30.142±0.030	10.4	0.1	0.1
26	30.68 ±0.02	486.5	0.1	32.3
27	29.94 ±0.27	1.2		0.4
		782.6		114.3

Evaluated Half-Lives

^{90}Sr	28.67	± 0.18y
^{137}Cs	30.11	± 0.08y

SUB-BARRIER NEUTRON-INDUCED FISSION OF ^{232}Th

B.I. Fursov, E.Yu. Baranoy, B.F. Samylin, Yu.M. Tuchin, and G.N. Smirenkin
Institute of Physics and Power Engineering, Obninsk, USSR

Abstract: The 232-Th/235-U fission cross-section ratio has been measured in the neutron energy range 0.16 - 1.6 MeV. Two different experimental techniques were employed. The results are compared with previous data. The low level of 232-Th fissility (ca. 10^{-6}) and its weak energy dependence in the energy range 0.1 - 0.6 MeV was confirmed.

Recently the fission of lightest actinides have been under active study relative to the problem of so-called thoriûm anomaly[1)]. Not long ago the ^{232}Th neutron-induced fission cross-section has been found to have a deep sub-barrier "plateau" on the level 3 - 10 microbarn in the energy range 0.1 - 0.7 MeV. Similar structure with the width up to 0.3 MeV in the sub-barrier fission cross sections were observed earlier in a series of Th and Pa isotops for various ways of excitation[1,3)]. The width of this structure for the case under consideration is at last twice as large. These salient features of the effect revealed in ref. 2 such a low level of fissility (ca.10^{-6}), its weak energy dependence could be explained by its backround origin. That is why it seemed reasonable to perform independent measurements of ^{232}Th/^{235}U fission cross-section with another experimental technique, fissile layers and a neutron source.

The ^{232}Th/^{235}U fission cross-section ratio was measured in the experiment. The measurements were carried out at electrostatic accelerator. The neutron source was T(p,n)-reaction on hard target. Fission fragments were detected with a twin ionization chamber with the efficiency about 100%. The absolute values of ^{232}Th/^{235}U fission cross-section ratio were determined in seperate experiments in the neutron energy range E_n = 1.5 - 3 MeV.

Even with insignificant impurities of ^{235}U in ^{232}Th being potential main source of backround in a deep-subbarrier ^{232}Th fission area in the energy rang 0.16 - 0.85 MeV, some additional measurements were conducted. An assembly of ^{232}Th layers on aluminium backings was adopted as well as that of metallic ^{232}Th foil 0.1 mm in thickness. The impurity of ^{235}U in a ^{232}Th foil had been predetermined in a special test to be < 2 . 10^{-7} (ref. 4). Fission fragments were detected by mica detectors, the number of tracks was evaluated visually under the microscope. The same ionization detector with ^{235}U was used to determine the neutron flux. The data obtained by the track technique were normalized for the $^{232}Th/^{235}U$ fission cross section ratio with E_n = 0.85 MeV. The employment of two independent experimental methods enabled us to increase reliability and statstical accuracy of measurements in the deep-subbarrier region of ^{232}Th fission.

In the addition, fission fragment angular distributions were measured in the cylindrical glass detectors technique.

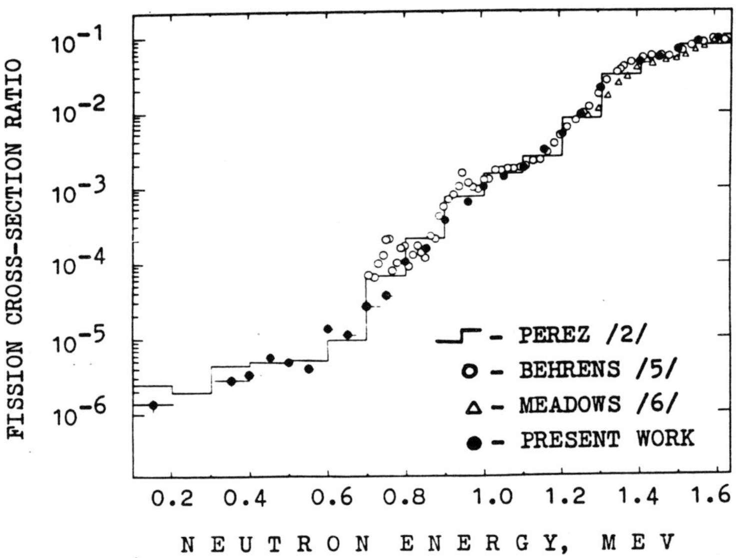

Fig.1 $^{232}Th/^{235}U$ fission cross-sections ratio

Our results of $^{232}Th/^{235}U$ fission cross-section ratio in the range $E_n \leq 1.6$ MeV are given in fig.1 together with the data of refs. 2,5,6. For the range $E_n \leq 0.85$ MeV the data obtained in the ionization chamber and mica detectors were averaged and their statistical accuracy is better than 10%.

The results obtained are in good agreement with the data of ref.2 and acknowledge the ^{232}Th fission cross-section in the range $E_n \leq$ 0.6 MeV to abruptly deviate from the exponentially droping dependence specifie to higher energies. Fig.2 illustrates angular distributions of ^{232}Th fission fragments for the energy values E_n = 0.6 and 0.8 MeV. It can be seen, that when passing to the relative ^{232}Th fission cross-section constancy region the angular distribution become

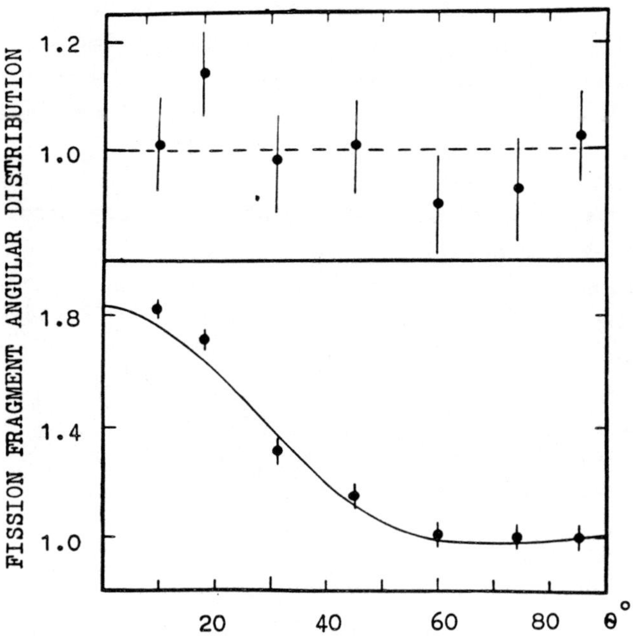

Fig.2
Fission fragment angular distribution W(θ) for En = 0.6 MeV (top) and En = 0.8 MeV (bottom)

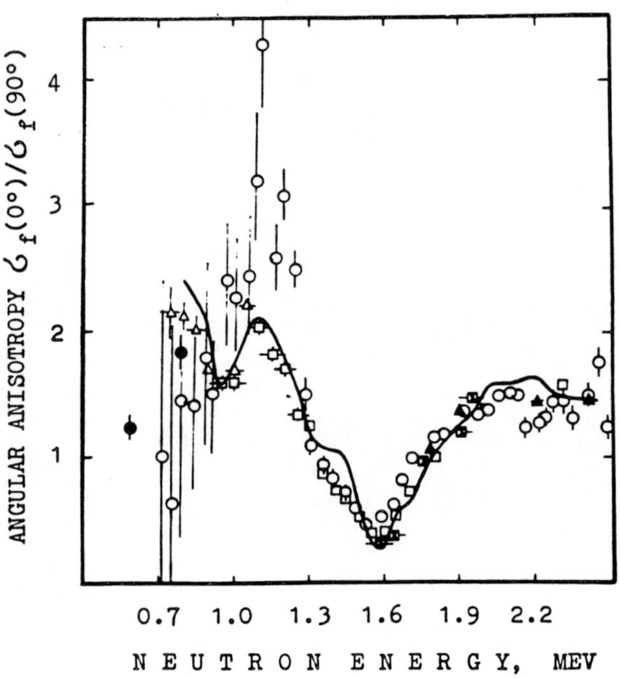

Fig.3
Angular anisotropy $\sigma_f(0^\circ)/\sigma_f(90^\circ)$ for $^{232}Th(n,f)$ [1].
The following points are added to the experimental data given in ref.1 and their theoretical description (ref. 10) shown in fig.3 by a solid curve:
o - ref. 9
o - the results of this work.

virtually isotropic. The experimental data treatment taking into account 2 and 4 degrees of polynom offers the angular anisotropy values A = 1.85 ± 0.18 and 1.23 ± 0.17, respectively, for E_n = 0.8 and 0.6 MeV. An assumption on fission fragment angular distributions in the range E_n < 0.8 MeV becoming more isotropic can be made in fig. 3 taken from ref. 1 supplemented by the recent data of ref. 7 and the results of present work. This is inconsistent with the tendency defined in the theoretical calculations[8)] and shown in fig.3.

Comparison of ^{232}Th fissility in the range E_n = 0.16 - 0.6 MeV and for the thermal neutrons offers the values (1-1.5) . 10^{-6} and (0.3-0.5) . 10^{-6}, respectively. The measurement results of thermal neutron-induced ^{232}Th fission cross-section (2.5-4 μbarn)[9)], thermal radiative capture cross-section $\sigma_{n\gamma} \approx \sigma_{abs}$ = 7.37 barn, compound nucleus formation cross-section in the range E_n = 0.16-0.6 MeV σ_{CN} = 2.82 barn[10)] were applied for calculation. It is worth mentioning, that if in the range E_n = 1.2-0.6 MeV the ^{232}Th fissility drops ~ 10^3 times then in the range from E_n= 0.6 MeV to thermal energy it changes only 2-3 times. This behaviour of fissility is anomalous and needs explanation along with the similar properties of some other light actinides. A satisfactory theoretical description within a three humped fission barrier for these extended (up to 0.6 MeV) structures has not been obtained yet. The main result of this work should be validation of data[2)], that allows the suspicion of relating the observable effect in a deep-subbarrier ^{232}Th fission cross-section to the experimental inaccuracies to be ruled out.

REFERENCES

1. A. Michaudon, Some Aspects of Nuclear Fission. Report CEA-N-2232 (1981).
2. R.B. Perez et al., Phys. Rev. **C28**, 1635 (1983).
3. B.B. Back et al., Phys. Rev. **C9, 1924** (1974).
4. V.E. Zhutcko et al., Yad. Fiz. **28**, 1185 (1978).
5. J.W. Behrens et al., Nucl. Sci. Eng.**81**, 512 (1982).
6. J.W. Meadows, Proc. Conf. Nucl. Cross Sect. for Technology, Knoxville (1979) NBS Spec. Publ. **594**,479 (1980).
7. J.A. Becker and R.W. Bauer, Phys. Rev. **C34**, 594 (1986).
8. H. Abou-Yehia et al., Report INDC (FR) 34L-NEANDC(E)204"L" (1979).
9. C. Wagemans et al., Nucl. Phys. **A259**, 423 (1976); **A285**, 32 (1977).
10. Ch. Lagrange, Report INDC (FR) 56/L-NEANDC(E) 228"L" (1982).

^{235}U FISSION IN (n, f) AND (n, n'f) REACTIONS

A.A. Goverdovskij, G.A. Kudyaev, Yu.B. Ostapenko, G.N. Smirenkin
The Institute of Physics and Power Engineering, Obninsk, USSR

1. Introduction.

The practical need constantly drow great attention to the research of neutron-induced heavy nuclei fission in the neutron range $E_n \leq 15$ MeV. Due to this fact a lot of experimental data on various characteristics of this reactions have been amassed. Nevertheless our knowledge of their energy dependence in actually limited by the first 'plateau' region about 6 MeV up to the threshold of the so-called emission fission, which means the activation of reactions with neutron preemission. As a result the fission of various excitation energy nuclei mixture is observed, and it turns an appreciable barrier in the recovery of reaction characteristics for the parent nucleus A, or so-to-speak, the first fission chance. It is one of the principal problems in nuclear fission physics.

In addition to scientific significance the solution of the emission ('chance') $\sigma_f(E_n)$ fission cross-section structure is indubitable practical interest. The energy distribution of neutron 'source' in nuclear reactors cannot be correctly specified without the reliable insight into it. One of the main difficulties of this problem lies in incompleteness and a great number of parameters required for specific calculations being unknown rather than in complication of theoretical description.

Due to the above-mentioned the search and measurement of fission process characteristics importing knowledge of the cross-section structure to be discussed aquires high priority urgency. The present work is focused on the energy dependence of symmetric mass distribution component of fragments $\sigma^2_{M,asymm}(E_n)$ featuring favourable propeties for the objective mentioned. The data on the chances of neutron-induced ^{235}U fission cross-sections in the second plateu region obtained in the analysis are adopted for the statistical description improvement.

2. Analysis of $\sigma^2_{M,asymm}$ dispersion energy dependence.

The value to be considered is

$$\sigma^2_{M,asymm} = \sum_M (M - M)^2 Y_{asymm}(M), \qquad (1)$$

where $Y_{asymm}(M)$ - is a normalized mass dustribution of the asymmetric component predominating in the total yield of $Y(M)$ over the whole neutron energy range of interest (further the index 'asymm' is omitted). Dispersion (1) is an integral characteristic and that is why it can be fairly easily measured with an accuracy better than 2% as evidenced by a high level of agreement of the data by various authors for ^{235}U in fig. 1. The values σ_M^2 are given ibid for $^{233,236,238}U$ in a narrower range of E_n. Unfortunately, no direct measurements of σ_M^2 for $^{234}U(n,f)$ were made and we would make use of an estimated dashed curve, which is in agreement with the value for the 'equivalent' reaction $^{234}U(d,pf)$ /7/.

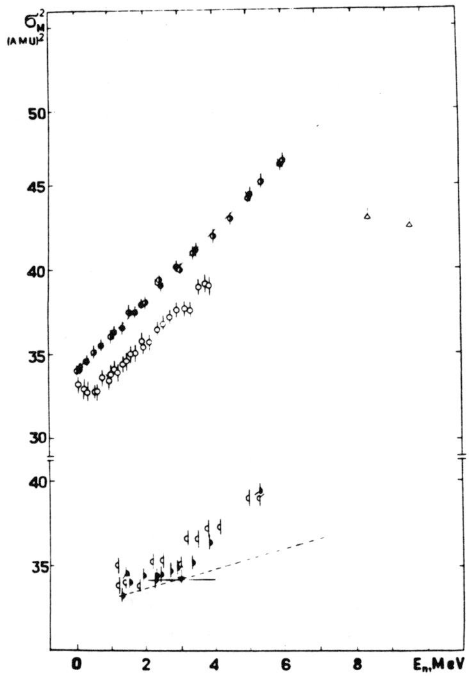

Fig.1 The dependence of fragments mass distribution dispersion on incident neutron energy. At the top: the data of Ref. /1/ - ◐ , /2/ - ● , /3/ - Δ for target- nucleus ^{235}U; /4/ ○ - for ^{233}U. At the bottom: ◗ - the data of Ref. /5/ for ^{236}U; - ◖/6/ for ^{238}U; ▼ - the data from reaction $^{234}U(d,pf)$ /7/; the dashed line - evaluation of the dependence for fissile nucleus ^{235}U.

The set of data in fig. 1 enables us to drow the following conclusion :

i) σ_M^2 for even-odd nuclei grows as E_n increases following a linear dependence with a slope $d\sigma_M^2/dE_n \cong 2$ (amu)^{2}MeV^{-1} with good accuracy;

ii) σ_M^2 for even-even nuclei-turgets in the vicinity of the threshold offers an extended region with weaker dependence $d\sigma_M^2/dE_n \leq 1$ (amu)^{2}MeV^{-1};

iii) σ_M^2 in the second plateau region for ^{235}U nucleus drastically changes the dependence in a reasonable compliance with the expected role of (n,n'f) reaction contribution, whose prefission neutron carries over on an average the 7-8 MeV energy.

A monotonic growth of $\sigma_M^2(E_n)$ in the first plateau qualitatively corresponds to the thermodynamic representations: for general reasons the dispersion is proportional to the temperature of nucleus at the "moment" of mass distribution formation, however, the observed dependence can be stronger due to shell effects attenuation /8/.

The second plateau dispersion can be represented by the ratio:

$$\sigma_M^2 = (\sigma_M^2)_A \frac{\sigma_f^A}{\sigma_f} + (\sigma_M^2)_{A-1} \left(1 - \frac{\sigma_f^A}{\sigma_f}\right), \qquad (2)$$

from which a disired fraction of the parent nucleus A in the total fission cross-section σ_f^A/σ_f is to be found. To this effect the mentioned properties of σ_M^2 are rather favourable: i) provides reliable extropolation of (σ_M^2) to the second platau region, and ii) provides stability of $(\sigma_M^2)_{A-1}$ estimation, that requires averaging over the spectrum of emitted neutrons, unsufficiently known due to uncertainty of non-equilibrium component contribution, iii) specifies a general rather high sensitivity of the method of σ_f^A/σ_f determination at issue.

Fig. 2 shows the total neutron-induced ^{235}U fission cross-section σ_f and its component σ_f^A. The analysis results σ_M^2 are in

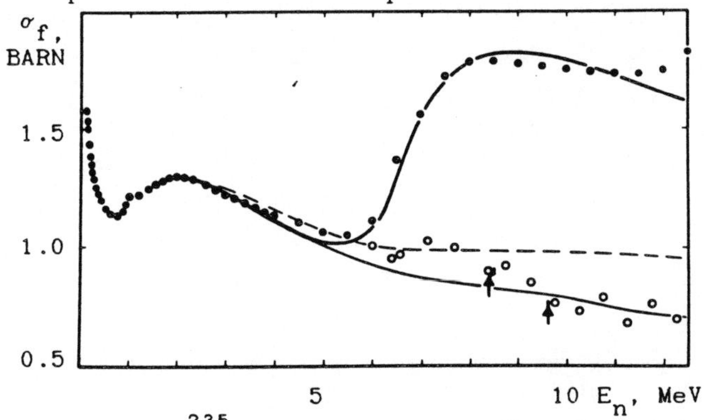

Fig. 2 Neutron-induced ^{235}U nucleus fission σ_f(● - /9/) and its component σ_f^A(O - /10/, ▲ - the present work). The solid curves - the calculation of the total fission cross-section and its component σ_f^A taking into account preequilibrium neutron emission; the dashed line - the calculation of σ_f^A without taking into account preequilibrium decay.

satisfactory agreement with the data of ref. /10/, where the ratio σ_f^A/σ_f was determined from the energy dependence of average prompt γ-quanta energy $\bar{E}_\gamma(E_n)$. In the latter case the utilization was initiated for the linear relationship between the values $\bar{E}_\gamma$ and $\bar{\nu}$ -

the average number of prompt neutrons per fission act - measured at a time, that was established in /11/. Up to now the nature of the mentioned relationship is not completely understood, particularly, the mechanism of spending more than 30% of energy introduced into the nucleus on the grows of $\bar{E}_\gamma$ is not clear to the last, and which is in this case the role of the bombarding particles mean angular moment increase with the energy. Finally, the results /3/ are indicative of essential changes in the mean kinetic energy of fragments $\bar{E}_k$ in the second plateau region without noticeable variation of the total fission reaction energy neglected in /10/.

3. Description of $^{235}U(n,f)$ fission cross-section.

The calculations of fission cross-sections, whose results are shown by the curves in fig. 2 have been perfomed by the statistical model using a code /12/. The values of neutron transmission coefficients have been taken from ref /13/. The preequilibrium reaction channel has been taken into account according to the recomendations of ref. /14/ for the value and energy dependence of a matrix element specifying life time of quasi-particle nuclear exitations. The fissile and residual nuclei level density has been calculated by the superfluid model with the phenomenological consideration of shell and collective effects /15/. The fission channel has been described within the approach given in Ref. /16/, in this case the idea of the energy surface of nuclei /17/ has been adopted. The dependence σ_f^A in fig. 2 has been calculated with the following values of the most essential parameters: 5.6 MeV and 5.8 MeV for the heights of inner and outer fission barrier humps; -2.0 MeV for the shell correction in the saddle point (on the outer hump B); 0.78 - 0.80 MeV for the ground state correlation function in neutron and fission channels.

The solid curve at $E_n > 8$ MeV gives a good description of the data on σ_f^A passing somewhat lower than the points at lower energies E_n up to 4 MeV. The reason of this deviation may well be related to inaccuracy of the preequilibrium decay channel description employed in /12,14/. This consideration is illustrated by a dashed curve, which describes σ_f^A without preequilibrium processes being taken into account and shows the agreement to be easily reached should their activation be required noticeably later - in the vicinity of 7 MeV.

The analysis of chance ^{235}U fission cross-section structure was also conducted in Refs. /14,18,19/. The phenomenological analysis results /18/ have extremely high errors, and only the conclusion on the drop of parent nucleus A fission probability as $\bar{E}_n$ grows follows from it, the data in fig. 2 being indicative of it with higher reliability. In the Refs. /14,19/ the coherent calculation of fission cross-sections and multiple neutron emission was attempted, however, their results are different and in both cases pass noticeably higher than the experimental points $\sigma_f^A(E_n)$. An essential negative shell correction in calculations /14,19/ was not considered to correspond to a mass-asymmetric saddle point, and this is the main reason of disagreements both with the experiment and the results of our calculations.

References.

1.Ch. Straede et al. Nucl. Phys., 1987, v. A462, p. 85-108.

2.V.G. Vorobjeva et al. Yad. Fiz., 1966, v. 4, n. 2, p. 325-331.

3.A.A. Goverdovsky et al. Atom. Energ., 1988, v. 64, p. 425-428.

4.N.P. Djachenko et al. Yad. Fiz., 1972, v. 16, n. 3, p. 475-480.

5.N.P. Djachenko et al. Yad. Fiz., 1979, v. 30, n. 4(10), p.904-908.

6.N.I. Akimov et al. Yad. Fiz., 1971, v. 13, n. 3, p. 484-492.

7.R. Vandenbosh et al. Proc. Int. Symp. Phys. and Chem. of Fission. Salzburg, 1965, v. 1, p. 547-555.

8.E.N. Gruzintsev et al. Yad. Fiz., 1987, v. 46, p. 1604-1613.

9.G.B. Yankov. The uranium-235 fission cross-section. In: Nuclear Data Standarts for the Nuclear Measurements. Technical Report Series n. 227, Vienna, IAEA, p. 39-45.

10.J. Frehaut. In: Proc. of the Third Advisory Group Meeting on Transactinium Isotope Nuclear Data, Upsala, 1984, IAEA-TECDOC-336, 1985, p. 105.

11.J. Frehaut et al. Proc. Conf. on "Nuclear Data for Science and Technology", Antwerpen, 1982, p. 78.

12.M. Uhl et al. Report IRK-76/01, Vienna, 1976; Addenda 1978.

13.Ch. Lagrange. Report INDC(FR) 56/L (NEANDC 228-L), 1982.

14.A.V. Ignatyuk et al. Yad. Fiz., 1988, v. 47, n. 2, p.355-362.

15.A.V. Ignatyuk et al. Yad. Fiz., 1979, v. 29, n. 4, p. 875-883.

16.G.A. Kudyaev et al. Voprosy atomnoi nauki i tehnici, Ser. Yadernye konstanty, M: TsNIIatominform, 1986, v. 3, p. 55; Yad. Fiz., 1987, v. 45, n. 6, p. 1534-1546.

17.V.V. Pashkevich. Nucl. Phys., 1971, A169, p.275.

18.D.G. Madland et al. Nuclear Science and Engineering, 1982, v. 81, p. 213-271.

19.A.B. Klepatsky et al. Voprosy atomnoi nauki i tehnici, Ser. Yadernye konstanty, M: TsNIIatominform, 1987, v. 3, p. 3-9.

^{236}Pu FISSION CROSS SECTION MEASUREMENT ON LEAD SLOWING DOWN NEUTRON SPECTROMETER

S.V. Antipov, V.V. Danichev, V.N. Dementyev, S.L. Isakov, V.S. Zenkevich
I.V. Kurchatov Institute of Atomic Energy, Moscow 123182, USSR
V.B. Funshtein, Yu.A. Selitsky, and V.A. Yakovlev
V.G. Khlopin Radium Institute, Leningrad 197022, USSR

Abstract: The ^{236}Pu fission cross section has been measured in the neutron energy range 0,1 eV - 50 keV on lead slowing down neutron spectrometer. The target contained 32,5 ng of ^{236}Pu. Simultaneously, measurements on ^{238}Pu and ^{239}Pu samples were made to introduce corrections for the impurity content in the target and ^{235}U for normalisation. 26-group cross sections and resonance fission integral are presented.

(plutonium-236, neutrons, fission, cross section)

The energy dependence of the ^{236}Pu neutron-induced fission cross-section is a matter of practical interest for an external fuel cycle of fast neutron reactors /1/. Moreover, unlike other even nuclei of Pu, this nucleus has no fission threshold /2/. The large value of the resonance integral (the calculated one is 960 barn /3/ and the experimental one is 970±60 barn /2/) gives grounds to expect that during some admissible time of the linac "Fakel" operation we could succeed in obtaining unvailable information on the ^{236}Pu fission cross section behaviour (having 32.5±0.2 ng of ^{236}Pu at our disposal) using a lead slowing down spectrometer (LSDS) /4/.

Indeed, under real experimental conditions, when an electron beam ($E_e \simeq 40$ MeV) with a 50-ns duration and 200-Hz repetition frequency initiates $5x10^{12}$ neutr./s in the e-γ-n converter, a statistical accuracy of 5-25% has been achieved in the energy range 0.1 eV-50 keV for a time of 100 hours. Certain difficulties were due to the presence of impurities in the sample used: ^{238}Pu (20.6±0.6 ng), ^{239}Pu (10.9±0.5 ng), ^{232}U (4.20±0.02 ng). For a correct application of corrections, measurements have been performed with isotopically pure samples of ^{238}Pu and ^{239}Pu. The measurements were made simultaneously in the same neutron flux by identical plate avalanch counters of fission fragments /4/, on one of which a sample of ^{235}U (197±4 μg, 99.99% enrichment) was attached for flux monitoring. Efficiency ratio of counters with uranium and samples of plutonium was determined by the counting rates of ^{252}Cf spontaneous fission fragments in the identical geometry. An experimental energy dependence of the ^{235}U fission cross section, obtained on the LSDS /4/ and normalized to the group cross sections of the BNAB-78 file, served as a reference cross section. The applicability of the measurement procedure to very small amounts of matter has been confirmed by the good agreement of experimental ^{235}U fission cross sections obtained with the samples of 175 ng and 197 h (Fig.1). Comparison of the latter with the evaluated fission cross section in the region of unresolved resonances (above 100 eV) /5/ indicated the normalization correctness.

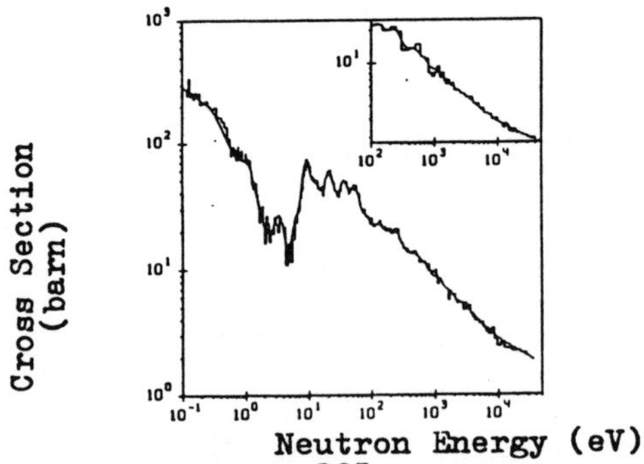

Fig.1. Experimental ^{235}U fission cross section. The histogram corresponds to a 175-ng sample; the smooth line - to a 197 g one. The histogram in the insert shows evaluated fission cross sections /5/

Fig.2 presents the experimental data from which the effects of fission of other Pu isotopes have been subtracted. They were processed in order to obtain average-over-the-1/E-spectrum fission cross sections within the intervals of 26-group constants. The treatment included the program of finding the energy dependence of the fission cross section corrected on the spectrometer-resolution-function /6/. In the energy range 1 eV - 1 keV this corrected cross section was used for calculation of average cross sections. A systematic error of average cross sections characteristiced this data processing procedure was estimated for ^{235}U and did not exceed 3% /4/. It should be noted that in the region of unresolved resonances and where the

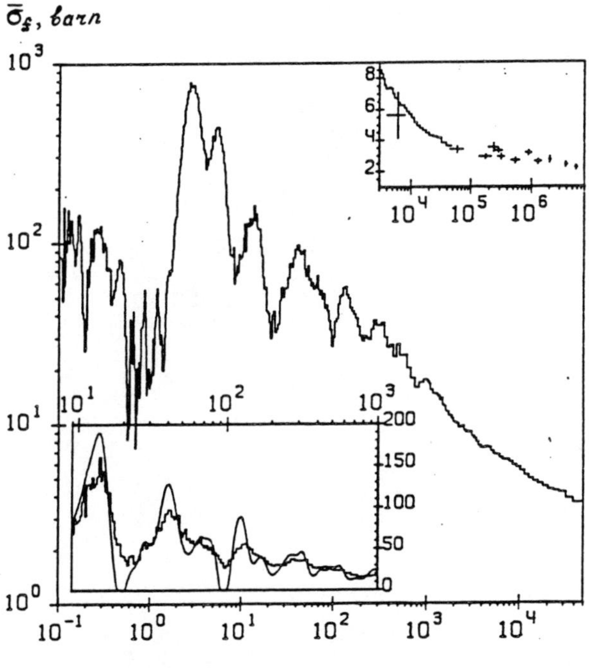

Fig.2 Experimental ^{236}Pu fission cross section (shown by the histogram). The smooth line in the lower insert corresponds to the corrected cross section. The cross marks in the upper insert show the results of ref. /2/.

cross section variation with energy has a monotone character the systematic error of the average cross sections calculated immediately from the experimental dependence, as has been shown earlier /4/, amounts to 3%. Therefore, at the sections below 1 eV and above 1 keV the group cross sections were calculated immediately from the experimental energy dependence. The results are tabulated. The same table gives statistical errors in each group. The normalization factor error being common to all the groups equals 3%. Then taking into account the contribution of the ^{232}U fission effect to the experimental group cross sections, some uncertainty appears which is due to the present-data status of the ENDF/B-V evaluations of the ^{232}U fission group cross sections. The average ^{236}Pu fission cross sections were obtained by subtracting the corresponding fraction of the latter.

The ^{237}Pu compound nucleus fissility, at the excitation energies close to the neutron binding energy, was determined by the $\sigma_f\sqrt{E_n}$ value averaged over the energy interval 0.1 eV-50 keV and being equ-

Table 1. Fission cross sections

Group No	Upper bound, eV	σ_f(exp), barn	$\sigma_f(^{232}U)$, barn	$\sigma_f(^{236}Pu)$, barn
10	46500	4.5±0.2	1.4	4.3
11	21500	5.4±0.3	1.6	5.2
12	10000	7.1±0.3	2.5	6.8
13	4650	9.5±0.4	4.4	8.9
14	2150	14.7±0.8	9.0	13.5
15	1000	22.7±0.8	13.2	20.9
16	465	35.9±1.5	20.0	33.3
17	215	49.9±1.3	30.0	46.0
18	100	52.8±1.5	42.5	47.2
19	46.5	81.5±2.0	73.7	72.0
20	21.5	116.6±2.0	195.4	91.0
21	10.0	267.2±3.0	131.9	250.0
22	4.65	616.0±6.0	4.2	615.0
23	2.15	25.5±2.0	4.3	25.0
24	1.0	36.0±3.0	7.2	33.0

al to 390±20 b.eV$^{1/2}$. The fissility value was equal to 0.9±0.05, which is characteristic of nuclei having no fission threshold. A satisfactory agreement of the ^{236}Pu fission resonance integral value, 980±30 barn, obtained by us, with the above-mentioned values should be pointed out as well as coincidence within the limits of the measurement result' errors at 60 keV (Fig.2), which in the final analysis, allows us to judge the character of the ^{236}Pu fission cross section in a broad range of neutron energies from 0.1 eV to 6.0 MeV.

References

1. L.N.Usachev: Yadrena Energ. 18, 3(1983).
2. P.E.Vorotnikov et al.: In: Neutron Physics, Int. Conf. on Neutron Physics, Kiev, (September, 1987).
3. Handbook on Nucl, Act. Data: 273 (1987), Vienna, IAEA.
4. Gerasimov V.F. et al.: See Ref. 2.
5. V.A.Kon'shin: Nucl.-Phys.Constants of Fissile Nuclei. A. Handbook, 98 (1984).
6. V.V.Danichev: VANT, ser. Obshch.Yadern. Fiz. 3, 75 (1986).

U-235 MEASUREMENT OF "ALPHA" VALUE IN THE THERMAL POINT

G. Georgiev, N. Tchikov, N. Yaneva
Sofia, Bulgaria
and **G.V. Muradyan**
Moscow, USSR

Abstract: U-235 "alpha" value was measured in the thermal point. The new method based on multiplicity spectrometry has been developed and the appropriate experimental set up was created with multisectional scintillation gamma detector "Romashka" as a main part. The experiment has been performed at the 2 MW research reactor in Sofia. The "alpha" value obtained is $\alpha = 0.169 \pm 0.004$. This result permits us to consider the discrepancy between the integral and the spectroscopic measurement of U-235 "alpha"-value in the thermal point resolved.

For the use of value "alpha" (the gamma-capture to fission cross sections ratio) in the calculations of nuclear physical and commercial characteristics of the reactors, the high precision data are needed. The requirement for the precision of U-235 "alpha" value determination in the thermal point is about 1%. α obtained with such a precision can be used as a constant for calibration of U-235 neutron cross sections and resonance parameters.

The evaluations available for this value, based on experimental data, differ in the limits of 17%. They can be grouped around the average value for integral experiments $\alpha = 0.169 \pm 0.003$ and those ones for spectroscopic experiments $\alpha = 0.155 \pm 0.003$ which themselves differ with about 9%.

We used the multiplicity spectrometry method[1)] to perform the direct spectrometric measurement of U-235 "alpha"-value in the thermal point on the 2 MW research reactor IRT-2000 in Sofia. The experimental set up has been created[2)] on the base of multisectional scintillation 4π gamma-ray detector of "Romashka" ("Daisy") type. The detector contains 12 NaI(Tl) scintillation crystals with total volume 16.6 l. The fission fragments are detected by two solid state detectors situated on both sides of the U-235 sample. The use of such a detecting apparatus with high efficiency (4π, 16.6 l total of the scintillator) on the stationary research reactor is facilitated by the application of the time-of-flight technique (on the base of the neutron chopper-monochromator) and the special system for beam collimation.

The separation of the fission and the radiative capture contributions in the multiplicity spectrum N_γ is well defined by the difference in gamma multiplicity of fission and capture events. As an additional procedure we perform the measurement of multiplicity spectrum $N_{\gamma f}$ in coincidence with the fission fragment registration by the solid state detectors. This spectrum $N_{\gamma f}$ contains only fission events.

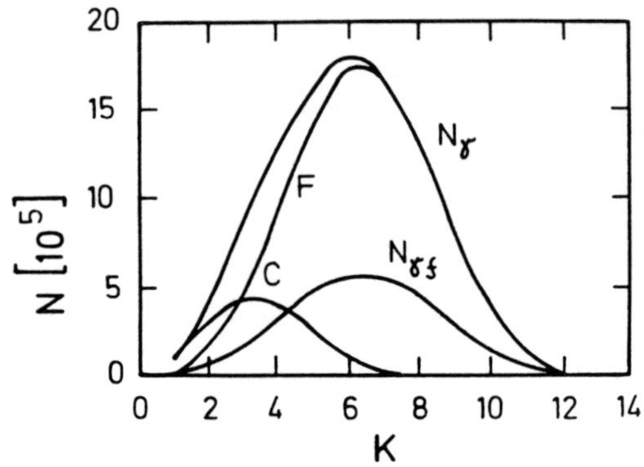

Fig. 1 Gamma multiplicity spectra - N_{γ} from "Romashka"; $N_{\gamma f}$ - in coincidence with fission fragments; radiative capture - C; fission - F

The comparison of these two curves at multiplicity K $\geq$ 8, when there are no more events of radiative capture, gives the calibration constant for calculating the total contribution of fission events F in the total multiplicity spectrum N_{γ}. Fig. 1 illustrates the separation of fission and radiative capture events. This permits to use the advantages of the multiplicity spectrometry method for high precision measurements, as at high multiplicity the background diminishes significantly. α is determined from experimental data by the following expression:

$$\alpha = \frac{\sum_{K=1}^{8} N_{\gamma}(K) - \beta \sum_{K=9}^{12} N_{\gamma}(K)}{(1+\beta) \sum_{K=9}^{12} N_{\gamma}(K)} \quad \frac{\varepsilon_f}{\varepsilon_c},$$

where

$$\beta = \frac{\sum_{K=1}^{8} N_{\gamma f}(K)}{\sum_{k=9}^{12} N_{\gamma f}(K)},$$

ε_c and ε_f - the efficiencies of radiative and fission registration are very high, close to 100%; ε_f = 99.1% and ε_c = 98.5%. The result for U-235 "alpha" value in the thermal point obtained in the described measurement is:

$$\alpha = 0.169 \pm 0.004.$$

This is an experimentally determined value with the use of new, perspective[4)] method different from all well known measurements and can be considered as an independent one. The high values for ε_c and ε_f permit to perform direct absolute measurement and to obtain an absolute α without influence of fission on the channel for registration of gamma-capture events. This is an absolute value, obtained in the direct spectroscopic experiment and it is in good agreement with the average value over the integral experiments. The numerical result obtained permits to consider the resolved discrepancy between spectroscopic and integral (over the Maxwellian neutron spectrum) measurements of α in the thermal point.

REFERENCES

1. G.V. Muradyan, Yu.V. Adamchuk et al., Preprint of Kurchatov Institute, Moscow-2634 (1976).
2. Yu.V. Adamchuk, A.L. Kovtun et al., Nucl. Instr. Meth. **A236**, 105 (1985).
3. Yu.V. Adamchuk, M.A. Voskanyan et al., Comptes rendus de l'Acad. Bul. Sci., **39**, 59 (1986).
4. L.N. Usachev, Neitronnaya fizika, Proc. All-Union Conf., Kiev, Vol. **4**, 348 (1980).

HIGH PERFORMANCE DELAYED FISSION NEUTRON MEASUREMENTS FOR SPECIAL REQUIREMENTS

N.N. Papadopoulos
National Research Center "Demokritos",
Aghia Paraskevi 153 10, Attiki, Greece

Nuclear fission is well known because of its use in military applications and electricity generation, but there exists also a variety of other applications. Thus the delayed fission neutron emission from the fission fragments serve as an analytical tool for the determination of fissionable material concentrations. The high accuracy required in certain cases, like in nuclear safeguards analysis, has been achieved by cyclic activation analysis in order to overcome the limited counting statistics due to counting rate restrictions because of pulse pile-up and radiation build-up effects. Modification of the pneumatic transfer tube terminal at the reactor core improved the performance of the nuclear analyzer extending its application range.

1. Introduction

After the discovery of nuclear fission and the extensive fundamental study of this phenomenon, the measurement of the delayed neutrons emitted from the fission fragments, being proportinal to the mass of the fissionable material, has been used to determine its concentration. There were however some restrictions in the accuracy of the method.

2. Improvements of the Technique

Since fission and delayed neutron emission are statistical phenomena, the accuracy of delayed neutron measurements depends on the counting statistics. In order to optimize the counting statistics the experimental conditions have to be adjusted to maximize the counting rate without exceeding the limit above which pulse pile-up and high radiation levels may be observed. By introducing cyclic activation analysis the total counts will be increased and thus the counting statistics improved while the partial counting rate of each cycle can be kept at the optimum level. The cyclic activation will also reduce the irradiation timing and capsule positioning incertainty. However, experience has shown that cyclic activation may cause capsule damage because of the capsule striking at the irradiation end position. Therefore, a new irradiation tube terminal has been designed (fig.1) and is being constructed based on a similar system at the pneumatic tube terminal at the counting position. During insertion of the capsule into the reactor for irradiation, the air is sucked through a new tube junction above the reactor core so that the capsule falls finally smoothly on the pneumatic tube end at the irradiation position, as well as at the counting position after irradiation. This precaution is particularly necessary when cadmium covered samples have to be analyzed because of the increased capsule weight.

The cadmium covers have to be used in case of isotopic enrichment or mixed uranium and thorium determination. Since then the neutron flux decreases drastically, the cadmium covered samples will be irradiated in a new pneumatic transfer tube closer to the reactor core than the bare ones. Since however for high fissionable material concentration, as in case of nuclear safeguards applications the neutron flux at the reactor core and thus the counting rate is relatively high, a new pneumatic tube system has been constructed with the tube terminal at a proper distance above the reactor core. Thus, by using simultaneously two pneumatic tube systems at different distances from the reactor core with different neutron fluxes for the bare and cadmium covered samples, similar counting rates can be achieved optimizing the accuracy of the measurements. After these modifications the home-made nuclear analyzer (1) can now be used for highly accurate isotopic enrichment determination of fissionable material with applications in nuclear safeguards analysis, certification of nuclear standards, sample homogeneity tests a.o.

3. Calculations

In order to determine e.g. isotopic Uranium-235 enrichment, bare and cadmium covered samples as well as two respective standards with different uranium enrichment are irradiated and counted. The calculations are based on equations similar to the ones used for thorium determination (2). The basic difference is that now, instead of the total uranium and thorium masses, the isotopic U-235 and U-238 masses and their ratios are determined. Preliminary measurements have been performed giving promising results (3,4).

4. Discussion

The delayed fission neutron counting method has not been used so far for these applications, because they require extremely high accuracy of the order of 0.1% relative standard deviation which only certain expensive highly sophisticated methods like mass spectrometry could offer so far. The presented method which is based on neutron activation of fissionable material in a nuclear reactor has moreover certain advantages over other methods. Due to the experimental improvements the method has now become competitive with other established methods offering significant additional applications of nuclear fission.

5. Conclusion

High performance delayed fission neutron measurements have been achieved by improving the home-made nuclear analytical system at the 5 MW swimming pool reactor of the National Research Center for Physical Sciences "Demokritos". By the cyclic neutron activation technique combined with timed intermediate storage of the fissionable material samples, non-destructive sample analysis with simple sample preparation and negligible matrix interference, high accuracy and sensitivity is assured, which make this method, based on nuclear fission, a powerful analytical tool with a variety of applications.

Acknowledgements

Thanks are due to the reactor and mechanical workshop staff of NRCPS "Demokritos" for their contribution to this work.

References

1. N.N. Papadopoulos, Development of activation techniques for extensive requirements in uranium and multielement analysis, J. Radioanal. and Nucl. Chem. Articles, Vol. 113, No 2(1987)351.
2. N.N. Papadopoulos, Improvements of neutron activation techniques for the determination of fissile material concentrations, J. Radioanal. and Nucl. Chem., Articles, Vol. 110, No 1(1987)253.

3. N.N. Papadopoulos, S. Synetos, Adjustment of nuclear analytical techniques to special applications, Isotopenpraxis 24(1988)7, 288.

4. N.N. Papadopoulos, Proc. 9th ESARDA Symposium on Safeguards and Nuclear Material Management, London,U.K., 12-14 May 1987, 449.

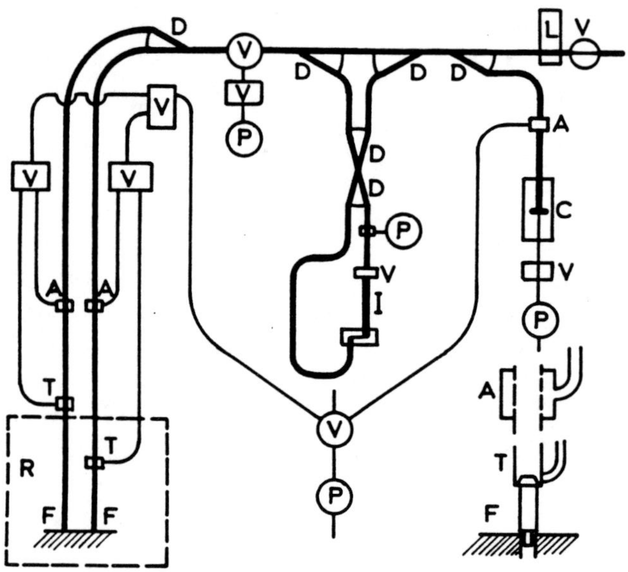

A = Air tube junction,C = Counting station
D = Diverter, F = Flexible tube end support
I = Intermediate storage, L = Loader, P = Pump
R = Reactor, T = Tube end junction,V = Valve

Fig.1. Cyclic nuclear analyzer

URANIUM FISSION FRAGMENT CONTRIBUTION IN THE ELEMENTAL DETERMINATION, USING INAA TECHNIQUE

S.M. Al-Jabori, A.K. Hussein, and M. Jalil
Nuclear Research Center, Tuwaitha, Baghdad, Iraq, P.O. Box 765

The Isotopes ^{95}Zr, ^{99}Mo, ^{103}Ru, ^{140}La, ^{141}Ce, and ^{147}Nd are commonly used in the activation analysis of corresponding elements. These isotopes are resulted due to the fission of uranium. The interference factor for these radioisotopes is defined as the activity produced by irradiating of 1 µg as pure natural uranium divided by the activity produced by irradiating of the chosen element.This contribution factor was determined experimentally by analysis of natural uranium standard using the same conditions as these for samples analyzed. The interfering factor for the isotopes ^{95}Zr, ^{99}Mo, ^{103}Ru, ^{140}La, ^{141}Ce, and ^{147}Nd as measured experimentally were 7.3, 1.65, 0.14, 0.018, 0.3, and 0.22 µgX / µgU, respectively.

INTRODUCTION

The INAA technique is recommended for REE and other trace elements analysis in geological samples. This technique is not free of serious analytical errors due to inferferences and unwanted contributions.As an example to the refered unwanted contribution are the fission products of U-235 which are composed of different radionuclie of REE and other elements. In the present work the contribution of product factors of U-235 are determined experimentaly and calculated theoretically using the formula proposed by Erdtmann [1] for deretmination the contribution factor of the fission products ^{95}Zr, ^{99}Mo, ^{103}Ru, ^{140}La, ^{141}Ce, and ^{147}Nd.

EXPERIMENTAL

Standard material of U,Zr,Mo,Ru,La,Ce and Nd are prepared using specpure oxides. These standard samples are weighed and sealed in high purity quartze vials for irradiation in the IRT-5000 nuclear reactor. The irradiation time was 5h, the neutron fluxes were measured using Indium flux monitor (with and without cadmium can) to datermine the epithermal and thermal neutron fluxes. The fluxes ratio were determined for different position in the reactor channel, it was found that

$$\phi_{epi}/\phi_{th} = 0.0015 \text{ and } 0.025$$

The gamma-ray activities measured with a 30cc Ge(Li) datector coupled to an HP-1000 computer through a 4096 channel ADC. The system resolution was 2.5 keV (FWHM) for the 1332 keV γ-ray of ^{60}Co.

The photopeak areas obtained for the individual element standards were compaered with the areas of the same photopeaks produced by fission products in the uranium standard. The same irradiation and counting geometry was used in both cases.

DETERMINATION OF THE EXPERINENTAL CONTRIBUTION FACTORS

In order to determine experimentaly the correction factors for U-235 contribution for Zr,Mo,Ru,La,Ce and Nd analyzed by INAA, the maesred amounts of fission products produced during irradiation in the uranium samples are given to the amount of total uranium as correction factors uX / uU. Table-I shows the obtaind results which are expressed in term of 1 μg X / 1 μg U.

CALCULATION OF THE THEORETICAL CONTRIBUTION FACTORS

For calculation the contribution factors of Zr, Mo, Ru, La, Ce, and Nd due to U-235 fission, the Erdtmann formula is used.

$$m = \frac{\sigma_f \; \mu}{\sigma_X} \; \frac{0.0072 \; m_{U_{nat}}}{238} \; \frac{M}{a}$$

m = Contribution factor

σ_f = Fission cross-section

μ = Accumulated fission yield for the respective radionuclies

$m_{U_{nat}}$ = amount of natural uranuim.

M = atomic weight.

a = isotopic abundance of the nuclide.

σ_x = activation cross-section, obtained by adding [the resonance integral (Ix) multipled by a ratio of epithermal to thermal flux (R)] to the thermal cross-section as follows:

$$\sigma_x = (\sigma_{th} + \frac{\phi_{epi}}{\phi_{th}} I_x)$$

The contribution factors were calculated using the parameters presented in Table -II. The obtained results are listed in Table-I.

Table-I. Theoretical and Experimental results of contribution factors expressed as μg of the respective element to μg of uranium

Isotope	Theoretical	Experimental
^{95}Zr	6.4	7.3
^{99}Mo	1.03	1.65
^{103}Ru	0.072	0.14
^{140}La	-	0.018
^{141}Ce	0.22	0.3
^{143}Ce	0.94	1.38
^{147}Nd	0.13	0.22

Table II. Nuclear parameters (n,γ) reaction products which have interferences with 235-U(n,f) reaction products.[5]

Reaction	M	a	σ [3] [barn]	I_x [4] [barn]	τ [d]	μ [5] [%]
$^{94}Zr(n,\gamma)^{95}Zr$	91.22	0.1739	0.075	0.38	64.0	6.371
$^{98}Mo(n,\gamma)^{99}Mo$	95.94	0.241	0.13	6.56	2.75	5.730
$^{102}Ru(n,\gamma)^{103}Ru$	101.07	0.316	1.3	4.2	39.35	3.264
$^{139}La(n,\gamma)^{140}La$	138.9	0.999	8.934	11.7	1.68	6.128
$^{140}Ce(n,\gamma)^{141}Ce$	140.12	0.885	0.567	0.28	32.5	6.008
$^{142}Ce(n,\gamma)^{143}Ce$	140.12	0.111	0.95	0.73	1.38	5.711
$^{146}Nd(n,\gamma)^{147}Nd$	144.24	0.172	1.4	2.91	11.0	2.100
$^{235}U(n,f)$	238.03	0.0072	584	280	-	-

DISCUSSION AND CONCLUSIONS

From the results indicated in Table I it is shown that the contribution factor for Zr is extremly large compared with the contribution factors La, Ce, and Nd. The correction factor are applicable for any decay time following the irradiation except ^{140}La which is special case because its correction factor was dependent on the length of time between irradiation and counting Td. This means that values in Table I hes been multiplied by the factor ($e^{0.35893Td}-1$). From the experimental results of La it´s seemed that the contribution factors independent on neutron flux and irradiation time.

It is shown that uncurrected results can potentially lead to uncorrect analysis conclusion, particulary if Zirconuim and barium concentration are used and that molybdenum assays may be deteriously affected by the presence of large amounts of uranium (Table I).

Ingeneral, it was found that the experimental results are systematically lower than the calculated values.

Comparison of measured and calculated correction factors of measured and calculated correction factors indicates accepted agreement. The only appreciable discrepancy is in the correction value for Ru-103 .

REFERENCES

1. G.Erdtmann, J.Radioanal Chem., 10 (1972) 137.
2. V.Reus, W.Westmeier, Atomic Data and Nuclear Data Tables, 29 (1,2) 1983.
3. JAERI-M., (Rep.) 86-030 , 1986 .
4. BNL - 325 (Rep.) 3 Rd. Ed. 1973 .
5. JAERI (Rep.) 1287 NEANDC (J) 91/U 1983 .

INVESTIGATION OF THE DIFFERENTIAL AND INTEGRAL NEUTRON EMISSION OF URANIUM-238 AT 14 MEV NEUTRON INCIDENCE ENERGY

T. Elfruth, T. Hehl, H. Kalka, D. Seeliger, K. Seidel, M. Toepfer, and S. Unholzer
TU Dresden, Sektion Physik, 8027 Dresden, Mommsenstr. 13, GDR

Abstract: Neutron leakage spectra were measured with time-of-flight and with proton recoil spectroscopy, activation and fission rates were determined for a metallic U-238 - sphere shell fed in its centre with 14-MeV neutrons. The results are compared with calculations based on recent data files.
Neutron emission cross sections have been measured with time-of-flight spectroscopy in a wide range of emission angles. The data obtained are compared with published results of experiments. The emission cross sections have been calculated without any parameter fit. Multiple chance fission, (n,inf) with i=0,1,2 and (n,jn)-processes have been taken into account where in the preequilibrium stage of the compound system single-particle and collective excitations have been included.

1. Introduction

Uranium-238 ist the main raw material used in nuclear power engineering. It is necessary to study the interactions of neutrons with uranium-238 nuclei from all points of view because of the varied role of this material not only in fast neutron reactors. The determination of the cross section for the total neutron production and of all partial processes must be performed with high accuracy [1], since one concept of an energy producing fusion reactor will be the so called hybrid reactor, which uses ^{238}U as a blanket component arranged at the plasma chamber wall for neutron multiplication, energy enhancement and plutonium breeding. By measuring the neutron leakage from an uranium assembly fed with 14 MeV neutrons and comparing it with neutron transport calculations the data of all neutron interaction and multiplication processes and the transport code are tested. Such a benchmark was carried out at the pulsed neutron generator of TU Dresden with an one-dimensional assembly, a spherical shell of 6cm thickness consisting of depleted (0.4 % ^{235}U) metallic uranium. Neutron leakage spectra were measured and calculations were carried out with transport codes using several data libraries.

In this work presented, also the double differential cross sections of the total neutron emission are determined with the TOF spectrometer. They are compared with data obtained by other groups [2]-[5]. Furthermore, the double differential cross sections of all neutron producing reaction channels are calculated using statistical models of direct reactions, compound reactions and fission. Comparing experimental and calculated double differential neutron emission cross sections with the library data used in the transport codes, one can discuss differences between measured and calculated neutron leakage spectra.

2. Benchmark Experiment

The experimental arrangement is shown in Fig. 1. The sperical shell consisting of metallic uranium has an inner diameter of 20 cm and an outer diameter of 32 cm. The neutron generator operated in pulsed regime with deuteron pulses of 2 ns f.w.h.m. and 5 MHz repetition rate, produces 2-5 * 10^9 neutrons per s. The neutron production is determined by counting the α-particles. The angular distribution of the neutrons per α-particle was calculated and measured with activation foils and with proton recoil spectroscopy.

The neutron flight path was arranged at 76° to the deuteron beam axis. At a flight path of 4.51 m and a time resolution of 2.5 ns the energy resolution was 6 % at 14 MeV and 1.6 % at 1 MeV.

The neutron detector is a stilben scintillator of 29 mm diameter and a length of 20 mm. Its neutron detection efficiency ε(E) is determined by time-of-flight spectroscopy scattered from hydrogen, of 14.1 MeV neutrons and of neutrons from a ^{252}Cf fission chamber. A Monte Carlo code NEUCEF is used to calculate ε(E) too.

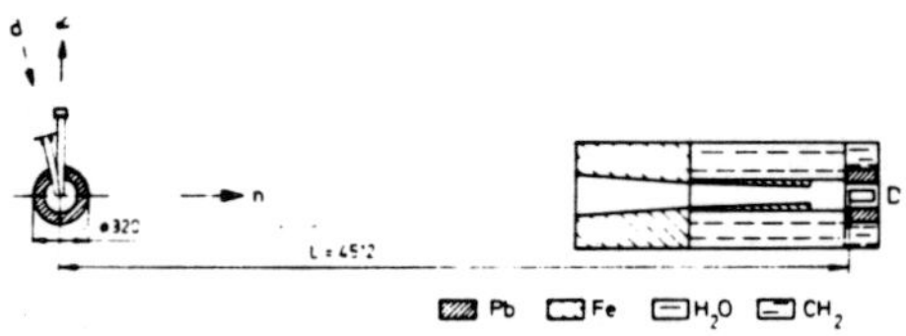

Fig. 1: Geometrical relations of the benchmark

To reduce the influence of experimental uncertainties on the experimental result, time-of-flight spectra were measured repeated at different bias with and without uranium shell. The data obtained were transformed into energy spectra after correction for background, dead time and differential nonlinearity.

Additional, the neutron leakage was measured in the lower energy-range with proton recoil spectroscopy. Eight activation and and fission rates were determined at the shell surface.[6]-[8]

Neutron transport calculations were carried out with the Monte Carlo codes BLANK [9] and MORSE [10] and the one-dimensional S_n code ANISN [11].The data used in the calculations with BLANK were taken from the libraries ENDL-75 and ENDF/B-IV. The group constants used were processed by the code LAVA.

For calculations the 100-group standard library DLC-37 (ENDF/B-IV) [12] and group constants generated with the code MINX [13] (JENDL-2) were used.

Some results are shown in Fig. 2. Remarkable deviations of the calculated leakage spectra from the measured distributions are observed in the high-energy-range, 6 MeV < E < 12 MeV. Here, the calculations predict less neutrons.

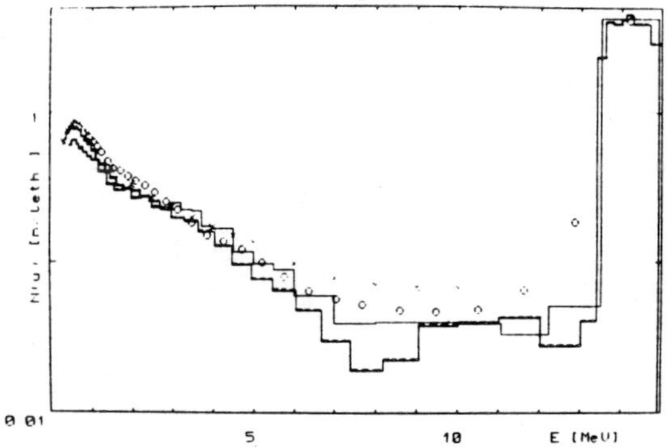

Fig.2: Neutron leakage spectra per unit of lethargy and source neutron measured with TOF (ooo) and PRS (xxx) and calculated with BLANK-ENDF/B-IV (——), MORSE-ENDF/B-IV (▬▬▬).

The neutron leakage in E = 0.1 - 14 MeV per one source neutron found in the measurement is 2.34 (Neutrons with $E < 0.1$ MeV contribute to the total leakage with 5 % only.) This value is different from the calculation with ENDF/B-IV data by 9 % (MORSE) and 17 % (ANISN), respectively.

3. Measurement Of Double Differential Neutron Emission Cross Sections

The experimental arrangement is shown in Figs. 3 and 4. Ring geometry with flight path arranged at 90° to the deuteron beam direction is used, so that the average incidence neutron energy E_0 is 14.1 - 14.2 MeV for all emission angles from 15° up to 165° (and symmetric to $\vartheta = 90^{\circ}$). The flight path is about 5 m.

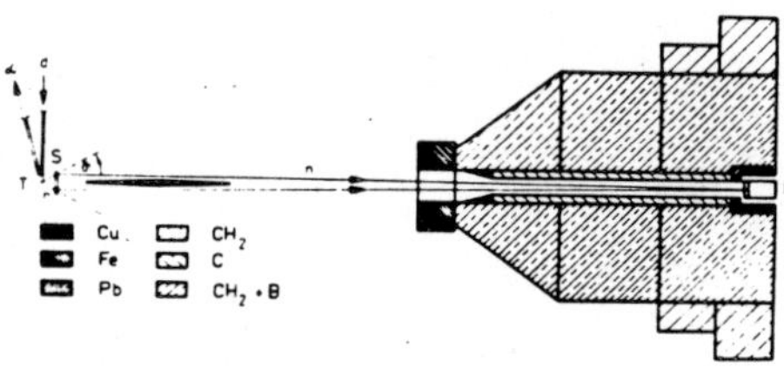

Fig.3: Geometrical arrangement of the time-of-flight spectrometer. T, tritium target; S, ring sample; D, neutron detector.

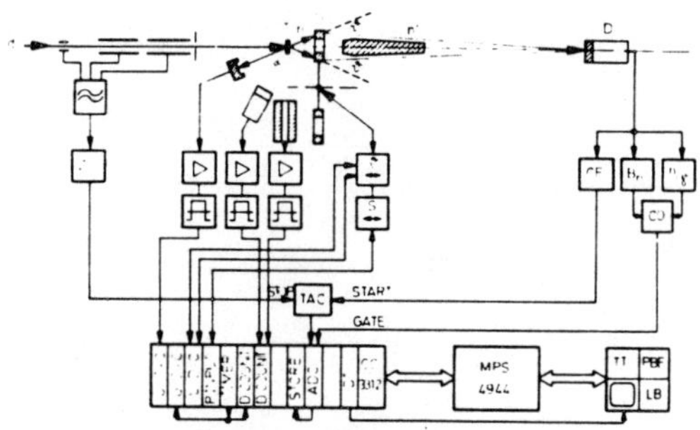

Fig.4: Block scheme of the spectrometer:

ZC, zero-crossing trigger; CF, constant fraction trigger; B_n, proton-recoil-energy discriminator; n/γ, neutron-gamma-discriminator; CO, coincidence; S, sample shifter; S, sample changer; TAC, time-to-amplitude converter; U/D-C, up/down counter; ADC, analog to digital converter; CC, controller of the CAMAC crate; MPS, microcomputer.

The neutron detector is a liquid scintillator NE 213 coupled with a XP 2041 photomultiplier. It is biased at 2 MeV neutron energy. The spectrometer is controlled by a microcomputer so that sample changing, varying of the scattering angle and monitoring is done automatically. This allows to subdivide the data acquisition in many short periods to check the spectra after each short-time run and to cover the chosen angles many times.

The data obtained are reduced to $\sigma_{nM}(E_o;E,\vartheta)$ and were transformed into center-of-mass system. The cross sections integrated over ϑ are shown in Fig. 5.

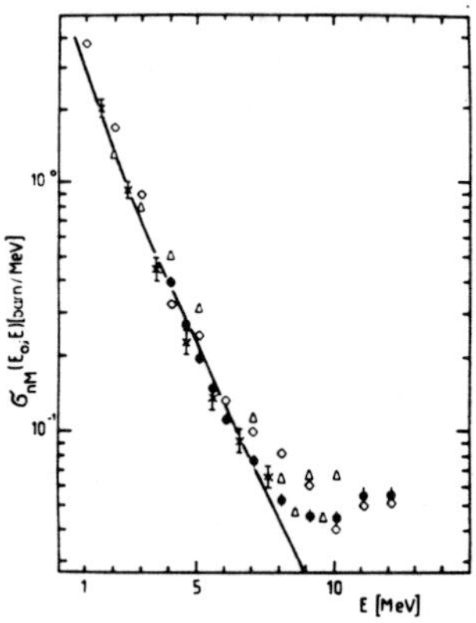

Fig. 5: Angle-integrated neutron emission cross sections of U-238 at E_o = 14.5 MeV.
◆ ,present work; ⌶,Ref. [2]; ◇ ,Ref. [3]; △ , Ref. [4]; +,Ref. [5]; —, ENDF/B-IV

The error bars represent the statistical errors only. The comparison of the $\sigma_{nM}(E)$ available in the literature, shows a spread larger than the required accuracy. The ENDF/B-IV evaluation underestimates at least the emission of high-energy neutrons. The angular distributions are in ENDF/B-IV assumed to be isotropic, with exception of the four pseudolevels. But, also in the middle energy-range they are forward peaked as shown in Fig.8 for E = 5.5 and 7.5 MeV and in Ref. [3],[5].

4. Calculations

The neutron producing reaction channels open at 14 MeV neutron incidence energy are shown in Fig.6. At each stage of a compound-system cascade coupled by neutron emission (n_1,n_2,n_3), the neutron emission competes with fission (leading to fragment neutron evaporation n_{fi}) and with γ-deexcitation. The $\sigma_{nj}(E,\vartheta)$ determines all following neutron emissions $\sigma_{nj+1},\ldots$ and the number of fission events in the channels f_j, $f_{j+1},\ldots$. Therefore, especially the n_1-emission should be treated carefully.

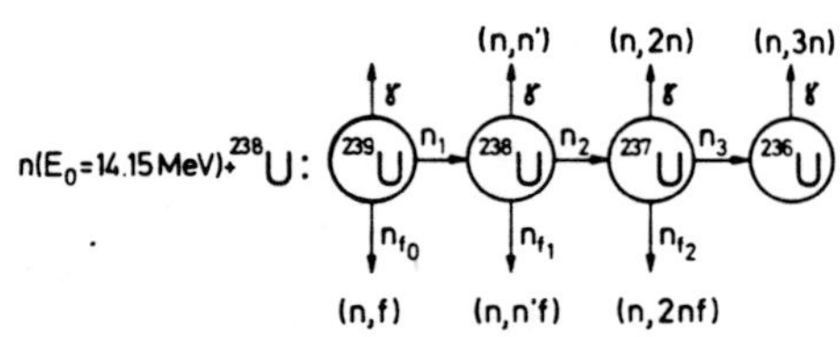

Fig. 6: Neutron producing reaction channels of U-238 bombarded with 14.15 MeV neutrons.

Contributions to n_1 do not arise from compound nucleus emissions only but also from direct and other precompound processes. In the present work, one- and two-step direct processes are treated statistically (SMD). They include both particle-hole excitation and excitation of direct collective vibrations (confined on two low lying phonon states $\lambda^\pi = 2^+, 3^-$) in a consistent way. The contribution of direct scattering from the ground-state rotational band is taken into account seperately using a calculation of Lunev [14].

The neutron emission from the other precompound and from the compound nucleus states (SMC) is calculated by solving the master equation beginning with n=5. The calculations of both SMD and SMC were performed with the computer code EXIFON [15],[16] using total absorption cross section and fission cross sections as input data.

The cross sections of the three fission channels are calculated by Maslov [17] using double humped fission barriers. The value obtained for $\sigma_{n,f}+\sigma_{n,n'f}+\sigma_{n,2nf}$ is in agreement with recent experimental data. The energy distribution of the fission neutrons $\sigma_{fi}(E)$ is assumed as neutron evaporation spectrum (Maxwellian) from the full accelerated fragments. The sum of all neutron producing components is compared in Fig. 7 with experimental data of angle integrated emission cross sections. The good description of the energy dependence shows that the main reaction mechanism is met.

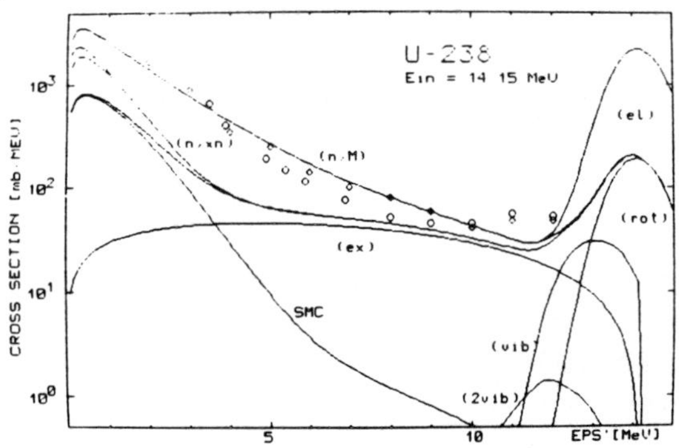

Fig. 7: Energy-differential neutron emission cross sections calculated as sum of the components indicated in Fig. 6 (n,M), and compared with experimental data (◆ [3]; o, present work). The n_1-emission consists of one-step direct scattering from rotational states (rot) with excitation of vibrations (vib) and with excitation of 1p 1h - states (ex), of two-step direct scattering and of statistical compound emissions (SMC) in the precompound and in the compound nucleus stage.

The angular distributions in the middle-energy range of the spectrum presented in Fig. 8, demonstrate that the forward-peaking is determined by the n_1-emission whereas n_f are symmetrically to $\vartheta = 90°$ (nearly isotropic) emitted. The ratio of these components seems also to be met by the calculations.

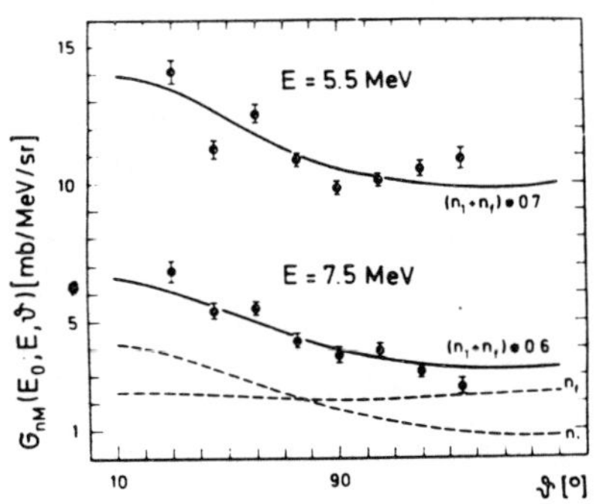

Fig. 8: Double-differential neutron emission cross sections at emission energies 5.5 MeV and 7.5 MeV calculated (——) as sum of the n_1 emissions (---) and all fragment neutrons ($n_{f0} + n_{f1} + n_{f2} = n_f$, ---), and compared with the experimental data of the present work. To discuss the relative angular distributions, the calculated values are adapted to the experimental by the factors inserted.

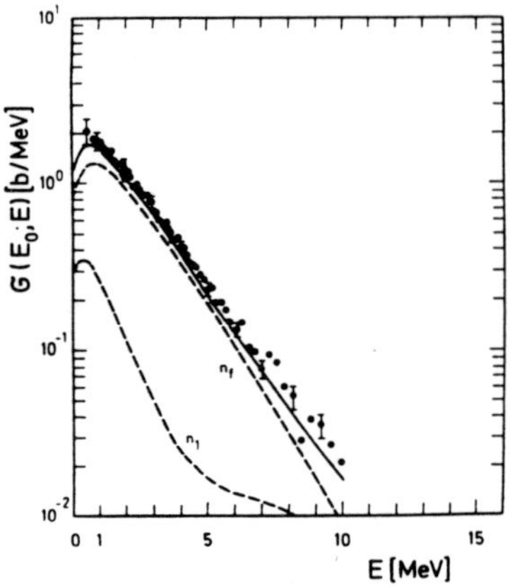

Fig. 9: Neutron emission spectrum from U-238, measured in coincidence with fission [18], compared with calculated spectrum, consisting of:
(--- n_f): $\sigma_{nfo}(E) + \sigma_{nf1}(E) + \sigma_{nf2}(E)$
(--- n_j):

$$\sigma_{n_1}(E) \cdot \frac{\sigma_{n,n'f} + \sigma_{n,2nf}}{\sigma_{n_1}} + \sigma_{n_2}(E) \cdot \frac{\sigma_{n,2nf}}{\sigma_{n_2}}$$

In Fig. 9, the theoretical calculations are compared with the experimental results of Baryba et al.[18], who measured the neutron emission from ^{238}U bombarded with 14 MeV neutrons in coincidence with fission events. A good description of the spectrum shape is given by the sum of n_j and n_f contributions. The visible constant factor between experimental and calculated spectra is caused by different neutron multiplicities , $\bar{\nu}$ = 4.445 used in the calculations and $\bar{\nu}$ = 5.07 obtained in the experiment of Ref.[18].

5. Conclusions

Neutrons emitted from ^{238}U at incidence energy 14 MeV in the preequilibrium stage and in direct reactions are not adequately represented in library data. A more established physical description of the input data for transport calculations must be the next step to clarify the differences between calculated and measured leakage spectra.

References

[1] World Request List for Nuclear Data, INDC(SEC)-88/URSF, IAEA Vienna, 1983
[2] J. Voignier et al., CEA-3503(9168)
[3] J.L. Kammerdiener, UCRL-51232(1972)
[4] A.P. Degtyarev et al., Yad. Fizika 34(1981)299
[5] Shen Guanren et al., AGM on Nuclear Theory for Applications, Beijing, 1987
[6] W. Hansen et al., Arbeitsbericht RPV-1/87
[7] G. Streubel, B. Doerschel, Kernenergie 20(1977)49
[8] W. Hansen, W. Vogel, ZfK-523(1984)
[9] S.V. Marin et al., Preprint IAE-2832(1977)
[10] S.K. Fraley, ORNL 1976
[11] W.W. Engle, Jr. A.E.C. Research and Development Report K-1693(1968)
[12] RSIC Data Library Collection, RSIC, Oak Ridge Nat. Lab., updated 11-7-79(1978)
[13] C.R. Weisbin et al., LA-6486-MS/ENDF-237(1976)
[14] V.P. Lunev, private communication, 1987
[15] H. Kalka et al., submitted to Phys. Rev. C, 1988
[16] H. Kalka et al., Proc. V. ISNIR, Smolenice 1988
[17] V.M. Maslov, private communication, 1987
[18] V.J. Baryba et al., FEI-947, Obninsk, 1979

X. SUMMARIES

THE CONFERENCE - A CONCERT*

R.W. Hasse

It was only after the Friday evening concert that Professor Seeliger asked me to give a summary of the theoretical talks of sessions II and III instead of Professor Greiner who left early. At first glance I thought that this must be formidable task because there were so many excellent talks. But then I realized that it is a great honor to me to be the substitute of (*or* second after) Professor Greiner. Furthermore, since my talk also fell into into these sessions and since it was common practice during the conference that the speakers also chaired (*or* the chairmen also talked during) their session, this fact facilitated the task because I can give a summary of my own talk. Finally I think that the concert already was the summary of the conference so that I can restrict myself to commenting on the concert. So I could not refuse.

At first I noticed that the piano concert had no conductor. From this one could conclude that the conference was not conducted. This is not so. The conductor of the conference, Professor Seeliger, rather stayed in the background and directed the concert in a great way.

The concert started with a tune by Johann Christian Bach, the son of Johann Sebastian Bach, played by four hands. One could think that the son Bach is second class after the great father like the first speaker in the session which happened to be me. This however, is not the case. Sebastian Bach continued the father's art of the fugue and brought it to perfection. In much the same way as the fugue is a seemingly continuous story which ends arriving at the beginning but with a different pitch and which opens new possibilities to the future, the art of dissipation and friction always continues with new ideas on the old subject. It was played by four hands but it could easily accomodate more. I hope that the auditorium got the message that one cannot always attribute discrepancies with theory to dissipative effects because, at least in low energy nuclear fission, they are small.

As in any concert there was an entracte (pause), at least for me because I did not listen to Dr. Flambaum's talk on parity violation in fission. Therefore I cannot refer to it.

Then came the big solo of the older pianist, a tune by Sergey Rachmaninoff which happened to be Vilen Strutinsky's talk. He promised six subjects with two amendmends, played two tunes and calmed down to a pianissimo. The first tunes played, however, were fortissimo; shells lying around everywhere, in particular in superdeformed nuclei, LiPa trees were growing, and during the grand finale he raped the nucleus its wall. From thereon the nucleus lay about naked to the eyes of the curious physicists subject to the decision whether it is a beautiful girl or a dirty old man. Everybody got the message that shells are always important.

The second solo, by Franz Liszt, was played by the younger pianist who could have been the pupil of the older. This happened to be Vitali Pashkevich's talk, played very soundly in adagio with much feeling in andante. He needed all his two hands and could have needed more, i.e. a fast computer (maybe at GSI). He collected all the debris left over from the previous talk, put together all the shells and formed a nice mosaic after he dressed again the nucleus with a shell model potential. Then he fissioned it into two pieces and looked at the sinuous landscape from the scission line and decided that the nucleus is beautiful from this aspect.

The last piece was by Antonín Dvořák, a Czech. But since his little black lake was put as the Black Sea I don't hesitate to locate him in the Soviet Union. This happened to be Professor Ignatyuk's talk. One really felt like walking through the Bohemian forest and one could imagine finding the spinning house where all the threads of different information on level densities were

*Summary prepared for the XVIII*th* International Symposium on Nuclear Physics - Physics and Chemistry of Fission, Gaussig (Dresden), GDR, Nov. 21-25, 1988

spun together to form a nice piece of linen. He ought to relize this because his book on level densities will be translated into English. His message could not be overheard, namely that at low excitation energy correlations (collective states) must be included in order to bring up the level density parameter from 1/15 to 1/8.

Since the concert was so good there was heavy demand for a Da capo - Zugabe - encore une fois - ещё раз. Therefore I don't hesitate to annex Dr. Brosa's talk although it was in session IV. The first encore was the Fantasy by Franz Schubert, very much in tune with the formulas presented by the speaker to fit mass distributions and *Nubar's*, the commonly used transcription of Greekless plotters for $\bar{\nu}$. Usually a theoretician should give simple formulas to experimentalists with much care because the latter may turn them upside down and apply them where not applicable. In this case, however, one need not worry because these formulas are not meant to have a physical background. Due to the speaker's popularity I also attribute to him the second encore, a Hungarian Rhapsody by Liszt which is so popular that everybody could understand it easily.

You might have noticed that I forgot Dr. Malev's talk on diabaticity. This is not the case. But I think that in the concert which was dominated by fortissimos something like études by Frédéric Chopin were missing which I could attribute to this speaker.

In conclusion I notice that although the piano concert was played on a short grand piano, it produced more vibrations than played on two or more full grand pianos. These vibrations of the 50th anniversary of the *discovery* of nuclear fission certainly will reach Berlin early next year when they celebrate the 50th anniversary of *publication* of the discovery of nuclear fission. These vibrations from Gaussig certainly will not be overheard there.

SUMMARY
BARRIERS - CHANNELS - VALLEYS

F. Gönnenwein

The present summary reviews the talks having been given in session I (Fission Fragment Distribution), session IV (Fragment Deexcitation) and session V (Ternary Fission). Looking for a common heading of the above contributions, the sub-title chosen by A. Michaudon for his introductory talk was found to be very appropriate: BARRIERS-CHANNELS-VALLEYS. Deliberately, some papers from other sessions, which may be subsumed under this heading, have also been included for the review.

1. FISSIONING NUCLEI LIGHTER THAN Ra

Until a few years it was generally accepted that fissioning nuclei lighter than (Ra, Ac) are characterized by symmetric fragment mass distributions. It is only upon approaching the Businaro-Gallone point for nuclear mass numbers ranging from about $A = 100$ to $A = 140$ that unstable asymmetric saddlepoint deformations and, hence, asymmetric mass splits were expected to occur.

The impressive bulk of experiments having been conducted at the Alma Ata cyclotron have led to a thorough revision of this picture. A summary of results from Alma Ata was presented by M. Itkis.

First, it was shown that purely symmetric mass distributions are only observed for nuclei definitely lighter than (Ra, Ac), namely mass numbers $A \lesssim 200$ and charge numbers $Z \lesssim 80$.

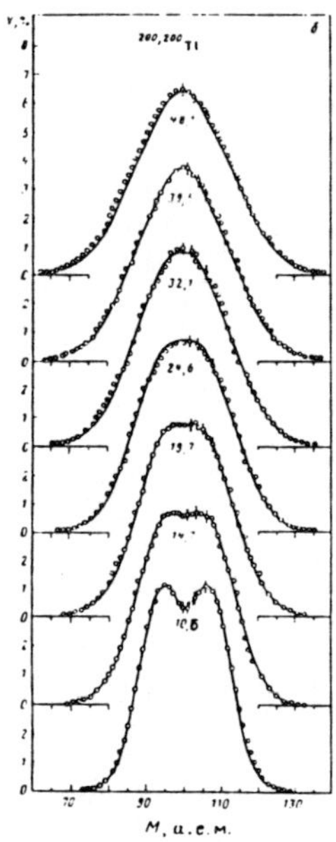

FIGURE 1 (left):
mass distribution of $^{200,201}Tl$ for various saddle point excitation energies

FIGURE 2 (right):
mass distribution of ^{205}Bi, ^{210}Po and ^{213}At at an excitation energy of ≈ 9 MeV at saddle

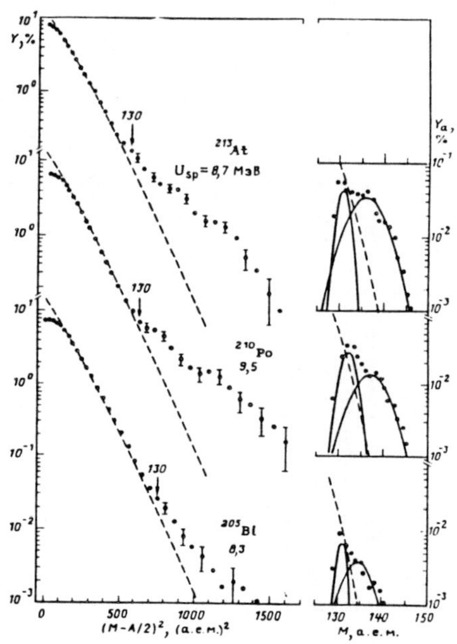

An example is given in Fig. 1 depicting the mass distribution for ^{200,201}Tl. The figures at each of the yield curves indicate the excitation energies at the saddlepoint. In fact, it is obvious from Fig. 1 that, strictly speaking, even for Tl the mass distribution is not perfectly symmetric: at low excitation energies a dip in the yield shows up right at symmetry. It is also evident from the figure that the mass numbers for the asymmetric mass peaks have nothing in common with the familiar peak mass numbers in the actinides. The rapid disappearance of this effect with increasing excitation energy is suggestive of a positive shell effect for masses around $A = 100$ being washed out at higher temperatures. One should also note the increase in width of the mass yield curve with excitation energy which is predicted from theory.

For nuclei heavier than $A \approx 200$ (but still lighter than Ra) we are faced with the surprising result that the mass yield curve exhibits distinct asymmetric components on the wings of a dominant symmetric mass distribution. This is shown in Fig. 2 for the three compound nuclei ^{213}At, ^{210}Po and ^{205}Bi at an excitation energy at the saddlepoint of about 9 MeV. These asymmetric components become even more evident when analyzing the kinetic energy distributions. The decomposition of the asymmetric yield into two parts as indicated on the right of Fig. 2 viz. a narrow component centred at the heavy mass $A_H \approx 132$ and a wider one centred around $A_H \approx 136$, receives further credit from the observation that the symmetric and asymmetric "modes" have quite different kinetic energies: in a mass interval $A = 130-135$, where the three modes overlap, the energy increases typically by as much as 15 MeV in going from the symmetric to the broad asymmetric mode, and by another 15 MeV when moving from the broad to the narrow asymmetric mode. It thus appears that the asymmetric mode controlling the fragment mass yield in the actinides extends down to $A \approx 200$. The common traits of the asymmetric fission mode for compound nuclei ranging in mass number from $A \approx 200$ up to $A \approx 250$ is demonstrated in Fig. 3: plotted as a function of the compound mass, the average mass of the heavy asymmetric fragment component stays constant while the light fragment mass shifts with compound mass. One should point out in passing that the gap in the data points for masses around $A = 220$ is simply due to the lack of suitable targets. The onset of asymmetric fission when moving from light to heavy nuclei across the mass number $A \approx 200$ may remind us of the change of the saddlepoint configuration from a dumbbell into a cylinder-like shape which is known from theoretical studies to take place right at the same mass number. For the moment it is not clear whether this coincidence is merely incidental.

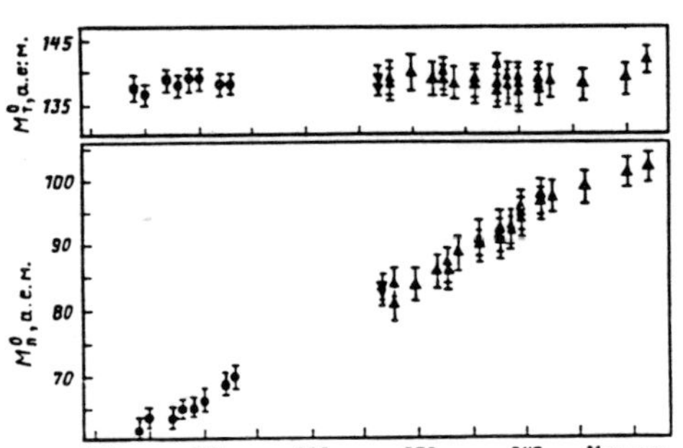

FIGURE 3:
average heavy and light fragment mass of asymmetric fission versus compound mass

As to be anticipated, the fragment mass distribution varies as a function of the compound excitation. But there is again a surprise: the asymmetric mass component vanishes both for high <u>and</u> low excitation energies, i.e. the mass distribution becomes strictly symmetric in both cases. For high excitation energies this feature is in line with what is known from the actinides and understood as washing out of shell effects. By the way, this also explains why the asymmetric component has that long escaped detection, since earlier experiments were performed at rather high excitation energies where the reaction cross sections are less frustratingly small. On the other hand, the disappearance of the asymmetric component at low excitation energies correlates with a similar observation for e.g. the ^{227}Ac compound nucleus, with the notable difference, however, that the roles of symmetric and asymmetric fission are just interchanged. This is brought about more clearly in Fig. 4 showing the excitation function for ^{213}At fission separately for symmetric and asymmetric modes (the insert gives the yield ratio of the two asymmetric components). Evidently, an offset in thresholds for symmetric and asymmetric fission is observed which is just the other way round from what is known for ^{227}Ac.

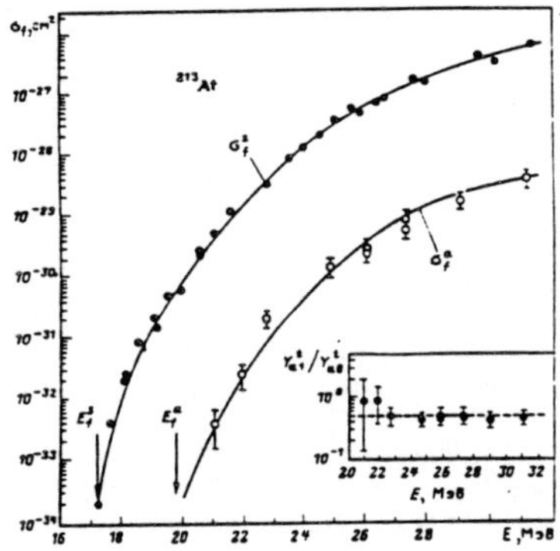

FIGURE 4:
excitation function for the symmetric (full points) and asymmetric (open circles) mode in the fission of ^{213}At

Looking for the theoretical background of the above experimental results one may go back to the year 1971, when V. Pashkevich published a study of the potential energy surface for the nucleus ^{208}Pb close to the saddlepoint. Fig. 5 gives a perspective view of this surface as a function of the collective coordinates elongation (to the right) and asymmetry (to the left). One recognizes two barriers, slightly shifted in energy: a lower symmetric and a higher asymmetric one. Obviously, this potential energy surface at least qualitatively explains the two shifted excitation functions shown in Fig. 4 for symmetric and asymmetric fission, provided the energy surfaces for ^{208}Pb and ^{213}At are not too different. At this conference V. Pashkevich reported on a new calculation of the potential energy surface for ^{213}At, with the computations being pushed down to the scission or — more precisely — the exit point. In the potential landscape indeed three valleys leading to scission were discovered. The predictions for the mass splits and kinetic energies corresponding to these valleys come remarkably close to the experimental findings. We have here a first example of how the phenomenological notion of fission "modes" may be given a firm basis by identifying the valleys being cut into an otherwise smooth Liquid Drop model surface with these modes.

2. FISSION OF THE ACTINIDES

In the fission of actinide nuclei the salient feature having come into focus these last 20 years has been the double-humped fission barrier (A. Michaudon reported on the experimental evidence for even a triple-humped barrier in Th nuclei). But it would be difficult to understand why the pairing and shell effects being responsible for the structure at the saddlepoint should not play also a role in the descent from the saddle to the scission point. This viewpoint has more recently

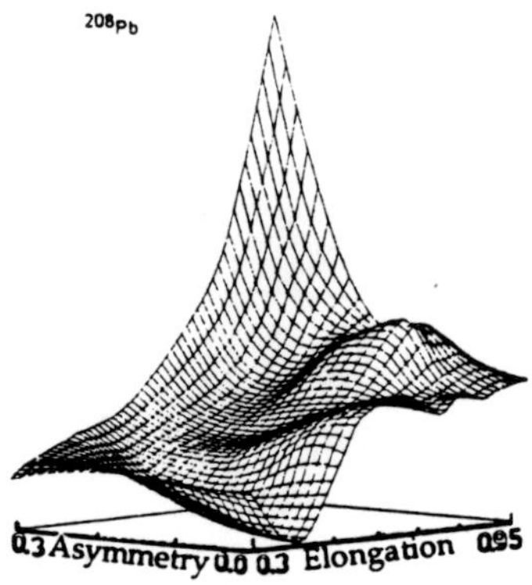

FIGURE 5:
potential energy surface of ^{208}Pb between the ground state and the saddle point as a function of the collective coordinates elongation (to the right) and asymmetry (to the left)

found fervent protagonists. For the standard thermal neutron reaction ^{235}U(n,f) A. Michaudon pointed in his talk to the excellent agreement between experiments having been conducted at Geel (H.-H. Knitter et al.) and rather detailed theoretical predictions by U. Brosa et al. Similar to the ^{213}At case (s. chapter 1) also for ^{236}U three valleys in the potential energy surface show up. The properties of these valleys (called channels by U. Brosa) lend themselves directly to an interpretation of even finer details in the mass-energy distributions of fission fragments.

There has also been considerable progress, reported at this Conference, in established fields of research in fission of the actinides. This progress is mainly due to refined experimental techniques. In the following only a few highlights can be given.

Let us start with the **fragment charge distributions.** There is first the remarkable result obtained by radiochemical methods (A.C. Wahl et al.) that the charge polarisation of the fragments $\Delta Z(A')$ changes sign when one moves from high-yield asymmetric to low-yield symmetric fission (s. Fig. 6 for thermal neutron fission of ^{235}U). This effect has been corroborated

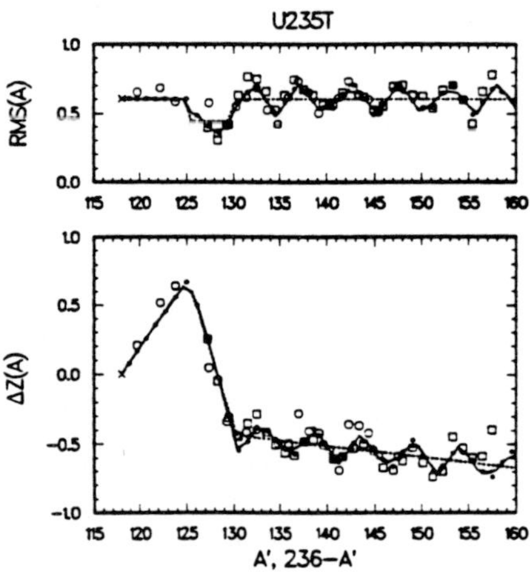

FIGURE 6:
average charge and standard deviation (designated RMS) of charge distributions for isobaric chains A in thermal neutron fission of ^{235}U. The average charge is plotted as deviation ΔZ from an unchanged charge distribution.

by physical methods at the Institut Laue-Langevin in Grenoble. There is so far no theoretical interpretation and one may wonder whether the channel theory of U. Brosa and others addressed above could be of some help here. Furthermore, one observes in Fig. 6 a pronounced fluctuation of both the charge polarisation $\Delta Z(A')$ and the conditional variance $\sigma(Z|A')$ which may be traced back to an odd-even effect of the element yields $y(Z)$. Experiments bearing on the comparison of higher even-Z and lower odd-Z yields have been discussed by N. Aras, T. Datta and F. Gönnenwein. The idea originally put forward by H. Nifenecker that the odd-even effect keeps track of the energy dissipated from collective into intrinsic degrees of freedom in the course of fission has been pursued further. An evaluation of all available data on charge yield odd-even effects from thermal and MeV neutron induced fission is summarized in Fig. 7 showing the smooth increase of dissipated energy as a function of fissility Z^2/A of the compound nucleus. It is noteworthy that in absolute value the dissipated energies are rather low, ranging from about 2 MeV for Th up to about 11 MeV for ^{250}Cf.

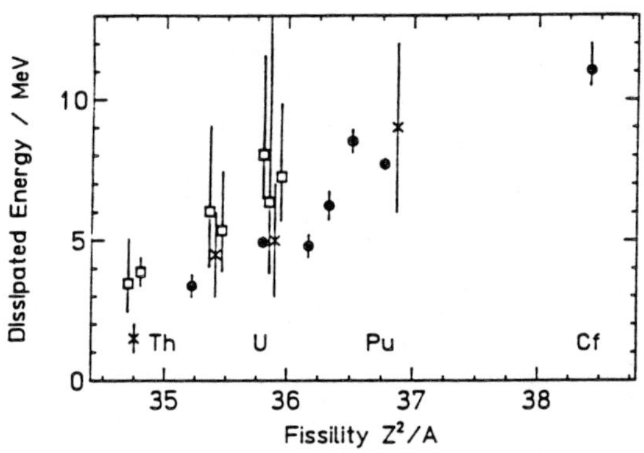

FIGURE 7:
energy dissipated between saddle and scission point for thermal (full points) and MeV neutron (open squares) induced fission versus fissility Z^2/A

There are also odd-even effects in the isotonic yields $Y(N)$ which were discussed by N. Aras. However, these effects are on the average much less pronounced than for the charge number and their interpretation is hampered by prompt neutron evaporation. Yet, A. Michaudon reported on recent measurements by P. Armbruster et al. having found huge neutron odd-even effects in the very low mass tail ($A = 70$ to 80) for fragments from thermal neutron fission of ^{235}U (s. Fig. 8). For the above mass bin the yield drops by 5 orders of magnitude. It appears that the neutron odd-even effects observed cannot be explained by an odd-even staggering in neutron evaporation.

Cold fission is a process where the available reaction energy is consumed by the total kinetic energy of the fragments. This topic was touched by A. Michaudon and discussed in the broader context of Cluster Radioactivity — Cold Fission — Bimodal Fission by D. Poenaru. May it suffice here to indicate that all of the cold fragmentation phenomena in binary decays of a nucleus are linked to specific fragment properties with shell effects being of paramount importance.

A longstanding effort in fission research has gone into the detailed study of **fragment kinetic energy** as a function of excitation energy of the compound nucleus. The recent experimental evidence that the total kinetic energy shows a fine structure, whenever the excitation energy

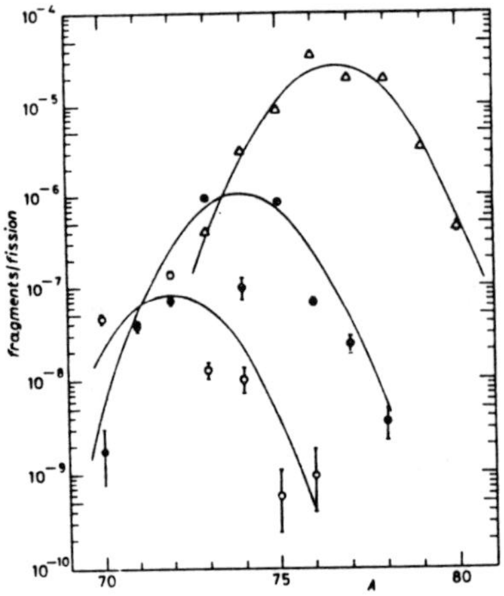

FIGURE 8:
isotopic mass distributions $Y(A|Z)$ for the elements Zn (triangles), Cu (full points) and Ni (open circles) in thermal neutron fission of ^{235}U

matches a vibrational resonance in the second minimum of the potential barrier, was analyzed by A. Ignatyuk. This interesting effect may give insight into the dissipation mechanism of a well defined collective state.

On the other hand, there is the intriguing puzzle which is well known since many years that for increasing excitation energy of the compound nucleus up to a few MeV above the barrier, the fragment kinetic energy increases for Th but decreases for Pu nuclei, while for U the situation is intermediate. In a talk by A. Ruben it was conjectured that these observations may be related to the barrier structure of the above nuclei: in going from Th to Pu the ratio of the second to the first barrier height E_B/E_A changes from $E_B/E_A > 1$ to $E_B/E_A < 1$. If, for the even-even compound nuclei considered, one now assumes that any excitation energy brought into the system, but which stays below the 2 quasiparticle excitations at the second barrier, remains collective in nature and adds to the kinetic fragment energy, while for still higher excitations one merely increases the heat energy, one may calculate the trend of fragment kinetic energy with compound excitation. The results of such a calculation in the framework of a simple two spheroid model at scission are given in Fig. 9. The experimental findings are surprisingly well reproduced. It should be stressed that the deformabilities of the fragments (including the dependence from temperature) are a crucial ingredient to the calculations. Likewise, the changes of the mass distributions with compound excitation have to be taken into account. Of course, it would be tempting to scrutinize whether the strange differences in the trends for the fragment kinetic energies from Th, U and Pu fission could not also be explained by the weights of different exit channels varying with compound excitation.

G. Petrov presented new experimental results obtained at Gatchina/USSR on **parity violation** in fission induced by polarised neutrons at thermal and resonance energies. Parity violation in fission manifests itself as a correlation between the direction of the incoming neutron spin $\vec{\sigma}$ and the momentum $\vec{p}$ of an outgoing fragment of given mass. Stated in other words, the average of the pseudoscalar $\vec{\sigma} \cdot \vec{p}$ does not vanish. The pioneering work in this field has been performed at ITEP/Moscow (G. Danilyan et al.) A non-zero effect has so far been disclosed for the target nuclei ^{233}U, ^{235}U and ^{239}Pu. The effect depends on the energy of the neutron inducing fission in a very sensitive way (at least for ^{233}U and ^{239}Pu), while appears to be independent from fragment mass or kinetic energy. The size of the effect measured as the up-down asymmetry α of fragment emission relative to the neutron spin is surprisingly large: $\alpha \approx 10^{-4}$.

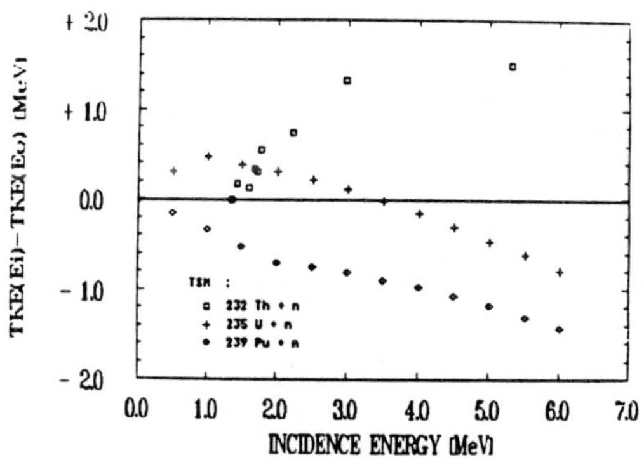

FIGURE 9: dependence of the total kinetic fragment energy on the incident neutron energy for neutron induced fission of ^{232}Th (squares), ^{235}U (crosses) and ^{239}Pu (diamonds)

The theory of parity violation in fission was expounded by V. Flambaum. The large size of α may only be understood by invoking a dynamical enhancement of the weak interaction at the compound nucleus stage, where the level densities are high. The mixing of levels of opposite parity induced by weak interaction at this stage is then shown to be transferred into a mixing of rotational states of opposite parity in the cold transitional stage at the saddlepoint. For fixed (J, K) of the transition state each of these rotational states is built up from states with opposite projection K on the symmetry axis of a pear-shaped nucleus. The parity mixing in the Bohr transition state finally leads to an angular asymmetry of the fission fragments. The magnitude of the parity violating effect predicted from theory is in good agreement with the experimental observation.

A special and even more ambitious type of parity violating study was also referred to by G. Petrov. Searching for a parity violating correlation between the spin polarisation of the neutron inducing fission and the momentum of ternary α-particles accompanying fission, no such correlation could be detected within experimental uncertainty. In contrast to this result, a recent experiment conducted by G. Danilyan at the ITEP/Moscow could demonstrate that the asymmetry measured for $\vec{\sigma} \cdot \vec{p}$ with p the fragment momentum is identical for both, binary and ternary fission. The two experiments taken together corroborate in a subtle way the generally accepted view that ternary fission is not settled at the saddlepoint. Instead, the ternary light particles emerge at scission from a basically binary process.

The topic of fragment deexcitation by **emission of neutrons** was covered by several authors: I. Kimura, U. Brosa, H. Märten, Li Anli and G. Petrov.

From a careful analysis of the fission neutron spectrum of ^{252}Cf in terms of a cascade evaporation model, it was concluded by U. Brosa that for neutron energies ε in the lab frame $\varepsilon \lesssim 0,5$ MeV there is a distinct deviation from an equilibrium (slightly generalized) Maxwellian distribution. In Fig. 10 the points with error bars come from experiment, while the thick continuous (and dashed) line is from theory. The difference between experiment and theory is attributed to scission neutrons. In a hydrodynamical picture the scission neutrons are thought to correspond to satellite droplets formed when the neck between the nascent fragments rapidly breaks apart. From a fit to the data it is deduced that only the fraction $p_0 = 0.01 \pm 0.003$ of all emitted neutrons are scission neutrons.

However, this low fraction was contested in a talk by H. Märten. Based again on a statistical model approach, the conclusion was reached that, within experimental and theoretical uncer-

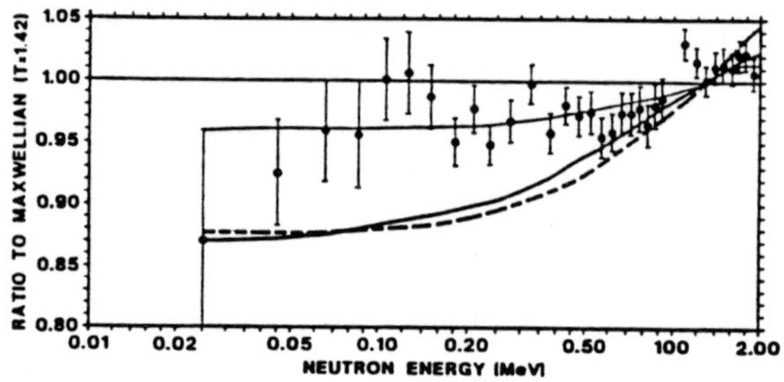

FIGURE 10:
neutron spectrum from ^{252}Cf(sf). Points with error bars: experiment. Thick continuous line: theory.

tainties, the energy and angular distributions of neutrons from ^{252}Cf(sf) are well reproduced with an estimated upper yield of scission neutrons of less than 5% . The question as to how many scission neutrons there are precisely thus appears to be still open.

The study of the evaporation model was extended by H. Märten to light charged particle emission. An interesting result was found: in contrast to α-particles the yield of protons observed in fission of ^{252}Cf(sf) may well be accounted for by evaporation from the exited fragments. Even the energy and angular distributions of the protons are well reproduced by the calculations. This is shown in Fig. 11. With θ the angle between the proton and the light fragment, the ratio of polar/equatorial emission (corresponding to $\theta \approx 0°$ or $\theta \approx 180°$ and $\theta \approx 90°$ respectively) is much higher than for α-particles. The ratio comes close to 0.1 for protons. On the other hand, rather similar to polar α-particles also polar proton emission from the light fragment is favoured as compared to emission from the heavy fragment. As is evident from Fig. 11, theory (continuous line) is in excellent agreement with experiment (points and histogram).

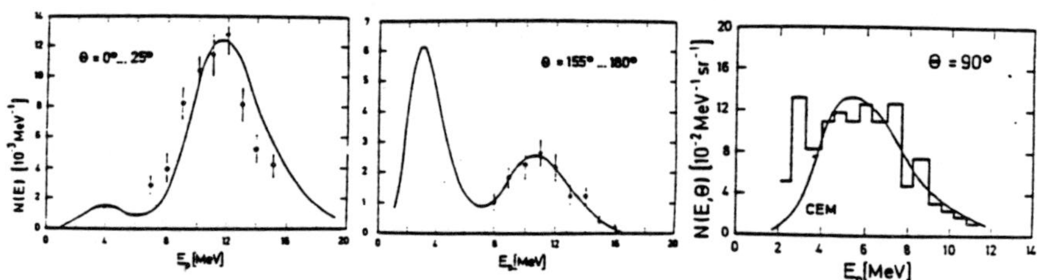

FIGURE 11: proton spectra from ^{252}Cf(sf) at various emission angles

As a last example for neutron emission studies let us dwell on a contribution by I.D. Alkhazov et al. which was presented by G. Petrov. In 1973 H. Nifenecker and C. Signarbieux submitted to the IAEA conference on fission physics in Rochester/USA two papers reporting on an experimental and theoretical analysis of the mean number, the variances and covariances of neutron emission from complementary fragments as a function of fragment mass and kinetic energy for ^{252}Cf(sf). The most puzzling result was that the covariance $cov(\nu_L, \nu_H)$ for neutron multiplicities ν_L and ν_H from the light and heavy fragment, respectively, vanished both for extremely large and extremely small total kinetic energies TKE.

These delicate measurements deserved to be repeated and the outcome of a new experiment by I.D. Alkhazov et al. is summarised in Fig. 12. The figure shows the moments of neutron multiplicity distributions for complementary fragments with masses $A_L = 108$ and $A_H = 144$ from ^{252}Cf(sf) as a function of the fragment kinetic energy TKE. One notices from the figure that both, $\bar{\nu}_L$ and $\bar{\nu}_H$, decrease almost linearly for increasing total kinetic energy TKE of fragments.

As to be expected, for the small neutron multiplicities close to the high energy limit of TKE, also the variances $\sigma^2(\nu_L)$ and $\sigma^2(\nu_H)$ of the neutron number distributions are small. Therefore, it is not surprising that the covariance $cov(\nu_L, \nu_H)$ being calculated from

$$2\ cov(\nu_L, \nu_H) = \sigma^2(\nu_l + \nu_H) - \sigma^2(\nu_L) - \sigma^2(\nu_H)$$

vanishes at the highest TKE measured. In contrast to this, at the low energy extreme of TKE where the multiplities ν and variances σ^2 are large, the observed vanishing of the covariance $cov(\nu_L, \nu_H)$ is not self-evident and, hence, carries new physical information.

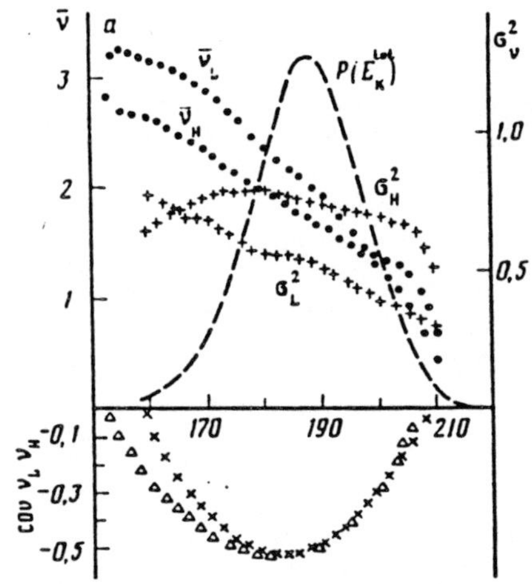

FIGURE 12:
average neutron number, variance and covariance of neutron emission in ^{252}Cf(sf) for the fragment mass pair $A_L = 108$ and $A_H = 144$ amu as a function of the total fragment kinetic energy. The dashed curve gives the kinetic energy distribution.

For the interpretation C. Signarbieux has argued that first, the covariance $cov(\nu_L, \nu_H)$ of neutron multiplicities is proportional to the covariance $cov(E_{XL}, E_{XH})$ of the excitation energies E_X of complementary fragments L and H, and second, for fixed mass split and energy TKE of the fragments (hence fixed total excitation energy) the covariance $cov(E_{XL}, E_{XH})$ is simply equal to the variance $\sigma^2(E_{XL})$ or $\sigma^2(E_{XH})$ in excitation energy of the light or heavy fragment (apart from a minus sign). One has therefore at given mass split and energy

$$cov(\nu_L, \nu_H) \sim cov(E_{XL}, E_{XH}) = -\sigma^2(E_{XL}) = -\sigma^2(E_{XH}) \quad .$$

The message then is that at the high and low borderline of kinetic energy release the vanishing of $cov(\nu_L, \nu_H)$ indicates a zero variance (or width) of the excitation energy distribution of the fragments. This means that in both cases not only the sum (E_{XL}, E_{XH}) but also the individual terms E_{XL} and E_{XH} are fixed. Again, this is trivial for the high energy limit of TKE with $(E_{XL} + E_{XH}) \approx 0$ and, therefore, necessarily $E_{XL} \approx 0$ and $E_{XH} \approx 0$, but rather unexpected for the low energy limit of TKE. Bearing in mind that generally the fragment excitation energy leading ultimately to neutron and gamma emission is at the scission point made up of two terms, viz. intrinsic excitation and fragment deformation energy, the frozen-in ratio E_{XL}/E_{XH} is only to be understood if at the scission point there is virtually no intrinsic excitation.

This view is corroborated by still another aspect of the data from I.D. Alkhazov et al.: not only for small but also for large total neutron multiplicities $\nu_{total} = \nu_L + \nu_H$ i.e. large total excitation energies, pronounced structures 5 amu apart do appear in the mass distributions $Y(A)$. The structures in the mass yield point to an odd-even effect in the charge yields. Recalling that odd-even effects indicate low intrinsic excitation energies at the scission point, one again concludes that the high excitation energies, when traced back to the scission point, stem from large fragment deformations. This then suggests that there are two different types of cold scission configurations: a most compact one leading to maximum kinetic energy release("cold fission"), and a most elongated one with minimum kinetic and large excitation energies ("cold-deformed fission"). Indications for the latter process have also been seen in other fission reactions. Note that in cold deformed fission one is tracing those events which happen to have small if not vanishing intrinsic excitation energies at the scission point and which, exactly for this reason, may reach the maximum deformations compatible with energy conservation and exhaust all of the free energy. In this way it is possible to reconcile minimum kinetic and maximum excitation energy (for fragments infinitely apart) with pronounced odd-even effects.

Finally, a further interesting feature from the work of I.D. Alkhazov should be mentioned: searching for events with the signature of cold-deformed fission (odd-even structure in $Y(A)$) it is found that not only the total excitation energy (neutron multiplicity) has to be large, but also this energy has to be shared in an unequal way between the two fragments. To be more specific, events with one fragment emitting no neutrons at all, while the complementary fragment emits, say, 4 or more neutrons, are among the best examples of cold-deformed fission. In these cases the mass distributions of fragments having emitted no neutrons exhibit furthermore a striking peak close to $A = 132$. This is very suggestive. It indicates that one may visualize the scission configuration of a typical cold-deformed fission event as composed of a heavy shell-stabilized spherical nucleus ($Z = 50, N = 82$) and a very strongly deformed lighter nucleus.

The field of **ternary fission** has much evolved these last years. Nevertheless, the answers given here by different authors to basic questions are still at variance with each other.

J.P. Theobald reviewed some of the more important results from the impressive data base having been accumulated by his group. For the discussion it is convenient to think separately of equatorial and polar ternary fission, with the light charged particle (in most cases an α-particle) being emitted either in the equatorial plane or along the polar axis defined by the momenta of the two main fission fragments respectively.

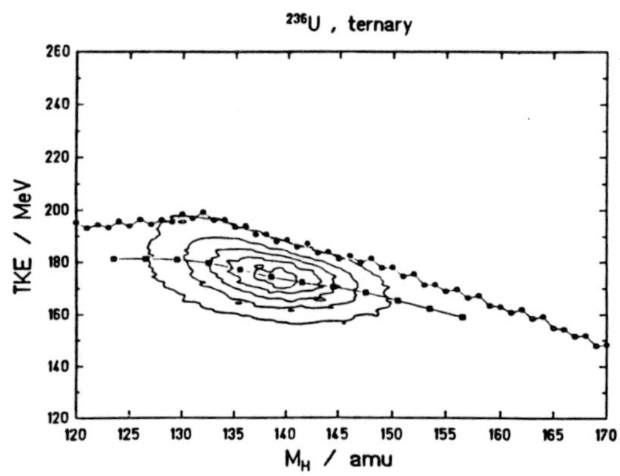

FIGURE 13:
contour plot of total kinetic energy release as a function of heavy fragment mass number in ternary fission (equatorial α emission) of thermal neutron fission of ^{235}U. For the symbols s. text.

For equatorial ternary fission of ^{235}U(n,f) at thermal neutron energies Fig. 13 shows a contour plot of the total kinetic energy TKE as a function of heavy fragment mass M_H. One should underline that in the present context TKE is the sum of the kinetic energies of both the fission fragments and the α-particle. Also given in the figure are the average $\langle$ TKE $\rangle$ (full squares) and the Q-value of the ternary reaction (full points) calculated for α-emission and fission fragment charges maximizing the Q-value. What catches the eye in Fig. 13 is the fact that in ternary fission the TKE release comes close to the Q-value for a broad mass bin. Stated otherwise, cold fission is not an exotic but instead a common feature of equatorial ternary fission. It is tempting to interpret this phenomenon with a naive picture. Imagine a fission prone nucleus with a neck developing between the two main fragments. In the standard case the neck will rupture at one single point and the separating fragments will have to reabsorb the two parts of the neck. Depending on the exact location of the rupture point the fragments will share the deformation in different amounts. Ultimately the deformation energy is converted into excitation energy becoming measurable through neutron and gamma emission. This is a well established model to explain the famous sawtooth of neutron multiplicity vs. fragment mass, having been reconsidered these last years by U. Brosa in the framework of exit-channel theory. In any case, it is seen that in standard binary fission excitation energies come into play and, hence, one will not observe a cold process. But nature may play a trick and cut the neck, almost at the same time, at two different points along the neck. In fact, a recent and rather successful theory of ternary fission has been put forward by V.A. Rubchenya et al. which is based on this idea. Now we may conceive that the nuclear matter from the neck is just used to form the third light particle, and we are then faced with a scission configuration with three virtually undeformed particles: two main fission fragments and the light ternary particle. Evidently, the Coloumb forces will push the neck particle into the equatorial plane. Thus, cold fission in the sense of cold compact fission with high TKE should not be an exotic process with equatorial ternary fission. This is in close agreement with experiment.

One should note in passing that, with the above picture in mind, an α-particle being created at rest on the symmetry axis of the main fragments should experience partly compensating Coulomb forces from the main fragments and, hence, end up with a small kinetic energy, while the fission fragments are efficiently accelerated and acquire large kinetic energies in this case. The anticorrelation in the kinetic energies is born out by experiment. At the same time, α-particles with small initial velocities should be well focussed and show narrow angular distributions. Also this feature conforms with observation. However, these simple arguments may be misleading and can certainly not replace full-fledged 3-body trajectory calculations. Yet even these calculations are not unequivocal. At the present Conference D. Seliverstov e.g. contested earlier claims by C. Guet et al. that from an analysis of ternary fission one may infer that the prescission kinetic energy of fragments is smaller than 10 MeV. D. Seliverstov argued that in order to reproduce consistently all the ternary fission data on angles, energies and masses (including correlations) he has collected over years, a prescission kinetic energy of 25 ± 5 MeV is mandatory. A major difficulty in all these trajectory calculations is obviously the proper choice of initial conditions.

Ternary fission with polar emission of α-particles differs markedly from the equatorial case considered so far. First of all, there is no indication for cold compact fission. The total excitation energies TXE calculated from TXE = Q − TKE are definitely higher for polar than for equatorial emission. The more is it surprising that the energy integrated mass distributions of the main fission fragments exhibit a fine structure in the polar case, but not in the equatorial one. Should these fine structures again hint to low intrinsic excitation energies at the scission point, but nevertheless unusually large deformations in order to bring about the large TXE observed? In order words, could polar emission of α-particles be a decay mode of cold deformed fission, in competition with the decay via strongly asymmetric neutron multiplicities discussed above? To help answer this question, the mass distribution of the main fragments for polar ternary fission of ^{252}Cf(sf) is plotted in Fig. 14 for the case where the α-particle was emitted by the light fission fragment. The mass distribution corresponds to events with larger than average energies and has been corrected for neutron emission. In the figure one discerns a conspicuous peak for the masses $M_L = 117$ and $M_H = 131$ amu. This then may be conjectured to mean that a favourable condition for the emission of a polar α-particle from the light fragment is indeed reached for a spherical heavy fragment close to $(Z = 50, N = 82)$ and a light fragment which has to be strongly deformed to make up for the large TXE. Though the evidence is certainly less clear-cut here, the situation resembles the one identified in neutron emission studies as cold deformed fission. Let us stress that cold deformed fission may only be one of several mechanisms leading to polar α-emission.

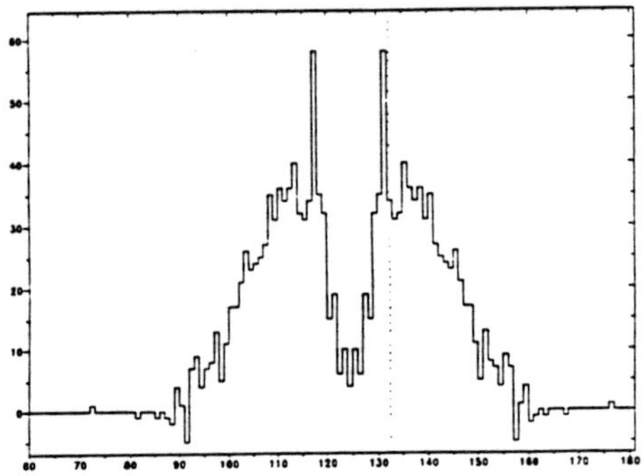

FIGURE 14:
mass distribution of ^{252}Cf(sf) conditioned by the emission of a polar α-particle by the light fission fragment

3. FISSION OF FERMIUM AND HEAVIER NUCLEI

Extending fission studies to very heavy nuclides it was discovered many years ago that in the spontaneous fission of ^{258}Fm a sharp departure from the behaviour of lighter actinides sets in: the mass distributions are symmetric and the kinetic energies are much larger than anticipated from the systematics. A detailed investigation of the fragment characteristics for spontaneous fission of nuclei with $Z \geq 100$ and $N \geq 156$ showed that the TKE distributions are markedly asymmetric, but may be considered as composed of two Gaussian shaped distributions centred at about 200 MeV and 233 MeV, respectively. Sorting events with energies larger or less than TKE = 220 MeV and plotting the corresponding mass distributions one finds that to the high-energy mode correlates a very narrow symmetric mass yield, while for the low-energy mode a still symmetric but broader mass distribution is characteristic. This phenomenon, so far explored for the isotopes ^{258}Fm, ^{258}No, ^{259}Md and ^{260}Md, has been called "Bimodal Fission".

K. Hulet, one of the pioneers and principal investigators of this field, reported to the Conference that a new isotope, ^{262}No, had been identified undergoing Bimodal Fission. The mass and TKE distribution of ^{262}No(sf) is represented in Fig. 15. Obviously, the mass distribution has a rather narrow central part with, however, sizeable tails for both low and high fragment masses, broadening the over-all distribution. The decomposition of the TKE distribution into the two Gaussions mentioned above is indicated in the figure by dashed and dotted lines.

A qualitative explanation of Bimodal Fission was offered by K. Hulet. Evidently, the high energy mode emerges from a compact configuration of two nearly identical spherical fragments at scission, while for the low energy mode one has more elongated nascent fragments. The compact configuration may be accounted for by fragment shell effects. In fact the high energy mode is only observed for fissioning nuclei in a narrow range of proton and neutron numbers

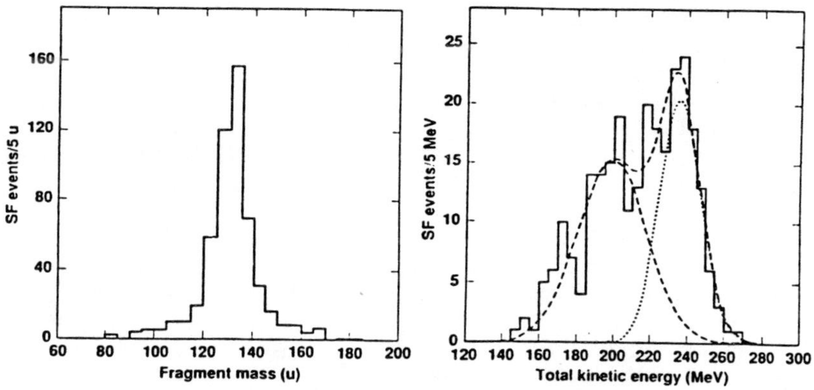

FIGURE 15: energy and mass distribution in ^{262}No(sf)

which may split into two doubly-magic spherical ^{132}Sn or nearby fragments. The ideal candidate for this compact scission mode would be ^{264}Fm. The narrow mass distribution of the high energy mode is fully in line with the stability of the magic fragments. Let us stress that the high-energy mode in the Fermiums and near-by nuclides on one hand, and cold compact fission in the lighter actinides or pre-actinides on the other, may be viewed as one single phenomenon. In both cases the process is triggered by magic fragment properties, a major difference being, however, the probabilities with which the processes occur.

For the interpretation of the low-energy mass-symmetric mode K. Hulet associates its appearance to the dropping of the second barrier (near the saddlepoint) below the ground state. This just happens in the region of the nuclide chart under study. From theory it is known that the first inner barrier is reflection symmetric and, thus, for the very heavy nuclides one comes back to the liquid drop model with one single symmetric barrier, predicting symmetric mass distributions and kinetic energy distributions consistent with the well known $Z^2/A^{1/3}$ systematics. Yet, on closer inspection this symmetric barrier turns out not to result from macroscopic liquid drop model terms but mainly from microscopic shell corrections to the potential energy surface of the fissioning nucleus. Therefore, both modes of Bimodal Fission are due to shell effects: either in the fragments (high energy mode) or in the parent nucleus (low energy mode).

The experimental discovery of Bimodal Fission has been a challenge to theory. Many authors have reinvestigated the potential energy landscape for nuclei around Fm and recognized that there are indeed several distinct channels or valleys leading from the saddlepoint region to scission (U. Brosa, V. Pashkevich, K. Depta, S. Cwiok, A. Sobiczewski, P. Möller, J.R. Nix, W.J. Swiatecki and others). The features of the individual modes are well reproduced by these calculations. However, as K. Hulet insisted in his talk, theory up to now is not capable to predict precisely the weights with which the different channels or modes are fed. In view of the sudden appearance and disappearance of modes in going from one isotope to a neighbouring one (a behaviour which is closely linked to the dominating single particle effects), the call for a more precise theory may prove to be a tremendous task.

4. CONCLUSION

At the present Conference fission phenomena for nuclei ranging from the preactinides up to Fermium and beyond have been covered. A newly discovered universal trait for all these nuclei is multimodal fission. Experimentally, a mode may be identified by its specific fragment mass and energy distribution. Yet, the notion of a mode has been in recent years put on firm theoretical grounds by realizing that the potential energy surface has a rich structure. Since 20 years or so interest has been focussed on the double-humped barrier. But structure is not limited to the immediate vicinity of the saddelpoint. Valleys (or channels), ridges, bifurcation points show up all over the potential energy landscape. Upon fission the flow of nuclear matter is channelled along these valleys, from the saddle down to the scission point. To each valley corresponds a characteristic mass and energy distribution and, therefore, these valleys may be considered as the modes showing up in experiment. Sometimes the modes are difficult to discern due to the overlap of fragment distributions from different valleys and this explains why the notion of modes has not always been felt to be compulsory in the past. Surprisingly, already the static properties of the potential energy surface in terms of channels allow for a satisfactory description of modes. For instance, the symmetry properties of the barrier are directly reflected in the basic character of the mass distributions, symmetric or asymmetric. As for the dynamics of the process, theory is in a less good shape. For example, there are no guidelines how to assess the rate of flow into the different channels. At the moment it is also not very clear how the link between the compound nuclear states, the well established transitional states at the saddlepoint (the A. Bohr channels) and finally the valleys (the exit channels) leading to scission really works. First steps to clarify this question have been taken (W. Furman, priv. comm.).

The unifying view offered by discussing fission in terms of the channel structure of the potential energy surface may help to disentangle the complexity of fission phenomena and has, therefore, to be considered as an important step towards a better understanding of the process. Future work has to be awaited to see whether also those limiting but highly instructive cases of cold compact and cold deformed fission, showing up in mass-charge distributions (with the odd-even effect as a fingerprint), in neutron emission and in ternary fission, can be accommodated within the proposed framework. Last but not least, parity violation studies in fission may serve as an important tool to clarify the relation between the Bohr transition channels at the barrier and the exit channels or valleys in the potential energy surface down to scission.

It is a pleasure to acknowledge helpful discussions and the supply of figures by many colleagues, but notably P. Heeg, K. Hulet, M. Itkis, H. Märten, A. Michaudon, V. Pashkevich, G. Petrov, A. Ruben and J.P. Theobald.

LIST OF PARTICIPANTS

Brazil

M.C. Khouri	Sao Paulo

Bulgaria

M.T. Matev	Sofia
N. Tchikov	Sofia

China, People's Republic

Li Anli	INS Beijing
Yan Yiming	University Beijing

Cuba

F. Guzman	Inst. Habana

France

A. Michaudon	ILL Grenoble

German Democratic Republic

T. Elfruth	TU Dresden
T. Hehl	TU Dresden
H. Helfer	TU Dresden
H. Kalka	TU Dresden
M. Kirchbach	ZfK Rossendorf
R. Kotte	ZfK Rossendorf
M. Matthies	ZfK Rossendorf
H. Märten	TU Dresden
P. Mädler	TU Dresden
K. Merla	TU Dresden
B. Milek	TU Dresden
R. Münze	ZfK Rossendorf
W. Neubert	ZfK Rossendorf
W. Pilz	ZfK Rossendorf
I. Rotter	ZfK Rossendorf
A. Ruben	TU Dresden
R. Schmidt	TU Dresden

List of Participants, cont´d

GDR, cont´d

K.-D. Schilling	ZfK Rossendorf
A. Schubert	ZfK Rossendorf
D. Seeliger	TU Dresden
F. Stary	ZfK Rossendorf
S. Unholzer	TU Dresden
G. Zwicker	ZfK Rossendorf

Germany, Federal Republic of

U. Brosa	KFA Jülich
P. David	Univ. Bonn
T. v. Egidy	TU München
F. Gönnenwein	Univ. Tübingen
W. Greiner	Univ. Frankfurt/M.
R.W. Hasse	GSI Darmstadt
D. Hilscher	Hahn-Meitner-Institut, Berlin/West
G. Herrmann	Univ. Mainz
H.-J. Krappe	Hahn-Meitner-Institut, Berlin/West
H. Machner	KFA Jülich

Greece

N.N. Papadopoulos	Aghia Paras. Attiki

India

T. Datta	Bhabha Atomic Research Centre

Iraq

S.M. Al Jabori	NRC Baghdad

Japan

I. Kimura	Kyoto Univ.

Poland

M. Jaskola	INS Warsaw

Romania

D. Poenaru	Centr. Inst. Bucharest

List of Participants, cont'd

Sudan

N.S. Shakir	Univ. Khartoum

Thailand

S. Chongkum	Bangkok

Turkey

N.K. Aras	Univ. Ankara

UK

T.D. MacMahon	Berkshire

USA

E.K. Hulet	Lawrence Livermore NL

USSR

G.E. Belovitski	INR Moscow
V.N. Dementyev	INP Moscow
V.V. Flambaum	INP Novosibirsk
V.I. Fursov	IPPE Obninsk
A.B. Ignatyuk	IPPE Obninsk
M. Itkis	INP Alma-Ata
S.P. Ivanova	INP Moscow
Yu.A. Lazarev	JINR Dubna
V.G. Nederezov	INR Moscow
Yu.T. Oganessian	JINR Dubna
V.V. Pashkevich	JINR Dubna
G.A. Petrov	NPI Leningrad/Gatchina
D.M. Seliverstov	NPI Leningrad/Gatchina
O.A. Yuminow	INP Moscow

Zambia

M. Nomai	Univ. Lusaka

Organizations

Wang DaHai	IAEA Vienna
V. Goulo	IAEA Vienna
V. Konshin	IAEA Vienna
J.J. Schmidt	IAEA Vienna

Index of Authors

Subject Index